OUR POLITICAL NATURE

OUR POLITICAL NATURE

THE EVOLUTIONARY ORIGINS OF WHAT DIVIDES US

AVI TUSCHMAN

59 John Glenn Drive
Amherst, New York 14228–2119

Published 2013 by Prometheus Books

Cover design by Grace M. Conti-Zilsberger

Inquiries should be addressed to
Prometheus Books
59 John Glenn Drive
Amherst, New York 14228–2119
VOICE: 716–691–0133
FAX: 716–691–0137
WWW.PROMETHEUSBOOKS.COM

17 16 15 14 13 5 4 3 2 1

Library of Congress Cataloging-in-Publication Data Forthcoming

Tuschman, Avi.
Our political nature : the evolutionary origins of what divides us / by Avi Tuschman.
pages cm.
Includes bibliographical references and index.
ISBN 978-1-61614-823-2
ISBN 978-1-61614-824-9 (ebook)

Printed in the United States of America

For Mark, Jana, and Eva

With special thanks to Paul, Frank, Bob, Cecilia, and Glenn

"*Our Political Nature* shows us that there are evolutionary underpinnings to our political attitudes, and that being liberal or conservative may reflect much deeper tendencies than we are inclined to think. This book is important reading for anyone trying to understand the sources of our present-day political world."

—Francis Fukuyama, *New York Times*–bestselling author of *The Origins of Political Order*

"In a remarkable interdisciplinary tour de force, evolutionary anthropologist Avi Tuschman integrates findings from social psychology, genetics, and neuroscience to provide a rich understanding of the polarization in politics throughout history, and of man's inhumanity to man. In *Our Political Nature* he makes clear that be it vote choice or the decision to go to war, our politics are the product of the passions that drive us, which are deeply rooted in humanity's evolutionary origins."

—Jerrold M. Post, MD, director, Political Psychology Program at the George Washington University, and author of *Leaders and Their Followers in a Dangerous World*

"At a time when unexpected political turmoil and economic crashes have exposed how feeble is our understanding of the forces that drive these crises, *Our Political Nature* provides a welcome respite from the intellectual confusion now reigning. In these pages Avi Tuschman offers a fascinating perspective on the deepest roots of the clashes that are changing our world."

—Moisés Naím, author of *The End of Power* and former editor in chief of *Foreign Policy*

CONTENTS

PREFACE

Political polarization in the United States has reached dangerous levels. On issue after issue, the US Congress is hopelessly paralyzed, and we have recently witnessed one of the most ideologically divisive presidential campaigns of our lifetimes. As frustrations deepen and rival accusations fly, "we the people" are left searching for answers. In this troubled time, *Our Political Nature* can perform a vital public service, for it examines two questions that are absolutely fundamental to the future of this country: Where do our political orientations really come from? And what are the root causes of our debilitating left-right divide?

Our Political Nature brings into the light the hidden dynamics of our most deeply held political convictions. It explains the factors that influence who we vote for, who we choose as mates, and whether or not we believe in God. The book also helps readers understand that the polarizing left-right divide is by no means unique to America; in truth, similar political spectrums run through almost every country around the globe.

This work is based on ten years of research at Stanford University and from my career advising heads of state on shaping public opinion. During this journey, I've gathered dozens of new insights from the diverse disciplines of neuroscience, primatology, genetics, and anthropology. Still, few of these studies have percolated up and been published by prominent authors or journalists.[1] So I wrote this book as a first effort to draw together this diverse research into one single, well-documented explanation of the biological foundations of our most important values. My aim here, in sum, is to paint a compelling and accurate portrait of our nature as political animals.

Today's political commentators would have us believe that we vote based on our views about the main issues of the day, on our economic circumstances, or on our longtime affiliations with this political party or that. That's wrong, or at least incomplete. Step by step, I will explain that our political orientations are not simply intellectual constructs, flowing from our upbringing, our schooling, our peer groups, or which newspapers we read. No, our political orientations are actually natural dispositions, molded within each of us by powerful evolutionary forces.

The substantiation comes from some surprising sources. For instance, I will explain why twins who have been separated at birth and nurtured in very different environments nonetheless grow up to share remarkably similar political attitudes. Likewise, I will highlight research that shows that a brain scan can accurately predict whether a person is more likely to be a liberal or a conservative. I will also reveal why a group's political spectrum shifts slightly to the left or the right depending on the homeland of its ancestors.

From different perspectives, I will show how *political orientations across space and time arise from three clusters of measurable personality traits*. The three clusters entail opposing attitudes toward tribalism, inequality, and differing perceptions of human nature. Together, these traits are by far the most powerful cause of left-right voting, even leading people to regularly vote *against* their economic interests. As I will explain, our political personalities also shape our likely choice of a mate and influence societies' larger reproductive patterns. Perhaps most importantly of all, the book tells the evolutionary stories of these crucial personality traits, which stem from epic biological conflicts.

The underlying science here is complex and often challenging. However, I have gone to great lengths to turn this material into a vivid and entertaining expedition through the worlds of hunters and gatherers, selfish and generous genes, and even viruses. Also, to make the book relevant to a wide range of readers, I have used current news stories and personalities to illustrate core concepts. For example, readers will meet Glenn Beck and Hugo Chavez and come to understand the hidden forces they represent. And they will understand why Ayatollah Khamenei goes on extended rants on state television to condemn miniskirts. In sum, by blending serious academic research with relevant contemporary examples, *Our Political Nature* will give readers an entirely new way to understand the events that are roiling our world today.

While this book covers extremely controversial issues, my intention here is not to take sides; it is to illuminate. I want to help people understand the origin of the painful ideological clashes that so dangerously divide and imperil our world. My intent is not to push people to the left or to the right, but to explain why our political personalities stretch across a conflictive spectrum.

Ultimately, I seek to provide readers with new insights into why the battles between moderates and extremists, and between the left and the right, will persist into the foreseeable future. Yes, our world is now in a period of profound change. News cycles are faster than ever. And social media, the Internet, and cable television have injected steroids into the muscle of public opinion. Still, most people remain largely uninformed about the evolutionary origins of political orientation. Through *Our Political Nature*, readers will gain deep insight into how evolutionary drivers and key demographic trends are right now transforming the future of our country and our world. In essence, this book provides readers with a pair of "evolutionary glasses," new lenses that will help them perceive the natural history of our species's political orientations, and how this history is intimately connected to today's news cycle and to our private lives.

It is my fervent hope that a deeper understanding of the hidden roots of our true nature and of our relationships with others will quiet the heart—and that the powerful yet elegant theory of evolution will help readers make sense of an otherwise perplexing world. Armed with that deeper understanding, perhaps we can raise the level of our political discourse and strengthen our noble democratic processes. That, at the highest level, is the mission and purpose of *Our Political Nature*.

INTRODUCTION

SEARCHING FOR THE ORIGIN OF POLITICAL IDENTITY

1.

THE BURNING MAN OF TUNISIA

On December 17, 2010, a Tunisian street vendor named Mohamed Bouazizi left his home in the countryside and walked twenty minutes into the town of Sidi Bouzid. Just as he did every day, Mohamed passed by groups of unemployed young men on the roadside, idling in plastic chairs and quietly smoking. When he reached the market, he stopped to set out his produce stand.

Mohamed was well liked in town. Sometimes he gave away free food to poor families. They called him Basboosa, which meant "sweet cake."[a] Yet Basboosa could scarcely afford his generosity, which made him more popular than it did rich.

Basboosa did not have an easy life. His father had died of a heart attack when his son was three. His mother toiled in the fields as a day laborer. As one of seven siblings, Basboosa started working odd jobs at the age of ten to support his family. Eventually he had to drop out of high school to work full-time.

Although he tried to join the army, and also applied to a number of other, more-promising jobs, Basboosa was rejected each time. At twenty-six, he barely managed to make ends meet by selling fruits and vegetables on the street. Basboosa's produce cart grossed the equivalent of five dollars a day, making him the breadwinner of the nine souls in his home. With these meager earnings, he dreamed of saving enough to buy a truck for his business.

On top of all this, Basboosa faced constant harassment from local authorities. There was no law that required Basboosa to have a license to sell his produce on the street, but that didn't stop the local police. They would routinely ask Basboosa for his license, just as a pretext to impose a "fine." Basboosa knew what this was: a shakedown. And sometimes the sum the police demanded amounted to a full day's earnings. Sometimes, too, the police simply confiscated all of his wares on a whim. This was infuriating and degrading, of course, but what could Basboosa do? This, alas, was the normal price of doing business in Sidi Bouzid.

Still, on this particular Friday, Basboosa had reached the end of his rope. The global economic crisis had sent food prices soaring, and he had just been forced to take out a $200 loan just to buy his produce. In every cell of his body, Basboosa burned with resentment.

At 10:30 a.m. that morning, a municipal inspector named Faida Hamdy approached Basboosa's cart to shake him down for a "fine." The amount she demanded was the equivalent of seven dollars, but to Basboosa it was pure blood money. And he refused to pay. This was a stunning act of defiance. The operative rule in town was that nobody ever said "no" to Hamdy; her father was a local police officer. But Basboosa no longer cared. He had no money—and he had stomached enough. When he refused to pay, a shocked Inspector Hamdy insulted Basboosa's dead father. Then she slapped him across the face, spat on him, and overturned his cart. Two of her aides proceeded to seize his electronic scale and beat him to the ground.

Basboosa was stunned and humiliated, but he was far from defeated. Immediately he marched to the governor's office to denounce the abuse. When officials there refused to even hear his complaint, Basboosa threatened to set himself on fire, right then and there. And he wasn't bluffing. A short time later, he returned to that spot with paint thinner to use as an accelerant. One way or another, he would have his say. Then, true to his word, right there in front of government headquarters, Basboosa doused his body with the thinner and set himself alight.

Help did not come quickly. By the time an ambulance could take Basboosa to the local medical clinic, the flames had burnt 90 percent of his body, and the heat had incinerated his lips and charred his clothes right into his body. Incapable of treating him, the clinic transferred the young man to the city of Sfax, eighty-four miles to the east.

Basboosa set more than himself on fire. Even though protesting was banned in Tunisia, angry demonstrations spread within hours through the town of Sidi Bouzid. Police tried to crush the protesters with violence, but their tactics backfired. By the following day, full-blown riots had broken out. While there were no conventional media to cover the unrest, the townspeople captured the violence with cameras and cellphones and uploaded images of police brutality onto Facebook and YouTube. That, in turn, gave birth to an Internet campaign to support Basboosa and the protesters in Sidi Bouzid. The "fire" was spreading.

Soon the country's lawyers' union went on strike and sent three hundred members to rally at the government palace in Tunis. In an attempt to defuse the crisis, the government had the severely burnt Basboosa transferred to a hospital in the capital, and President Zine El Abidine Ben Ali personally went to visit him at the trauma center. The president even promised the Bouazizi family that he'd be sent to France for treatment, and the government also offered a $15 million economic-aid package to pacify Sidi Bouzid. None of that worked; the civil unrest continued to escalate.

Basboosa's example soon inspired others. Lahseen Naji, a protestor in his twenties, railing against "hunger and joblessness," climbed up an electricity pylon and electrocuted himself. Then Ramzi Al-Abboudi, a desperate man overwhelmed by micro-credit debt, publicly committed suicide. As the protests swelled, the international hacker group "Anonymous" shut down over a dozen websites of the Tunisian government. This helped shine the global spotlight on Tunisia and its corrupt regime.

But it didn't come fast enough for Basboosa: he died from his wounds the following day. To millions of Tunisians, though, he died a hero, and in a final tribute, more than five thousand people attended his funeral procession.

Still, the final triumph belonged to the humble produce vendor. The protests against President Ben Ali and his repressive regime did not subside; they grew—with the popular anger this time serving as the accelerant. As the demonstrations spread across the country, their focus shifted from the injustice in Sidi Bouzid to the gaping economic disparities in Tunisia. The unbearable unemployment rate contrasted sharply with the opulent lifestyle of the elite. An especially intense hatred centered around the family of Tunisia's first lady, Leila Trabelsi, whom a lecturer at the University of Exeter had once called an "Imelda Marcos incarnate. But instead of shoes, Madame Leila collects villas, real estate and bank accounts."[1]

Protestors then directed their rage at the tight restrictions on media and political freedoms, which President Ben Ali had smothered during his twenty-three years in power. The resentments ran deep. His security forces had persecuted the socialists, on the left; on the right, they had driven Islamist leaders into exile. Ben Ali's government also had forbidden the wearing of headscarves in public institutions. He called the hijab a "sectarian dress," perhaps pejoratively referring to the Qur'an's mandate (33:59) that women draw a garment over their heads so as to avoid harassment.

The end was not pretty. With a mass of people now shouting for him to step down, President Ben Ali and his imperious wife Leila were forced to flee Tunisia on January 14, 2011. France denied them asylum while they were in midair. Ultimately, the Ben Ali family was granted exile in Saudi Arabia, one of only four states in the world that does not even claim to be a democracy. Within only weeks of Ben Ali's ouster, his people gained several freedoms; the new government in Tunisia relaxed dress codes and legalized three banned political parties.

* * *

The story of Basboosa does not end there. Scarcely a day after the overthrow of Ben Ali, a man in Algeria set himself on fire when a local mayor refused to meet to discuss his unemployment. Then a second Algerian self-immolated . . . and a third. Arab men set themselves on fire in Egypt, Saudi Arabia, Mauritania, Syria, Morocco, and Iraq.

Following the flames, protests broke out in each of these countries. In addition, a wave of civil unrest rolled across Bahrain, Djibouti, Jordan, Oman, Yemen, Kuwait, Lebanon, Sudan, Western Sahara, Iran, and Pakistan. Echoing the "Twitter Revolution" of the 2009–2010 Iranian election protests, the young organizers of the "Arab Spring" used social media to declare "days of rage." Like Basboosa's Tunisia, many of these other states were suffering from massive unemployment, government corruption, and long-entrenched dictatorships.

By early February 2011, the fire had spread to Egypt. The regime in Cairo attempted to counter immense demonstrations by disabling access to the Internet—and also by unleashing brutal violence on the protestors. In only eighteen days, however, President Hosni Mubarak had arrived at the point where he had no choice but to end his thirty years in power. And within the year, the country at the heart of the Arab world had its first free elections in three generations. In fact, the ensuing parliamentary and presidential races were the most genuinely reflective polls of the popular will in the civilization's five-thousand-year history.[b]

Next, armed conflict erupted in Libya, which the notorious Col. Muammar Gaddafi had ruled for forty-one years. Eight months and thirty thousand casualties into the Libyan Civil War, rebel soldiers fished a terrified Gaddafi out of a drainage pipe and lynched him in a frenzy of gunshots, beatings, and stabbing. A thoroughly dead dictator then went on display in an industrial freezer. Libyans flocked from hundreds of miles away to see the corpse with their own eyes.

The underlying currents driving the Arab Spring then reached Syria and the regime of Bashar al-Assad. The Syrian people rose up, Assad responded with ruthless killings, and the situation spiraled into civil war. When the Arab Spring arrived in Yemen, President Saleh negotiated his resignation in exchange for immunity from prosecution—after suffering severe injuries from a bomb attack on his compound. To escape the rising tide, the leaders of Sudan and Iraq announced their decisions not to seek re-election, and the kings of Jordan and Morocco implemented legal reforms in the hope holding on to power.

In one of the world's most authoritarian regions, the increasing force of the public voice is eroding the power of autocrats. The appearance of free elections in several countries represents one step toward democracy, although the outcomes have been decidedly less liberal and secular than many optimistic Western pundits had expected. And it remains to be seen whether countries like Egypt will develop the other elements of democracy (protecting minority rights, balancing powers, respecting term limits, and so forth). But whatever the future holds, one thing is clear: public opinion will impact the political destiny of the Middle East more than ever before. Dictating just isn't what it used to be.

And where did all this start? Many observers traced the roots of the Arab Spring right back to Mohamed Bouazizi. The news program *60 Minutes* was one of many that attributed the entire transformation of the Middle East to, in their words, "the desperate act of one single man . . . a poor fruit vendor who decided that he just wasn't going to take it anymore."

* * *

The story of Basboosa illustrates one of the key tenets of this book: that there are certain hot-button realms inside the human psyche that lie just below the surface, realms which even government leaders often fail to perceive. And we ignore them at our own peril. Inspector Hamdy tripped right over one of these hidden triggers, and in doing so, she lit the match that set an entire region ablaze.

In the pages ahead, we are going to dig deep into the personality traits that divide us into left and right, helping lock and load these hidden triggers. We will discover how these universal character traits make some people prone to conform, and others prone to rebel. Likewise, we will learn why some people have conservative dispositions, while others are liberal. As the underlying science here becomes clear, we will understand what was ticking inside Basboosa on that fateful day that drove him to commit the ultimate protest. What was it, specifically? An unbearable sense of indignation.

Yes, it was precisely Basboosa's moral *rejection of inequality* that activated one of these universal hot buttons residing within him, and within so many of his compatriots. And this is why his story resonated with a critical mass of people in the Middle East, for whom Basboosa symbolized the humble, well-meaning common man systematically abused by government fiat and corruption.

Now, the story of Basboosa might seem rather remote to many American readers. But the same hidden trigger at play in the Middle East underlies the concurrent transformation of the political landscape in the United States. As the Arab Spring was unfolding, the "Occupy Wall Street" movement emerged on the far left of the US spectrum. The demonstrators in New York shouting, "We are the 99 percent!" were railing not against Middle Eastern dictators, but rather against bankers and large corporations. And yet the same issue that ignited the Arab Spring had inspired and galvanized their movement: a moral rejection of economic and social *inequality*.

Despite the fact that the Occupiers claimed to represent 99 percent of the population, the American public was evenly split in favor of (35 percent) and against (36 percent) their cause. This division didn't simply pit the haves against the have-nots; rather, it polarized America based on people's *attitudes* toward inequality. And these attitudes stemmed from the *personality differences* at the core of partisanship: Democrats were almost five times more likely than Republicans to support the Occupy Movement.[2]

At the same time, the Tea Party emerged on the far right of the American political spectrum. This movement's central aims included reducing taxes and government spending (especially on social programs, like Obamacare). Its underlying ideology opposed expensive government interventions designed to force an equality of outcomes. Tea Partiers were instead committed to incentives based on market forces, regardless of the social and economic inequalities that result from free competition. Compared with the Occupiers, adherents of the Tea Party had a much higher *tolerance of inequality*.

Still, as a group, the Tea Partiers' ideology didn't come primarily from the self-interest of big fish at the top of the food chain; in terms of their income bracket, employment status, educational background, race, gender, and age, Tea Partiers were quite similar to the general population. Rather, their tolerance of inequality reflected a *personality disposition* that lies at the heart of political partisanship. That's why Republicans were over six times more likely than Democrats to support the Tea Party.[3]

Attitudes toward inequality are only one group of hot-button traits that separate people on the left from people on the right. In the coming chapters, we will also examine the two other clusters of measurable personality traits that give rise to political orientations across space and time. One emerges from opposing attitudes toward tribalism, and the other concerns differing perceptions of human nature.

Step by step, we will unearth the evolutionary roots of these three personality clusters, to learn why a left-right political divide runs through every country on earth. And the excavation begins in the next chapter. In this age of public opinion, it is high time we understand at a much deeper level why we love and hate in such different ways.

2.

THE UNIVERSAL POLITICAL ANIMAL

Among the first scientists to dig for the roots of political orientation—be it for people here in the United States or in a rural village in Tunisia—were a couple of pioneering psychologists in California named Jack and Jeanne Block. Back in 1969, the Blocks asked themselves two challenging questions: How deep do our political leanings run? And how early in life do these leanings begin to form within each of us?

In search of answers, the Blocks devised a very unusual study, and they began it with kids who were still in nursery school. On the face of it, the premise of the study seemed absurd: what did nursery school kids know about Democrats or Republicans, or about the complicated, hot-button issues of the day? Still, the Blocks were serious researchers from UC Berkeley, and they were determined to break new ground.

For their experiment, the two professors placed a group of 128 nursery-school children under the close observation of several teachers for a period of seven months. Then the Blocks had each of these caretakers measure the three-year-olds' personalities and social interactions, using a single standardized test. The same children then underwent this process again at age four, with a different set of teachers at a second nursery school. The Blocks tabulated the scores for each child and then locked the numbers away in a vault.

The test scores then sat in the vault for the next two decades, while the children from the study went their separate ways in life. They grew up, completed their educations, and turned into young adults. After twenty years had passed, the Blocks succeeded in tracking down 95 of their original 128 subjects, in the hope of measuring how liberal or conservative each of them had become. This time they asked the young adults, now age twenty-four, to situate themselves on a five-point political spectrum. They also asked them to express their opinions on a number of highly partisan, hot-button issues. In particular, several of these questions measured their *tolerance of inequality* between the genders and between different racial groups. In addition, they were asked to describe any political activism they might have participated in during the intervening years.

The results, published in 2006 by the *Journal of Research in Personality*, were astonishing.[1] In analyzing their data, the Blocks found a clear set of childhood personality traits that accurately predicted conservatism in adulthood. For instance, at the ages of three and four, the "conservative" preschoolers had been described as "uncomfortable with uncertainty," as "rigidifying when experiencing duress," and as "relatively over-controlled." The girls were "quiet, neat, compliant, fearful and tearful, [and hoped] for help from the adults around."

Likewise, the Blocks pinpointed another set of childhood traits that were associated with people who became liberals in their mid-twenties. The "liberal" children were more "autonomous, expressive, energetic, and relatively under-controlled." Liberal girls had higher levels of "self-assertiveness, talkativeness, curiosity, [and] openness in expressing negative feelings."[a]

The Blocks's experiment suggested that the roots of our political orientations emerge as early as the fourth year of life. But it begged further, essential questions: Would children from different regions or socio-economic backgrounds diverge into similar personality groups? And how much deeper are the origins of these crucial personality traits?

When the Blocks's findings are connected with other relevant studies in genetics, neuroscience, and anthropology, a composite image of our political nature begins to emerge, which few people have ever contemplated. As this portrait of ourselves comes into focus, we will learn what made some of those nursery-school children grow up to become liberals, and some of them, conservatives. We will discover exactly how deep the roots of our political proclivities extend, and why they have a similar influence on children in America, the Middle East, and most everywhere else.

The reason for these crosscutting commonalities, as we'll see in this chapter, is because *political orientations are natural dispositions that have been molded by evolutionary forces*. Taken together, those deeply ingrained political orientations form what could be called "The Universal Political Animal."

THE DEEPEST ORIGIN OF POLITICAL ATTITUDES

To begin with, exactly how far down *do* the roots of preschoolers' political orientations extend? Do these roots somehow spring from the hard wiring they were born with? Perhaps W. S. Gilbert, the nineteenth-century English dramatist and poet, was onto something when he mused:

I often think it's comical
How nature always does contrive
That every boy and every gal
That's born into this world alive
Is either a little Liberal
Or a little Conservative[2]

In the early 1990s, the University of Minnesota's Center for Twin and Family Research set out to test Gilbert's humorously posited theory. Armed with a valuable list of ten thousand twin pairs and their family members, these scholars had a unique way to determine whether our genes—as opposed to our environment—exert any effect over our political convictions.

The Minnesotan scientists began their study with a left-right political-orientation test called the RWA scale, which we'll learn about in great detail in the following chapters. They handed it out to over fourteen hundred identical and fraternal twins. Each pair was raised in the same family. In a parallel study, psychologist Thomas Bouchard gave the same test to a very special subset of siblings: eighty-eight identical twins and forty-four fraternal twins *raised in completely different environments*.

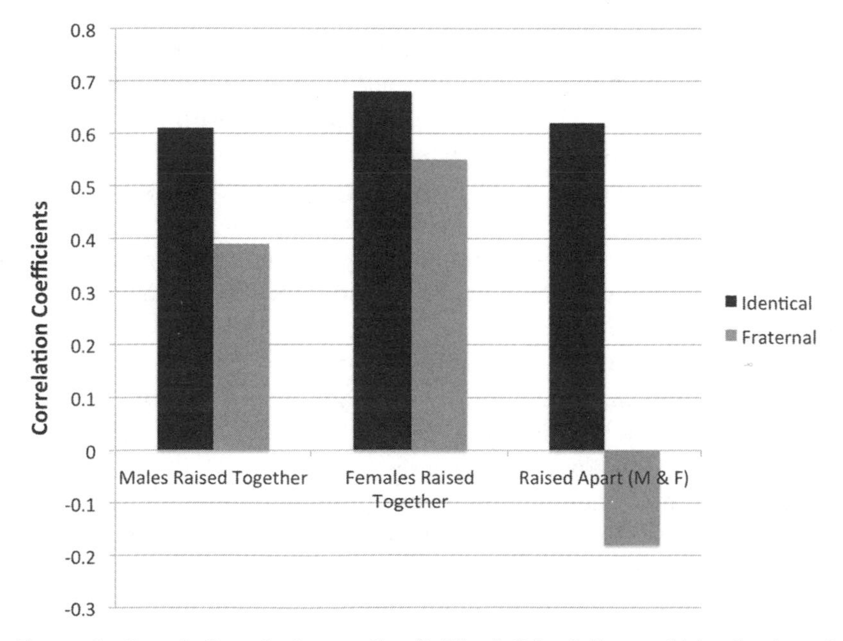

Figure 1. Correlations between the Political Orientations of Identical and Fraternal Twins, Raised Together and Apart.

Comparing both types of twins is crucial: in the case of twins raised together, the shared environment (such as family dynamics, interactions between twins, and social perceptions) could influence the development of the twins' personalities. But for twins separated at birth, most of these confounding factors drop away; this makes it easier to chalk up their personality similarities to genetics.

Figure 1 shows the correlations between the left-right orientations of twins raised together and apart (Appendix A explains correlations and the simplified notation used in this book). The black bars correspond to identical twins, and the gray bars to fraternal twins. The first two clusters show the twins raised together. Identical twins (who share 100 percent of their genes) had more similar political orientations than fraternal twins (who share only 50 percent of their genes, like normal siblings).

The third cluster shows the amazing finding of Bouchard's survey: identical twins reared apart had a strong correlation between their political orientations; but the scores of fraternal twins raised separately didn't correlate significantly.[3] These results suggest that genetics plays a decisive role in determining political attitudes. In other words, identical twins are more likely than fraternal twins to agree on divisive issues, precisely because they're more closely related to one another.

To make this idea more concrete, let's consider a few specific hot-button issues: "capitalism," "segregation," and "immigration." These three words come from the twenty-eight items on the Wilson-Patterson Conservatism Scale, which political scientist John Alford has researched at Rice University. These terms are guaranteed to set off the hidden triggers of political polarization. In the first case, conflicting attitudes toward "capitalism" are essentially what have divided Tea Partiers from Occupiers. Why? Because unfettered free markets reflect and create inequalities, to which people have varying levels of tolerance, depending on their political personalities. Words like "segregation" and "immigration," on the other hand, evoke opposing attitudes toward tribalism—another of the three universal personality clusters.

In 2005, Alford asked nine thousand identical and fraternal twins to agree, disagree, or express their uncertainty toward these three words and twenty-five others like them. Positive responses to half of the twenty-eight items raise one's conservatism score, and positive responses to the other half lower one's score. Negative reactions do the opposite.

Alford's twins considered these terms and wrote down their answers. Amazingly, for every single item, the identical twins' political orientations correlated more strongly than they did for the fraternal twins. And in every case, the difference in strength was significant.[4] Alford's experiment and others like it consistently show that 40 to 60 percent of the variation in our political attitudes comes from genetic differences between individuals.[5] The remaining differences come from environmental factors.[b]

Alford's findings suggest that the Blocks's nursery-school children weren't inevitably predestined to become Republicans or Democrats. Nonetheless, their adult opinions on controversial issues ran even deeper than their childhood personality traits; a substantial proportion of their political dispositions stemmed from their genetic makeup.

It's only a matter of time before scientists identify specific parts of our genetic makeup that may influence political orientation. In fact, they're already beginning to do so: in 2011 researchers from the Virginia Institute for Psychiatric and Behavioral Genetics joined hands with colleagues from the Queensland Institute of Medical Research, and they took DNA samples from over thirteen thousand individuals who had taken the same kind of liberalism-conservatism test given by Alford. Then, the scientists conducted the first genome-wide linkage analysis to search for genes that correlated with political attitudes. The genomic regions they identified accounted for up to 13 percent of the variation in their subjects' political orientation.[6] Chief among these candidates was a gene on Chromosome 4 for a neurotransmitter receptor called NARG1. This particular receptor binds an amino acid derivative called NMDA. And NMDA has been associated with fear conditioning, as well as prosocial, antisocial, and aggressive behaviors.[7]

Including this work, six published studies thus far have explored over a dozen genes that may affect political orientation. And additional articles are currently in press. The early study mentioned here is undoubtedly an important milestone. But based on current trends in genetics, it's likely that future research using more powerful techniques (such as genome-wide association studies) and sample sizes ten to twenty times larger, will soon make much more definitive discoveries.

It's important to bear in mind that a complex trait like political orientation will undoubtedly be influenced by an extremely large number of genes. Although a tremendous amount of work remains in order to find and understand them, we'll have a chance in the coming chapters to learn about a few specific cases where we can already see the influence of genes on political behavior.

Because our political attitudes have such deep roots, we don't normally undergo radical ideological shifts later in life. As people pass through their thirties, forties, fifties, and sixties, their identities stay quite stable; when measured from one decade to the next, the average individual's political orientation correlates very strongly with itself (at .80).[8] Political attitudes *do* in fact change over time. But they change at predictable ages, and in foreseeable directions. We'll learn much more about this natural tendency in chapters 18 and 21.

THE UNIVERSALITY OF LEFT AND RIGHT

When we talk about our specific political convictions, we think of them as lying somewhere on a spectrum between a "left" and a "right." These terms come from the 1791 French Legislative Assembly, where monarchists sat on the right and antimonarchists sat on the left.[9] But exactly how far do these concepts transcend their origin? Are notions of left and right more relevant to Californians than they are to Tunisians? Does *everyone* have a left-right orientation?

The National Election Studies, which are the leading survey of voters in US presidential polls, have found that over three quarters of Americans feel comfortable pinpointing their views on a two-dimensional political spectrum. Over 90 percent of college students will do the same, even when they're allowed to not participate. Collecting people's liberal-conservative self-placements is useful because they correlate very strongly with their actual voting habits.[10]

How does the United States compare to the rest of the world? For the past thirty years, a global network of social scientists has researched basic beliefs across every habitable continent. Periodically, they produce a colossal report called the World Values Survey. Between 1981 and 2008, these surveys asked almost three hundred thousand people from ninety-seven countries if they were willing and able to place themselves on a left-right spectrum. Only 3 percent chose not to answer, and 18 percent chose "Don't know." But nearly eight out of ten people in the world identified with a particular political orientation.

Since the surveyors operate in many authoritarian countries, many of the nonresponders probably considered it safer *not* to express a political opinion. Indeed, people in (historically) less democratic countries are not as willing or able to tell a researcher their ideological orientation. For example, only 46 percent of Algerians would do so, compared with 77 percent of Eastern Europeans, and 88 percent of Western Europeans.[11]

Of the great majority of people around the world who *will* express an ideology to a surveyor, we find an intriguing pattern in their responses. The World Values Survey always asks: "In political matters, people talk of 'the left' and 'the right.' How would you place your views on this [ten-point] scale, generally speaking?" Figure 2 shows how over a quarter million people answered this question.

This chart approximates a bell-shaped curve.[c] Measures of our bodies (such as height, weight, and blood pressure) frequently form a similar, natural shape; graphs of income distribution, however, seldom do.

If we were to graph the political orientations of every country separately, each population wouldn't necessarily produce the same exact left-right curve on the same segment of an absolute scale. Indeed, public opinion in certain countries skews further to the right (as in Communist Vietnam) or to the left (as in Mugabe's Zimbabwe). In addition, some countries have narrower or broader spectrums; we'll learn what determines a spectrum's breadth in the next chapter. The important point, though, is this: ordinary people everywhere use the concepts of "left" and "right" to describe their political orientations. And the experts agree with them about the existence of this spectrum. A survey once asked fifteen hundred professional political analysts if they could easily place political parties on a left-right scale, in their forty-seven countries of specialization. Virtually *none of them* had any problem in doing so.[12]

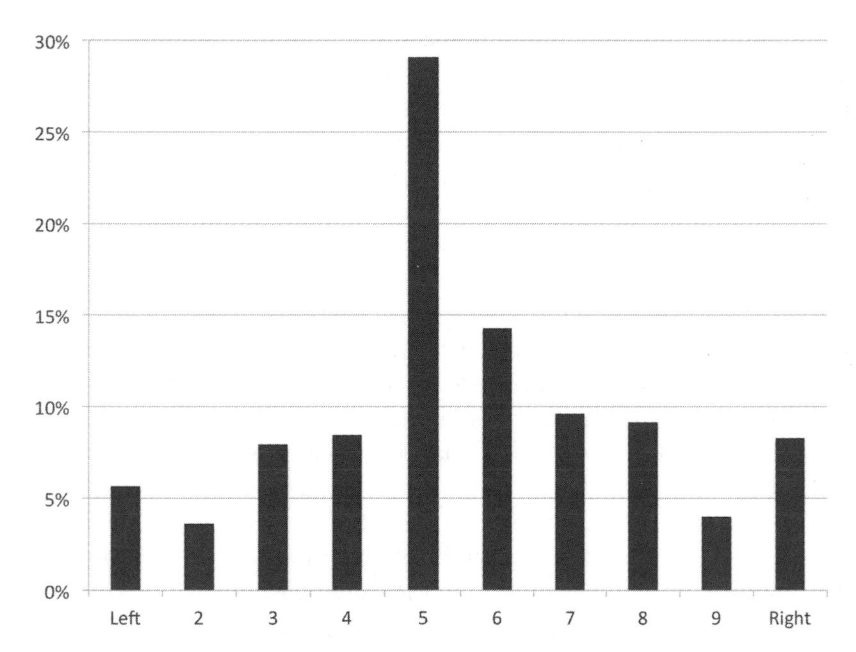

Figure 2. Global Self-Positioning on the Left-Right Political Spectrum (251,724 people in 97 countries, 1981–2008).

So the left-right political spectrum is universal. It forms a natural, bell-shaped curve. Some countries have wider or narrower spectrums. But how can we know whether the left-right orientations of groups as a whole compare with one another? In other words, are the Blocks's children as conservative or as liberal as their counterparts in Tunisia?

In psychology, there are a few standard personality traits that have been measured across truly diverse human groups. They belong to a well-tested and widely accepted inventory called the "Big Five" personality dimensions. Specifically, these traits are Openness, Conscientiousness, Extraversion, Agreeableness, and Neuroticism (it's easy to remember them because they spell out OCEAN). The first three dimensions (O, C, and—to a lesser degree—E) correlate fairly well with left-right voting. Therefore, these traits can serve as a universal yardstick for measuring the dispositions of disparate cultures, which would otherwise be difficult to compare in reference to specific political issues.

Psychologist Robert McCrae, with the help of his colleagues from numerous countries, has collected measures of these Big Five dimensions from nearly twenty-eight thousand people of thirty-six distinct cultures around the world. The participants represented the Indo-European linguistic family, as well as the Uralic (Finland, Hungary, Estonia, etc.), Dravidian (South India), Altaic (Turkic, Mongolic, etc.), Malayo-Polynesian, Sino-Tibetan, and Bantu (Sub-Saharan Africa) ethno-linguistic groups.

Variation in these Big Five personality traits was greatest *within* cultures. This finding makes intuitive sense, since a given population has a bell-shaped distribution of left-right political orientation. Moreover, Big Five dimensions such as Openness and Conscientiousness also form bell-shaped curves within a population.

In addition to the large variation within cultures, however, McCrae, together with his colleague Jüri Allik, also discovered small variations in personality traits *between* groups. Conscientiousness, which predicts conservative voting, moderately correlated with proximity to the equator. And Extraversion, which is associated with liberal voting, correlated strongly with greater distance away from the equator.

Temperature also mattered. Hotter climates correlated strongly with Conscientiousness. So even though a political spectrum runs through all populations, groups closer to the equator (and in hotter environments) have a more conservative average group disposition than groups living further away from the equator (and in cooler climes).

What happened when groups of very different genetic backgrounds live in the same environment? In this case, each group's average personality scores differed according to the *origin of their ancestors*. For example, the personality traits of white South Africans clustered closer to the Swiss, while black South Africans had personalities more similar to Zimbabweans. Likewise, groups that have traditionally lived in geographically adjacent territories have more similar average personalities than groups separated by large distances.

There is a chance that these facts could be explained by culture or by the direct effect of the environment. However, they suggest that populations around the world have personality distributions that are genetically adapted to their ancestral environments.[13] We'll unravel this mystery much further in chapter 10.

POLITICAL ORIENTATIONS IN THE BRAIN: WHAT AN MRI CAN SHOW

A skeptic might question whether it's really possible to study something as fuzzy and subjective as personalities in a scientific way. Is there any concrete, physiological evidence that could explain the apparent differences in our political personalities?

On a BBC program in 2010, the Oscar-bound actor Colin Firth laid down a tongue-in-cheek challenge. He asked scientists to find out what was "biologically wrong" with people who didn't agree with him on political matters.[14] In response, researchers at University College London recruited ninety students, and had them confidentially place themselves on a five-point political spectrum, just as the Blocks's twenty-four-year-olds had done. Their choices could range from "very conservative" to "very liberal." Then neuroscientist Geraint Rees used magnetic resonance imaging (MRIs) to scan each of their brains.

The results were stunning. From the MRIs, the scientists were able to accurately predict which of those individuals was more likely to be a liberal or a conservative. The more conservative students had a larger right amygdala; greater liberalism, on the other hand, was associated with a larger anterior cingulate. Figure 3 shows the correlations between political self-placement and the volume of these brain regions.

Someone who only had the measurements of these two brain regions would be able to correctly guess whether an individual was "conservative" or "very liberal" about 72 percent of the time (no student identified as "very conservative"). Aside from the right amygdala and the anterior cingulate, no other regions showed a significant and independent correlation with political orientation. An additional study later replicated the same findings.

The researchers noted that the amygdala has an emotion-processing function, which could explain why conservatives are more sensitive to threatening facial expressions than liberals are.[15] As we'll discover in chapter 19, this variation in the amygdala likely corresponds to differences in our *perceptions of human nature*, and these perceptions constitute one of the three personality clusters that underlie political orientation.

In the course of this book, we'll learn about fascinating findings from neuroscience, and other physiological differences between liberals and conservatives.

Some of these studies, such as this one, cannot determine whether political attitudes originate in the brain, or whether learning attitudes from others changes the structure of the brain (we know that the brain *can* change through training and gaining new skills). But other experiments *do* clarify the direction of the causation.

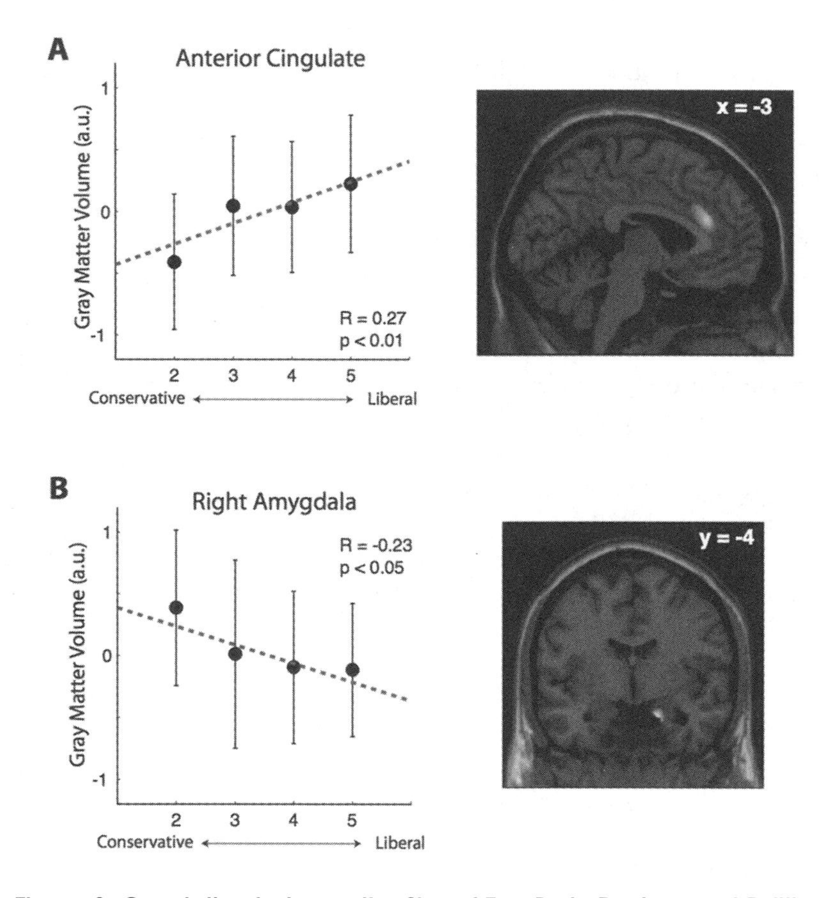

Figure 3. Correlation between the Size of Two Brain Regions and Political Orientation.

FINDING OURSELVES IN THE BIG PICTURE

All right, so perhaps there are physiological differences between liberals and conservatives—although we'll definitely want to learn more to be convinced.

Does this mean that it's possible to think about political conflict from a biological perspective? Surely politics is a uniquely human phenomenon that our species has developed long after we evolved fully modern "hardware," right?

Not so, according to one of the world's most renowned primatologists. Academy of Sciences member Frans de Waal has written a book called *Chimpanzee Politics: Power and Sex among Apes*. His book describes the complex political life of our closest living relatives. De Waal recounts in detail how chimpanzees form alliances between small groups of high-status individuals. And he tells of their betrayals and realignments, and of the brutal murders of chimpanzees fighting for social status (and mating opportunities) within a troop.

In one case, political rivals went far beyond simply killing an alpha male named Luit; they even ripped off Luit's toes and fingernails, and pounded out his testicles. Most chimp violence occurs between males, however females indirectly support male candidates for high positions by cheering and even intervening during conflicts.[16]

In addition to rudimentary political factions within their troops, chimpanzees have personalities that are meaningfully similar to ours. Groups of human caretakers have separately scored large numbers of chimpanzees on the Big Five factors, and their independent ratings of each dimension coincide with each other for individual apes.

The importance of this discovery is not just that chimps have variable personalities; rather, it's that they are the only animals known to have traits analogous to *all* of our Big Five factors (in addition to a sixth factor of their own called Dominance). Biologists have found personality traits similar to Agreeableness, Neuroticism, and Extraversion across some twenty primates and other mammals. But Openness is less common; for example, it's only partially present in the semi-social orangutan. And full Conscientiousness has only been found so far in humans and chimpanzees.[17]

Openness and Conscientiousness, as we've just learned, are the personality dimensions that best correlate with left-right voting in human beings. But before we focus on particular personality traits and their evolutionary roots, it's vital to make sure that we aren't overlooking shallower conflicts that could more easily explain our political battles. What about simple self-interest? Does your income bracket explain how you feel about paying taxes? Does having or lacking health insurance determine what you think about President Obama's controversial healthcare-reform bill? Economics does in fact influence politics, as we'll see in the next chapter—just not in the way that most people expect!

3.

WHAT ECONOMICS CAN AND CANNOT PREDICT

Around the time of the 2010 mid-term elections, President Obama's healthcare-reform bill remained an extremely divisive political issue. The British Broadcasting Corporation, commenting on the controversy from the perspective of a country with socialized medicine, couldn't understand why so many potential beneficiaries of the Patient Protection and Affordable Care Act so strongly rejected it. One perplexed BBC reporter wrote:

> It is striking that the people who most dislike the whole idea of healthcare reform—the ones who think it is socialist, godless, a step on the road to a police state—are often the ones it seems designed to help. In Texas, where barely two-thirds of the population have full health insurance and over a fifth of all children have no cover at all, opposition to the legislation is currently running at 87 percent.[1]

In a country where a majority of personal bankruptcies involve medical bills, healthcare reform seems like it would divide Americans along socio-economic lines. But it doesn't.

In this chapter we'll take a deeper look at how economics *actually* influences political attitudes. We'll also test how well economics can compete with personality traits in predicting political orientation.

VOTING AND SELF-INTEREST IN THE UNITED STATES

In general, do Americans vote according to their self-interest on economic issues? No. Although there are substantially more poor people than rich ones in the United States, every National Election Study has shown that the poor are no more likely than the rich to vote in favor of redistributive economic policies that would clearly serve their financial interests.[2] To put the irony of this in perspective, consider that the richest 1 percent of the country owns 42 percent of America's financial wealth, while the bottom 80 percent of the country owns less than 5 percent (as measured in 2010).[3]

On the other side of the coin, many wealthy individuals support redistributive economic policies that *reduce* their bank accounts. Americans sometimes refer to such people as latte liberals or limousine liberals (the British call them champagne socialists, and in Peru they've been called twenty-four-karat liberals).

American journalist and historian Thomas Frank wrote about this paradox in his 2004 bestseller *What's the Matter with Kansas?* Frank traces the history of his home state, which turned from a hotbed of left-wing activism at the end of the nineteenth century into a staunch supporter of right-wing populism by the end of the twentieth century—with most Kansans apparently voting against their economic interests. Frank weaves a well-told narrative in which mid-twentieth-century American politics once centered around economic and class issues. However, the socially tumultuous cultural revolution of the late 1960s, with its sex, drugs, and rock and roll, eventually gave way to a "Great Backlash": in reaction to the rapid liberalization of cultural norms, right-wing elites used explosive social issues such as abortion and sacrilegious art to gain popular support for a platform that ultimately benefited the extremely wealthy at the expense of the poor.[4]

Are economically irrational politics truly a new phenomenon of the Great Plains and red-state America? And how does modern Kansas compare with other trends in US history? According to Harvard economists Edward Glaeser and Bryce Ward, moral issues have actually played the central role in motivating political trends across the country's history—much more so than any economic logic. For example, the main issue that bound together late nineteenth-century Republicans was the Democrats' "unholy" trinity of "Rum [alcoholic indulgence], Romanism [Catholicism], and Rebellion." In more recent decades, Republicans have been concerned about "Guns [preserving the right to bear arms], God [keeping Christian morality in politics], and Gays [opposition to the civil rights of homosexuals]." These three "G"s served as the 1994 campaign slogan of Jim Inhofe, whom Oklahomans elected to represent them in the US Senate. In America, religion easily out-competes income in predicting Republicanism, Glaeser and Ward conclude, and "economic opinions don't appear to respond to economic interests."[5]

The two Harvard economists agree with Frank that there was a time in the middle of the twentieth century when politics centered on economic issues; Glaeser and Ward, however, consider this brief rational era to be the "true aberration" of American history.[6] But even though there was a short time when economic issues stood at the center of American politics, people's economic status wasn't driving their views. We know this thanks to Harvard political psychologist Daniel Levinson, who published a key study in 1950 in which he reported that "there is no simple relation between income and ethnocentrism."[7] As we'll learn in chapter 6, ethnocentrism is a core element of political conservatism.

Today's top political scientists definitely agree with Levinson that financial self-interest fares poorly in predicting political affiliation. UCLA's David Sears, for one, has worked extensively to understand how people make decisions about economic policy issues and pocketbook voting. "The conclusion," he writes, "is quite clear: self-interest ordinarily does not have much effect upon the ordinary citizen's sociopolitical attitudes." Sears does find occasional exceptions, but even then, the effects of self-interest tend to be small and unreliable.[8]

Even noneconomic forms of self-interest barely affect Americans' political preferences. For instance, whether or not affirmative action or busing to desegregate school districts directly impacts an individual has little to no influence on the individual's support or rejection of these policies.[9]

In this respect, the United States is not a political anomaly. In other ethnically diverse countries such as India, ethnocentrism easily outweighs income in predicting political affiliation. In fact, it takes an unusually secular and homogeneous country like Sweden for income levels to correlate with political affiliation better than religion does.[10]

THE ECONOMICS OF POLITICAL EXTREMISM AND MODERATION IN LARGE POPULATIONS

In most environments, people's economic self-interests do not cause them to espouse liberal or conservative political views. That is, economic data cannot readily "see" the difference between left and right on the spectrum, and people vote against their economic interest as often as not. Yet when we think about socio-economic class, we have an intuition that it must make a difference in politics. It does make a difference—just not necessarily in the way that most people think it does.

What economics *does* predict is the level of political moderation and extremism in large populations. Extremism, of course, comes in the two basic flavors of left and right. Although the middle and upper classes often blame the poor for supporting the "wrong" side of the spectrum, a nation's poverty actually nurtures *both* of the extremes.

Political Extremism and Moderation in Peru

Forecasting Peruvian elections is notoriously difficult for many reasons. One of the only reasonable a priori assumptions to make is which social classes are more likely to support moderate candidates, and which social classes tend to support more extreme ideologies.

Let's look at which socio-economic quintiles of the Peruvian people supported each type of candidate in the run-up to the 2011 presidential election. Figure 4 shows the results of a national, urban survey.[11] The letters A through E represent the five socio-economic classes of Peru, with A representing the wealthiest, and E representing the poorest.

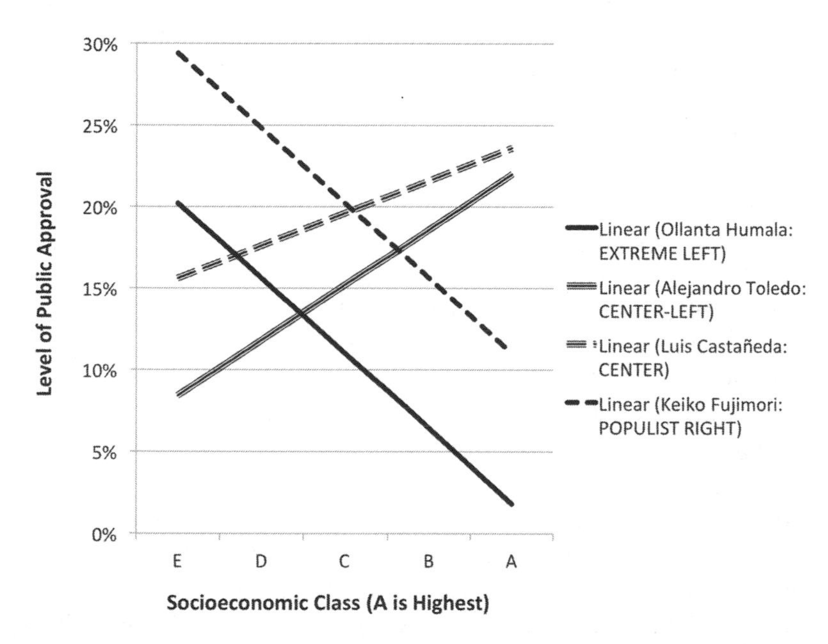

Figure 4. Percentage of Support for Presidential Candidates in Peru, by Socio-Economic Class (A is highest) and Candidates' Ideology (trend lines calculated in July 2010).

The two radical candidates on the extreme left and right of the ideological spectrum are shown by the two lines running "downhill" from left to right. The solid black line on the bottom represents the candidate on the far left, who was Lieutenant Colonel Ollanta Humala. Humala's father was a card-carrying member of Red Fatherland, the Communist Party of Peru. The candidate Humala had also aligned himself with the far-left Venezuelan leader Hugo Chavez. The downhill line on top indicates the approval ratings of the candidate on the populist far right, Keiko Fujimori. Peruvians associated Keiko with the dictatorship of her father, who dismantled democratic institutions during an internal war against left-wing terrorist groups like the Shining Path.

Despite coming from opposite ends of the spectrum, both Humala and Fujimori found their highest levels of support among the poorest people; in contrast, the centrist candidates on both sides (shown by the trend lines running "uphill" from left to right) appealed most to the richest voters. Therefore, it's not surprising that Keiko celebrated the launch of her candidacy by dancing to "popular" music in one of Lima's shantytowns; the center-left candidate Alejandro Toledo, on the other hand, announced his candidacy through Twitter.

In Peru, income quintiles can clearly distinguish extremist ideologies (the downhill, negative slopes) from moderate ideologies (the uphill, positive slopes). But economics cannot determine left from right.

Political Extremism and Moderation in Latin America

Just as the richer segments of Peruvian society tend to vote for more moderate candidates, the growing wealth of Latin America over the past decade has expanded political moderation within the region. Between 2002 and 2008, Latin America's real gross domestic product (GDP) per capita increased by almost 19 percent; over the same period, the percent of Latin Americans who identified as centrists increased by 13 points, up to 42 percent. In other words, every 1.46-percent increase in real GDP per capita bought a 1 percent increase in political moderation.

Although ideological policy differences still abound to the left and the right, the breadth of Latin American political spectrums has shrunk with the swelling of the middle class. Michael Shifter, the president of the Inter-American Dialogue think tank, has commented that predictability and pragmatism have also increased during this period—especially in Colombia, Chile, and Brazil. In the case of Brazil (which is by far the largest country in the region), almost a sixth of its population escaped from poverty between 2003 and 2009![12]

On the other hand, the popularity of the then de facto leader of the hemisphere's extreme left, Hugo Chavez, fell to a seven-year low by 2010. Chavez's popularity sunk largely because of his mismanagement of the Venezuelan economy—at the same time as the region's market economies were growing.[13]

Latin America's economic rise appears to have dampened enthusiasm for political extremism. As prosperity has lifted people into the middle class, it has pushed them toward the center of the political spectrum. However, knowing the socio-economic status of a Latin American tells us absolutely nothing about whether she is more likely to support a liberal or a conservative party. Very solid numbers back up this assertion: in 2008, AmericasBarometer carried out an eighteen-country survey that interviewed nearly thirty thousand people across Latin America. Social class had *no significant effect* on whether or not people voted for the left.[14]

Political Extremism and Moderation at the Global Level

How do Peru and Latin America fit into the global picture? A couple of Scandinavian economists have made it quite easy for us to grasp the worldwide relationship between political moderation and wealth. Remember the ten-point political spectrum used by the World Values Survey (in figure 2)? Carl-Johan Dalgaard and Ola Olsson calculated the percentage of respondents in each country who *did not* place themselves at the very extremes of the political spectrum (anyone who did not answer "1" or "10"). They called this proportion of moderates "political cohesion."

Then Dalgaard and Olsson compared this measure of political moderation with the GDP per capita of seventy-one countries. The two measures strongly correlated with one another (figure 5 shows what this .67 correlation looks like visually). In the average country, 84 percent of people do not select "1" or "10"; in the high-income OECD countries, the proportion of political moderation jumps up to 93 percent.[15]

The Dutch political scientist Jos Meloen has also found a similar relationship between GDP per capita and measures of "state authoritarianism." As a country grows richer, not only does public opinion become more moderate, but there is also a lower chance that the country will have an authoritarian government.[16]

The relationship between wealth and political moderation makes intuitive sense in historical perspective. Nearly two thousand five hundred years ago, Aristotle proposed that a greater distribution of wealth improves the likelihood of democracy. Most historians of the ancient world have cited unsustainable economic disparities as a key cause of the decline in Rome's democracy—and the fall of the Republic.[17]

The Great Depression of the twentieth century provides a modern, textbook example of how a severe downturn ferments political extremism. As the German economy plummeted between 1929 and 1932, the ultra-right-wing Nazi party increased its votes from eight hundred thousand to 17 million. Political scientist John Jost points out, however, that the very same economic depression spawned a significant *left-wing* movement in the United States, led by Franklin

D. Roosevelt.[18] It's vital to note that Communism also formed an important part of the fractured political spectrum that the Nazis eventually monopolized. Likewise, the economic shock generated some fascist activity in the United States and England; nevertheless, neither country fell to a Hitler, even though America suffered higher unemployment than Germany in 1932.[19]

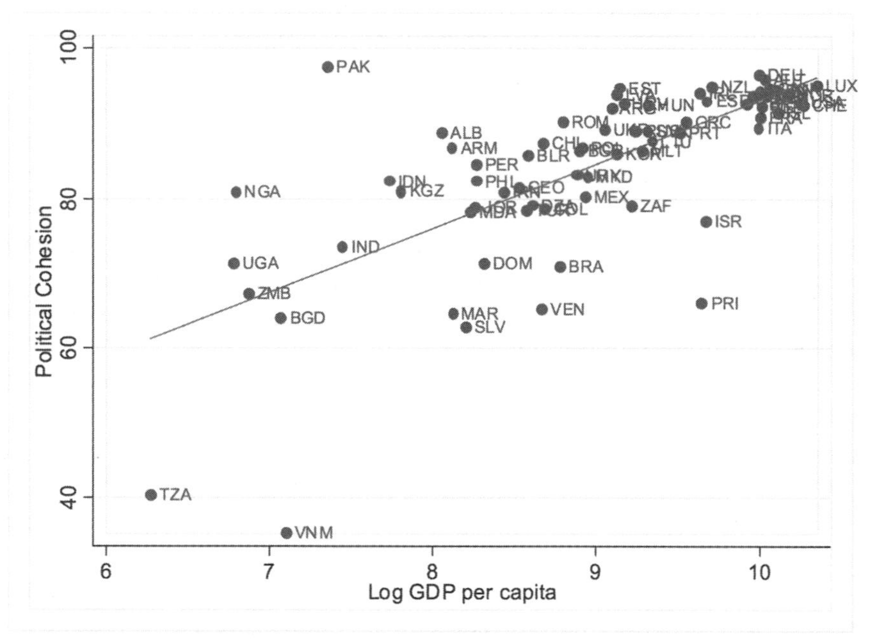

Figure 5. Correlation between Political Moderation and GDP per Capita for Seventy-One Countries (1981–2000).

The global economic recession of 2008 also broadened the American political spectrum in both directions. On the left, a forty-seven-year-old Barack Obama ran on a platform of "change." And the American electorate—of which less than 13 percent was black—elected the country's first black president. In response to Obama's election, the new Tea Party movement emerged, further to the right of the mainstream Republican Party. And then by 2011, Occupy Wall Street emerged on the far left.

Why do people under greater economic pressure tolerate more extreme ideologies? Perhaps having less financial freedom pushes people to look for help and answers beyond themselves, making the notion of strong political authority more palatable. Indeed, the results of a Latino National Political Survey taken at the turn of the 1990s seem to support this hypothesis. As poverty increased among the Mexicans, Puerto Ricans, and Cubans surveyed, so did their faith that "government officials could be trusted to do what is right."[20]

Based on these findings, one could speculate that greater portions of the respondents in the lowest income brackets would be more susceptible than the middle and upper classes to trust strong authorities. However, we shouldn't expect income to predict leftist political preferences from rightist leanings.

THE INCOME AND PERSONALITIES OF RED STATES AND BLUE STATES

Although wealth can predict moderation and extremism in large groups, it is all too often useless in understanding left-right political attitudes. So what *can* outperform economic data in predicting liberal versus conservative voting?

Take a look at the measurements displayed in figure 6. They come from three consecutive US presidential elections: (1) Bill Clinton vs. Bob Dole in 1996, (2) Al Gore vs. George W. Bush in 2000, (3) and John Kerry vs. George W. Bush in 2004. For each election, Cambridge psychologist Peter Rentfrow gathered measurements of personality traits and family income from approximately two hundred fifty thousand diverse Americans from every state. Then he took the median family income and the average personality traits for each state and fed them into a regression model to predict votes for Democratic candidates (shown by the striped bars) and votes for Republican candidates (shown in shades of grey).

The size of each bar shows how strong the relationship was. When a black line outlines a bar, it indicates that the relationship was significant.[a] Let's start with median family income. Since none of these six bars is outlined, there was no significant relationship between median family income and voting for the Democratic or the Republican candidates over the three elections. There was a slight tendency for the rich to vote Republican, and the poor to vote Democrat; however, there's a greater than 5 percent possibility that this measurement occurred by chance. For a sample of a quarter million people (which is one hundred times greater than a large poll), these economic results aren't impressive or useful.

In stark contrast, the relationship between Openness and left-right voting was about seven times greater. Moreover, the finding that more Open states voted liberally and less Open states voted conservatively was extremely significant (the odds are less than one in one thousand that this strong trend occurred by chance). Conscientiousness also significantly predicted left-right voting in each election, to varying degrees.[21]

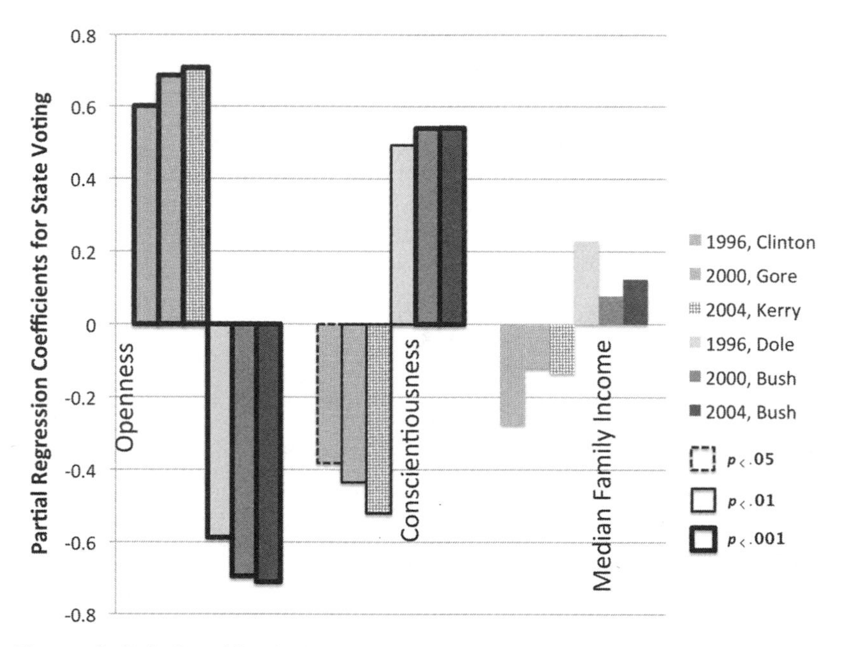

Figure 6. Relationships between Voting Records and Personality Traits vs. Median Family Income (calculated for each US state for the 1996, 2000, and 2004 elections).

Finally, there is something we can measure that's able to distinguish between left and right—at least across three American presidential elections. These two personality traits belong to a well-known model in psychology called "the Big Five" factors. In addition to Openness and Conscientiousness, they consist of Extraversion, Agreeableness, and Neuroticism. Where did these traits come from?

The Big Five actually began as the number 17,953. In the 1930s, psychologist Gordon Allport was busy researching personality. He was dissatisfied with psychoanalytic models, which he viewed as too "deep"; yet behavioral models were too "shallow" for his taste. Instead, Allport hoped to use an empirical approach to discover the root factors underlying personality. The figure 17,953 was the number of personality-describing words that he and his collaborators were able to find in the English language. They later reduced this figure to some 4,500 words that corresponded to permanent, observable traits.

Over the years, more and more psychologists began working on the problem. They conducted personality tests, and used computers to perform factor analyses for boiling down the interacting parts to their underlying structure. United States Air Force researchers conducted a meta-analysis of multiple samples, and were able to distill everything down to five personality factors. Since then, many social scientists have replicated the results and confirmed the five factors across diverse world cultures. These five personality factors correlate with behavioral differences and gender. Various studies in Europe and North America have also determined that the Big Five dimensions have a genetic component. For example, the average heritability of Openness is 57 percent; that of Conscientiousness is 49 percent.[22]

The discovery of these five traits took place through a tremendously empirical, data-driven process. In fact, social scientists still don't even know *why* the five traits exist—only that they do. The important point is that the Big Five represent significant facets of the human personality.

Although the Big Five were developed completely outside of political psychology, figure 6 shows that two of these personality traits predict left-right voting far better than an economic variable commonly used by pollsters to stratify their samples (Extraversion works better than income as well; like Openness, Extraversion usually predicts leftist voting, but the effect is weaker and less predictable).

What is Openness? Rentfrow's findings mean that people in "blue states," on average, are more creative, imaginative, curious, have many interests, and are more tolerant of differences than their average counterparts in "red states." Conscientiousness, in which red states score higher, entails reliability, organization, dependability, self-discipline, orientation to achievement, and adherence to tradition and rules.[23]

Political scientists agree that Openness is the Big Five dimension that most powerfully predicts left-right voting. John Jost and his colleagues have found that Open people tend to vote liberally in Australia and the United States.[24]

How do the Big Five fare in predicting an election in a non-Anglo country? In the general Italian election of 2001, Silvio Berlusconi ran for prime minister with the center-right House of Freedoms party. His rival, Francesco Rutelli, represented the center-left Olive Tree party. Again, voters who scored higher in Openness and lower in Conscientiousness were much more likely to vote for the center-left party. And vice versa. The typical demographic strata of political polls—income, education, age, and gender—hardly differed between supporters of the Olive Tree and the House of Freedoms.[25]

In South Korea, Openness correlates with participation in political protests. But South Koreans high in Conscientiousness are less likely to do so (with the same being true in Uruguay and Venezuela).[26]

We've reached a point where personality factors have overtaken economic indicators in explaining the liberal-conservative spectrum running through public opinion. This conclusion makes sense: Openness and Conscientiousness have a bell-shaped (normal) distribution, just like left-right political orientation; most economic variables, in contrast, don't (especially not income distribution).

We'll refer back, from time to time, to the two political elements of the Big Five personality dimensions, because they serve as convenient universal reference points. But the Big Five are the most general of personality dimensions, and we have distinctly political questions to answer. From here, we must explore the controversial origins of scientific research on the *political* personality.

4.

THE INVENTION OF THE POLITICAL LITMUS TEST

DEFEATED NAZIS AND "THE AUTHORITARIAN PERSONALITY"

Interest in political extremism peaked with the rise and fall of Nazi Germany. During and after World War II, a group of psychologists at the University of California, Berkeley, wanted to know why some individuals readily supported Nazism, and why others objected. After analyzing the speeches of anti-Semitic agitators, and interviewing highly ethnocentric subjects, the research group hypothesized that a constellation of nine personality traits predisposed individuals to favor fascism.[1]

As for theory, the Berkeley researchers' approach was influenced by the dominant psychological paradigm of the era: psychoanalysis. As such, they attributed the fascist personality to punitive parenting. Repressed hatred toward harsh parents, they believed, was displaced onto scapegoat out-groups. Thus, the famous book that they published in 1950 was titled *The Authoritarian Personality*.

Although today the word "authoritarian" refers to governments and ideologies on both the extreme right *and* the extreme left of the political spectrum, the Berkeley professors used this term for—and focused on—only the extreme right. We'll use the word "authoritarian" in the Berkeley sense of the word throughout this book only when referring to psychological tests.

The Berkeley researchers claimed that nine traits defined the authoritarian (i.e., the radical right-wing) personality. Seven of these nine traits are fairly straightforward:

(1) Conventionalism (a rigid adherence to conventional values);
(2) Submission to authority figures;
(3) Aggressiveness (especially toward unconventional members of out-groups);

(4) Superstition (fatalistic belief in mystical determinants of the future) and "stereotypy" (rigid categorical thinking);
(5) Power and toughness;
(6) Destructiveness and cynicism; and
(7) Sex (an excessive concern with the sexual activity of others and violations of sexual norms).

Two of the nine traits, however, were especially psychoanalytic and a bit harder to grasp:

(8) "Projectivity" (the outward projection of unacceptable impulses, such as sexual content); and
(9) "Anti-intraception" (a disapproval of emotionality, imagination, intellectualism, and subjectivity)[2]

The Berkeley group created a series of statements designed to test each of the nine traits. For example, statements about sex included: "*The wild sex life of the old Greeks and Romans was tame compared to some of the goings-on in this country, even in places where people might least expect it*," and "*Homosexuals are hardly better than criminals and ought to be severely punished.*"[3]

The long list of items became known as the F-scale. "F" stood for pre-Fascist, since those who scored high on the test ostensibly had the potential to develop into a full-blown fascist. Together with the F-scale, the Berkeley group developed an Anti-Semitism scale, an Ethnocentrism scale, and a Politico-Economic Conservatism scale.

The Authoritarian Personality, a massive thousand-page tome, detailed the results of these tests, taken by a wide sample of the American public. The respondents were university students, members of service clubs (Rotary, Kiwanis, Lions), psychiatric patients, inmates, military officers, veterans, various types of working-class union members, PTA members, and church members.[4]

The book is filled with countless correlations between scores on the various scales and demographic data. Most impressively, however, the F-scale in particular successfully predicted various types of *known* political extremists. For instance, volunteer activists who had joined the Nazi party before Hitler's rise to power scored significantly higher on the F-scale than did rank-and-file members of the German Armed Forces.[5] Former members of the SS (Hitler's paramilitary organization that perpetrated the worst atrocities) also scored significantly higher than men who'd been enlisted in the regular army.[6]

In the early 1970s, Yugoslavian researchers Rot and Havelka used the F-scale to test high-school students in a couple of Serbian cities. The students scored so highly that the researchers wrote: "It is difficult to accept that these results signify . . . subjects' extraordinary readiness to accept an antidemocratic and even fascist ideology." A decade and a half later, Serbian leader Slobodan Milošević and his regime committed the worst war crimes in Europe since the Nazis.[7]

On the other end of the spectrum, Communists in the United States and England scored lower than average on the F-scale.[8]

The F-scale seemed to produce a magic score capable of predicting a whole range of seemingly unrelated attitudes, tied together only by political ideology. In the first seven years after the publication of *The Authoritarian Personality*, at least 230 articles made reference to this study.[9] A handful of books were even written all about the research process itself, such as *Studies in the Scope and Method of "The Authoritarian Personality"* (1954).[10]

Despite its enormous success, *The Authoritarian Personality* was eventually beleaguered by criticism. In short, the work faced three kinds of problems: political, theoretical, and technical.

Hardly any of the social scientists working on political personality criticized the political bias of the F-scale. Perhaps few people called attention to its political bias because left-leaning academic psychologists and sociologists dominated the research area, or because the test's creators worried predominantly about preventing a resurgence of Nazism. But the political problem was this: the F-scale claimed to measure pre-fascism, but in reality it was measuring conservatism (and indeed, the entire political spectrum). This bias in the test's name would be equivalent to calling a test that quantifies left-leaning views a "pre-Communism" scale. To make matters worse, the Berkeley group considered "pre-fascism"—which in reality was normal conservatism—to be an undesirable personality syndrome.[11]

In addition, *The Authoritarian Personality* (*AP*) study was beset by a couple of rudimentary technical problems. For one, the authors used a "convenience sample"; this means that they simply recruited subjects from readily available groups, clubs, and organizations. As a result, all the subjects belonged to a social group; there would be no way of knowing how representative the sample was of the national population. The Berkeley psychologists likely excluded many individuals such as Groucho Marx, who "would never belong to a group that would accept someone like [him] as a member."

The other technical issue was that several of the questionnaires, including the F-scale, suffered from an "acquiescence response bias." This simply means that every time a subject answered "yes" to a question, it raised his score (and, consequently, his level of "pre-fascism"). A proper survey would have had "constraint" questions worded in the opposite direction to prevent acquiescent test-takers from producing false positives.[12]

As for the theoretical problems, the *AP* was far too complicated. The thousand-page book contained endless measurements, relationships, and tables. Sometimes items within a scale would correlate with items in another scale better than they would with each other. This mismatch threw into question the validity of the concepts that the authors claimed to measure.[13]

The last major issue that later discredited the study in the eyes of many was the difficulty in disproving the *AP*'s psychoanalytic logic. According to the authors, people who scored highly on authoritarianism grew up in homes dominated by a strict father who employed harsh physical punishment to discipline his children. The interviewers expected these children to repress these traumatic memories, and instead to over-idealize their portrayals of family life. Yet at the same time, the *AP*'s interviewers also expected negative attitudes toward punitive fathers to seep out occasionally.[14] Thus, in the study's qualitative research interviews, sometimes the Berkeley psychologists understood the subjects' responses as literal and realistic, and other times they interpreted responses in terms of unconscious conflicts (e.g., glorifying an abusive parent).

For instance, one of the study's main authors, Elsie Frenkel-Brunswik, correlated ethnocentrism with her subjects' "conception . . . of a *distant, stern father*." Her impressions of attitudes toward fathers could have easily been susceptible to subjectivism.[15] This is not to claim that Frenkel-Brunswik was necessarily wrong—merely that a reasonable skeptic would correctly judge her methodology to be problematic. To scientifically correlate adults' ethnocentrism scores with the type of childhood discipline they received, it would take a good twenty-five years of research and a substantial invasion of privacy. Not surprisingly, hardly any research at the time had linked child-rearing practices with adult political attitudes.[16] And so the skeptics "won."

Despite these flaws, the seminal research undertaken in *The Authoritarian Personality* remains as important as ever, as testified by summaries in introductory psychology textbooks.[17] Today, a Google Scholar search shows that the original book has been cited in over ten thousand publications.

Back in the 1950s, though, *The Authoritarian Personality*'s flaws eventually began to seem insurmountable; what doomed the project more than anything was the theoretical mess it created through excessive complexity, and the impossibility of disproving its psychoanalytic assertions. Researchers began to give up on the problem. The **first theoretical era of modern political psychology**, defined by the *AP* and psychoanalytic approaches to ideology, faded in the 1960s. With the entrance of a new era, very different questions were asked.

Yet the holy grail of political psychology (the origin and nature of left-right attitudes) had slipped away without full redemption.

Second Era: 1960s–1970s. During this period, the mass media emerged and dramatically changed politics. Communications theorists gained great influence in the field, and political psychologists began assuming that people's political nature was highly impressionable. A prominent group of political scientists at the University of Michigan claimed that people lacked the cognitive sophistication for public opinion to truly exist in any stable, coherent form.[a] Political psychologists shifted their focus to behavior. They researched "voting habits." Many of them even identified themselves as "behavioralists" rather than as "psychologists." Anthropologists and psychiatrists were no longer needed during this "end of ideology" era, since notions of personality, enduring political attitudes, and even human nature were in crisis.[18]

Third Era: 1980s–1990s. Then computers began to change the world. Political psychologists grew exuberantly enthusiastic to theorize about information processing. The metaphor of the political brain as a computer dominated their thinking. They wrote about information storage, memory types, inputs, and outputs. Even their flowcharts looked like logic boards. Computer scientists and cognitive scientists gained a greater share of the multidisciplinary field of political psychology. Voting decisions became merely a function of accessing data about candidates. Many information-processing models even ignored emotions![19] In addition, economists and rational-choice theorists left a large mark on the field.

When these approaches produced little fruit, a theoretical malaise set in. Political psychologists even began to openly express disheartenment.

Fourth Era: 2000s–2010s. Finally, the fourth era of political psychology has returned to the problem of ideology—back to the roots of the First Era, sans psychoanalysis. Events such as the rise of Islamist terrorism and the ideological polarization of the United States have made the original questions impossible to ignore—especially since eras two and three never advanced or supplanted the grand topic of era one.[20]

THE "RIGHT-WING AUTHORITARIANISM" TEST

One man, however, was completely out of sync with the four eras of political psychology. In 1965 Robert Altemeyer was a self-described lazy Ph.D. candidate at Carnegie Mellon University. Altemeyer, in a fateful moment, failed a question about *The Authoritarian Personality* on a test. He recalls:

> I had to write a paper to prove I could learn at least something about this research, which had gotten itself into a huge hairy mess by then. However, I got caught up in the tangle too. Thus I didn't start studying authoritarianism because I am a left-winger (I think I'm a moderate on most issues) or because I secretly hated my father. I got into it because it presented a long series of puzzles to be solved, and I love a good mystery.[21]

By the beginning of the 1980s, when political scientists were just getting hooked on computers, Bob Altemeyer had become a psychology professor at the University of Manitoba, where he was still obsessed with the original puzzle. To cut through the Gordian knot of *The Authoritarian Personality*, Altemeyer began by discarding the Berkeley school's psychoanalytic premise. Thus, unconscious struggles, repressed hatreds, and projected hostilities no longer mattered. An adult attitude toward parental or societal authority still did count, but the virtually untestable origin became irrelevant.

Next, Altemeyer sheared the Berkeley school's nine personality traits down to the three attitudinal clusters that regularly and substantially covaried. These three personality traits formed the core of the new authoritarianism, which Altemeyer renamed "Right-Wing Authoritarianism" (RWA). They were:

(1) "**Authoritarian Submission**—a high degree of submission to the authorities who are perceived to be established and legitimate in the society in which one lives;
(2) **Authoritarian Aggression**—a general aggressiveness, directed against various persons, that is perceived to be sanctioned by established authorities; and
(3) **Conventionalism**—a high degree of adherence to the social conventions that are perceived to be endorsed by society and its established authorities."[22]

In his new RWA test, Altemeyer cleaned up the technical problems that had snagged the F-scale. To avoid acquiescence, he worded an even number of questions to express opposite sentiments. Lazily answering "yes" would no longer automatically raise one's score. The RWA test also used a standard, seven-option Likert scale. All of these changes helped produce more reliable results, measuring more valid concepts, in a more unilinear way.

Unfortunately, the name of the test stuck—as did a couple of accusations of ideological bias.[23] Instead of calling it the "Right-Wing Authoritarianism" scale, Altemeyer could have called it a conservatism scale. But let's not throw out the baby with the bathwater.

Altemeyer began testing students in his introductory psychology course, and their parents. As more and more scores accumulated, he discovered that the curve formed a normal distribution: the average score was around 150 points, with a top quartile 20 points higher and a lower quartile 17 points beneath average. Altemeyer also noticed that, as the national political mood grew more conservative between 1973 and 1987, the entire curve of RWA scores shifted to the right (i.e., increased) by 14 percent.[24]

In 1986, Altemeyer's work received the American Association for the Advancement of Science's Prize for Behavioral Science Research. As members of his field heard about Altemeyer's work, numerous independent researchers around the world were able to successfully replicate his results with the RWA scale. They proved the test to be both reliable (it logged similar results consistently) and valid (meaning that respondents answered the test questions relating to each personality trait in an expectedly coherent way).

By 1996, over fifty researchers had carried out thirty-three studies of the RWA scale among tens of thousands of Canadians, Americans, Spaniards, South Africans (in Xhosa), Israelis, Palestinians, Australians, Russians, and Estonians.[25] In each case, the results showed a high degree of internal consistency, despite the diversity of cultures and languages tested. When the numbers were all in, the RWA scale had outperformed the F-scale (and five other minor competitors) in internal coherence and unidimensionality.[26] That is, the RWA scale had reached the purest core of political bias, and it had succeeded in measuring it on a single dimension with the least amount of interference from peripherally related concepts.

RWA scores also covaried with the politically charged elements of the Big Five personality traits: When RWA scores (i.e., conservativeness) rise, Conscientiousness goes up and Openness goes down.[27]

Much more importantly, however, RWA correlated brilliantly with left-right voting. Altemeyer first discovered this in Canada. On the left end of the Canadian political spectrum stood the New Democratic Party (NDP), with its socialist roots. In the middle was the Liberal Party, while the Progressive Conservatives held down the right wing. Among his sample of students and their parents, supporters of this right-wing party always had a significantly higher average RWA score.

Altemeyer then sent his RWA test to provincial legislators in Canada, requesting their anonymous participation on a survey about "a variety of social issues." Figure 7 shows the results from four provinces. Not every party was represented in each state legislature or survey responses, which is why some bars are missing. But the results were astronomically significant—indeed, stronger than the relationships between height and weight, race and income, or gender and aggression.[28]

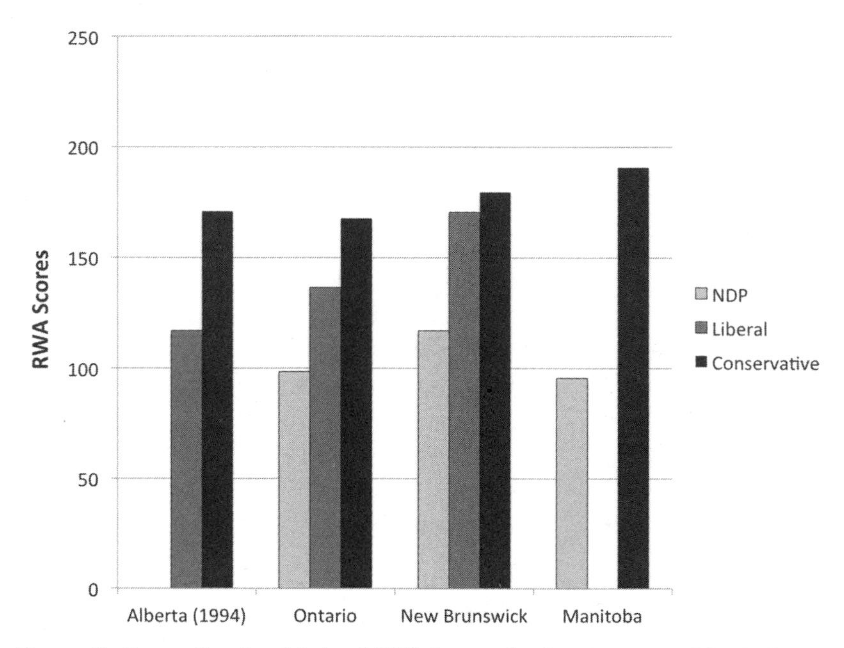

Figure 7. Canadian Legislators' RWA Scores by Provinces and Party Membership, 1983–1985 (Alberta, 1994).

In America, college students from half a dozen states that Altemeyer visited tested like the Canadians. Republicans scored significantly higher than Democrats on the RWA scale. The same held true for 1,200 politicians from forty-eight state legislatures who took Altemeyer's test between 1990 and 1993. Republicans scored higher in forty-seven out of forty-eight states (the exception was a small group of politicians from Louisiana).[29]

The RWA scale certainly seems like a useful tool for predicting left-right partisanship in the United States and Canada. But how well would the test work in an extremely different region of the world?

Gidi Rubinstein, an Israeli researcher who is leading the study of authoritarianism in his country, had the RWA scale translated into Hebrew and Arabic in the mid-1990s. Then he gave the test to 708 Jewish Israeli university students. About 29 percent of the students' parents were Ashkenazim who had returned to Israel from the Diaspora in Europe, the Americas, Australia, and South Africa; 20 percent had parents who were Sephardim or Mizrahim (from the Middle East and Asia); 28 percent had parents of mixed backgrounds; and the remaining 23 percent had parents who were native-born Israelis. Rubinstein also gave the Arabic version of the test to 120 Muslim Palestinian students in the West Bank.

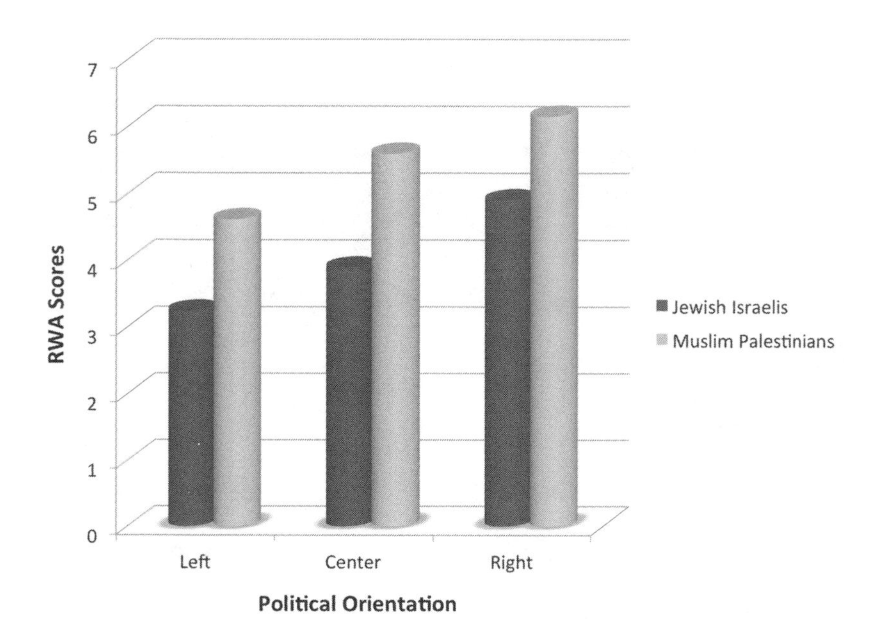

Figure 8. RWA Scores of Israeli Jews and Palestinian Muslims, by Partisanship (1996).

Figure 8 shows the results of the RWA scores, according to the ideological orientation of the political parties that the respondents supported. The Jewish Israelis with the lowest RWA scores supported left-wing parties; those with intermediate scores supported the centrist Labor Party; and the highest-scoring students were affiliated with right-wing and religious parties (including parties with a "greater Land of Israel" ideology).

On the Muslim Palestinian side, respondents with the lowest scores supported left-wing parties; intermediate scorers supported the centrist Fatah party; and those who scored highest supported Hamas and Islamic Jihad (which have "destroy Israel" ideologies). In all cases, the differences between scores were extremely significant.[30] The fact that the lighter bars reach substantially higher than the darker bars suggests that the Muslim students were much more conservative than their Jewish counterparts.

For those who may be skeptical of which parties classified as left-wing and right-wing, figure 9 shows the remarkable breakdown of Jewish Israeli RWA scores by specific political party:

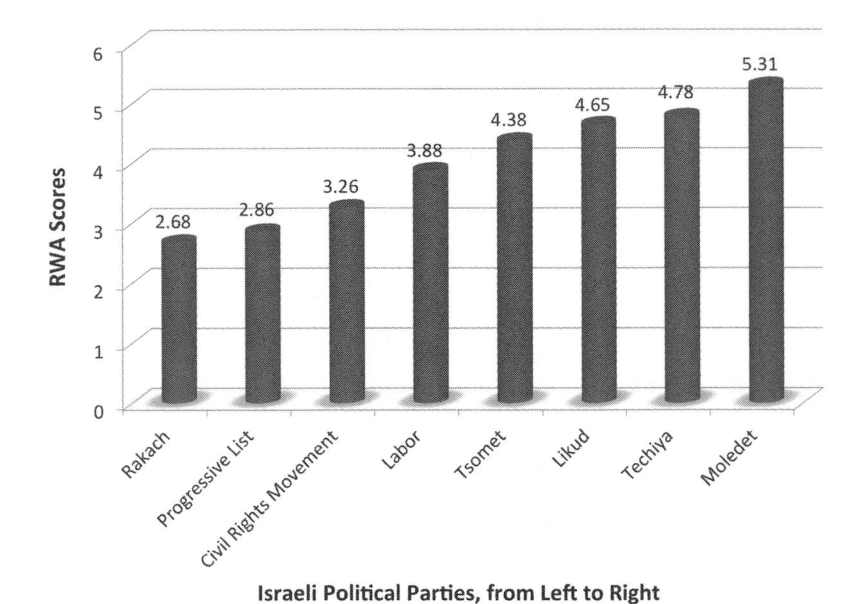

Figure 9. RWA Scores of Israeli Jews, by Specific Party Affiliation (1996).

On the far left, the lowest RWA scores belong to supporters of the Rakach Party (which became Maki, the Communist Party of Israel) and the Progressive List for Peace. Both of these parties had a mixed Jewish-Palestinian membership, and they both demanded Israel's total withdrawal to the state's 1967 boarders. Ever so slightly to the right stood the Civil Rights Movement (which later merged with Meretz), whose platform sought to grant full autonomy to Arab territories. On the mainstream center-left was the Labor Party, which advocated territorial compromise. Tsomet was a right-wing—but emphatically secular—party. To the right of Tsomet, as calculated by average RWA score, was the mainstream right-wing Likud party. Likud argued that Jewish Israelis should get along with Arabs, but without stopping the building of Jewish settlements in the West Bank and Gaza. Further to the right stood Techiya, which opposed giving up any land to Arabs. On the extreme right of the 1996 Israeli political spectrum was the small Moledet (Homeland) Party. Not only did Moledet oppose surrendering land to Arabs, but the party even advocated for the "voluntary transfer" of Arabs from the West Bank and Gaza to the neighboring Arab countries of Jordan and Egypt.[31]

RWA scores are far more effective in predicting left-right voting than many conventional demographic variables, such as income brackets or gender. Sometimes race, religion, region, or socio-economic class has some predictive power, but much less than RWA. Although it's less convenient to gather RWA scores, their superior predictive power could help to improve the accuracy of important polls.

The RWA scale works brilliantly. But how did Altemeyer himself explain why this is so? The position that Altemeyer took on this question was more an argument for how RWA does *not* work than a theory of why it does. That is, he demonstrated beyond a doubt that RWA could still work "without" psychoanalysis.

In any case, Altemeyer's key technical corrections and years of experimental refinements did produce a far better metric than the F-scale. Social scientists around the world have replicated his results and proven the test's reliability and validity. As in the case of the empirically developed Big Five factors, however, the mystery remains as to why individual differences in human personality exist.

Altemeyer's answer, which he mentions only cursorily, is that individuals acquire their political attitudes through "Social Learning Theory." This model postulates that children pick up their social attitudes not from parental discipline shaping innate inner conflicts, but rather through cognitive mimicry of their parents and other societal influences during their development. Altemeyer's RWA scale itself, however, has drawn exceedingly little—if at all—on this social-learning explanation. The improvements that he *has* implemented are actually technical ones, and the theoretical makeover is more a disavowal than a content-transforming shift.

Even if children *do* pattern their political attitudes after the influences of their family and greater environment, questions remain: Where did these attitudes come from to begin with? Why is there a normal distribution of self-placement on the political spectrum of traits like Openness and Conscientiousness, and of RWA scores? Why is it that RWA scores show such a high level of heritability—even in twins separated at birth?

Our mission is to answer these questions—to understand, in the deepest sense possible, why Altemeyer's test works so well around the world. In the following chapter, we'll break down the content of the RWA scale and lay out a roadmap for the rest of this book.

5.

UNEARTHING THE THREE ROOTS OF POLITICAL ORIENTATION

The RWA scale is the best cross-cultural predictor of left-right voting and party affiliation. The test can calculate the relative positions of political parties along the spectrums of truly diverse populations. So what specific topics does the test cover? Let's give the RWA scale a new analysis with fresh eyes.

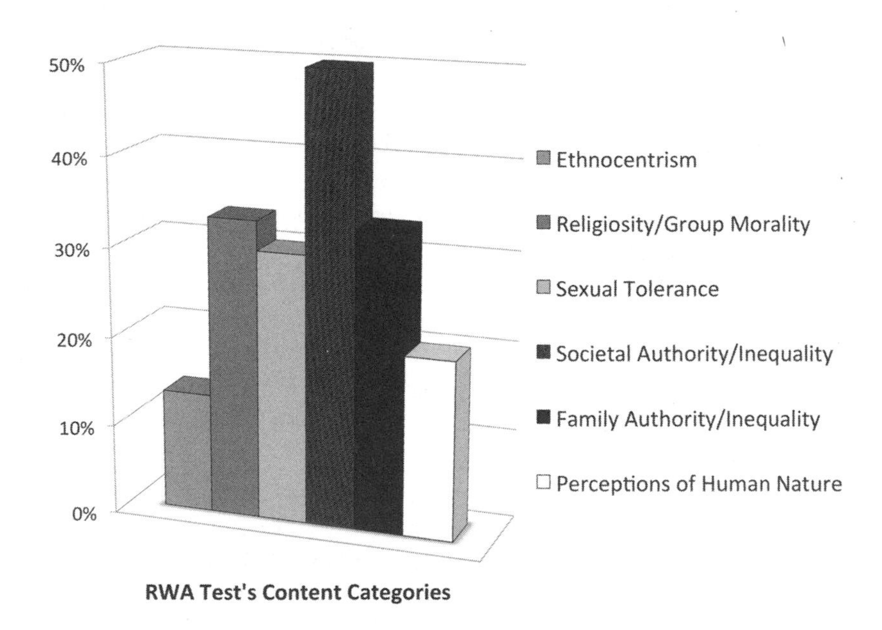

Figure 10. Underlying Content of the RWA Test (as percentage of thirty questions).

Altemeyer's test consists of thirty controversial statements. Figure 10 breaks down the content of these statements into six categories. Each bar represents one of these content categories and shows the percentage of the thirty statements that makes reference to it.[a]

The six content categories, in turn, can be lumped into three larger groups: the grey cluster, the black cluster, and the white cluster. The three categories within the grey cluster are ethnocentrism, religiosity/group morality, and sexual tolerance. These are the three elements that comprise the "tribalism" cluster of personality traits.

The two categories in the black cluster measure tolerance of inequality: the first concerns attitudes toward inequality and authority in society, while the second category pertains to inequality and authority within the family.

The white personality cluster has only one category, which measures perceptions of human nature.

This breakdown of political ideology coincides very well with independent research conducted by a multidisciplinary team from Stanford University, UC Berkeley, and the University of Maryland, College Park. These professors carried out a meta-analysis of studies on political orientation in North America, Europe, the Middle East, Africa, and New Zealand. The eighty-eight studies they evaluated covered nearly twenty-three thousand individuals who took various types of political tests between 1958 and 2002. These tests included the F-scale, RWA, several other political-conservatism scales, economic-conservatism scales, self-reported ideological positions, issue opinions, and voting records.

After analyzing all of the data, the team concluded that the "two relatively stable, core dimensions that seem to capture the most meaningful and enduring differences between liberal and conservative ideologies" are: (1) "attitudes toward social change versus tradition" and (2) "attitudes toward inequality." The analysis showed that, although these two factors are "often related to one another," they are "obviously distinguishable."[1]

The meta-study's first dimension clearly corresponds to the grey cluster (tribalism) in figure 10. The second dimension is the same as the black cluster. What about the white cluster? The debate over the nature of human nature is the ancient subject of political philosophy, which stretches back for millennia. The problem of human nature has also been studied extensively in its own right by political psychologists and evolutionary biologists, who have made valuable discoveries about this core political topic. Moreover, the white cluster overlaps substantially with both the grey and the black clusters; we'll explore the relationship between them later on.

These three clusters relate directly to this book's two arguments:

The Personality Argument: Human political orientation across space and time has an underlying logic defined by three clusters of measurable personality traits. These three clusters consist of varying attitudes toward tribalism, inequality, and different perceptions of human nature.

These three factors correspond, of course, to the grey, black, and white color groups in figure 10. To go into slightly greater detail:

- **Tribalism**. Tribalism breaks down into ethnocentrism (vs. the opposite force, xenophilia, which means an attraction to other groups), religiosity (vs. secularism), and different levels of tolerance toward nonreproductive sexuality.
- **Tolerance of Inequality**. There are two opposing moral worldviews toward inequality; one is based on the principle of egalitarianism, and the other is based on hierarchy.
- **Perceptions of Human Nature**. Some people see human nature as more cooperative, while others see it as more competitive.

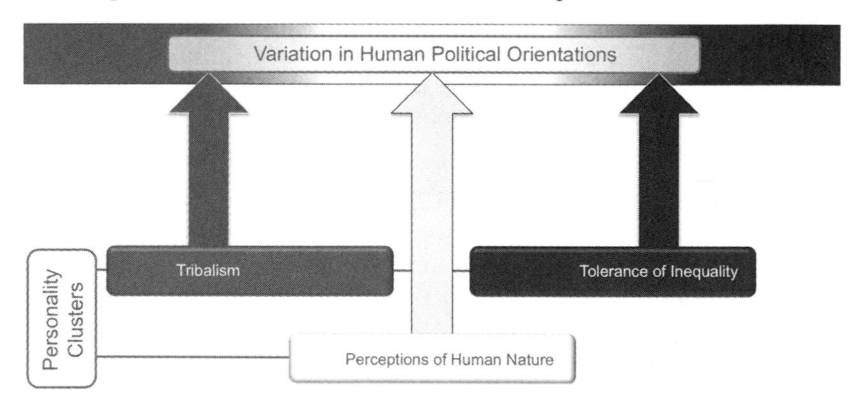

Figure 11. The Three Universal Personality Clusters Underlying Political Orientation.

Figure 11 shows these three personality clusters, which underlie variation in human political orientations. The horizontal positioning of the elements in figure 11 does *not* relate to the left-right spectrum; rather, the important dimension is depth, which shows different levels of analysis, focused on by different disciplines.

The top bar, which stretches all the way across the diagram, represents the variation in a population's political orientations. These orientations are manifested by different attitudes and behaviors toward controversial issues (for example, members of the Tea Party feel differently about taxation and social spending than supporters of Occupy Wall Street).

Whatever the time or place, specific political attitudes are reflections of one or more of these three underlying personality clusters, which are represented by the rectangles in the diagram (for instance, Tea Partiers have greater tolerance of inequality than Occupiers). The upward-pointing arrows show that the personality clusters are the greatest cause of political orientations, on the top level. The grey and the black rectangles are at the same level because they suffice to explain most political phenomena, and because this pair of clusters has been intensely studied by the modern social sciences. The white rectangle is slightly deeper down because the topic has a much more ancient intellectual history within the field of philosophy. In addition, the white cluster's influence heavily overlaps with both the grey and the black clusters. This overlap means that the clusters are related to one another (for example, people who think human nature is essentially cooperative are more likely to be xenophilic).

The following is the book's second and most important argument:

The Evolutionary-Origins Argument: Each of the three personality clusters (the grey, black, and white rectangles) has roots that reach even deeper down into evolutionary origins. In other words, *political orientations throughout time and space systematically and predictably reflect much deeper biological conflicts*.

Below are the three evolutionary roots of the three personality clusters (figure 12 shows how the six components connect to one another). Each of these deeper roots pertains to an area of evolutionary theory. Don't worry if you haven't heard of these terms before; the relevant chapters will fully explain each of them:

- **The Biology of Tribalism** concerns pushes and pulls between populations, which primarily occur due to tradeoffs between inbreeding and outbreeding. Ethnocentrism and other tribalistic personality facets have evolved to influence mate choice and encourage "optimal outbreeding." The book will explore these and other tribalistic political phenomena that impact the evolution of populations, including gender inequality, warfare, and genocide.
- **The Biology of Family Conflict** (Parent-Offspring Conflict) is the field of evolutionary theory that explains why the interests of the most closely related individuals do not always align, and thus why different family disciplinary strategies exist. The two opposed disciplinary models are based

on egalitarian and hierarchical moralities. These conflicts are linked to the variation in people's tolerance of inequality.

- **The Biology of Altruism and Self-Interest** is the area of evolutionary theory that describes how and why people cooperate with and betray one another; this field sheds light on why some people perceive human nature so differently than others.

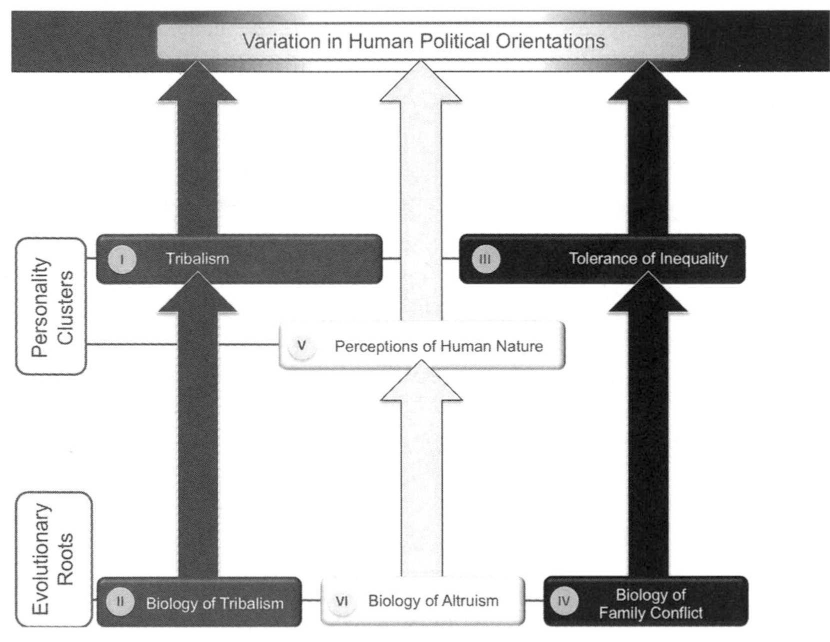

Figure 12. The Evolutionary Roots of the Three Personality Clusters Underlying Political Orientation (in this map of the book, the white circles show the order of the sections [I, II, III, IV, V, VI]).

This book is divided into six parts, which correspond to the six roman numerals and their rectangles in figure 12. Part I tells how different levels of tribalism influence people's placement on the political spectrum, and part II explains the biological roots of tribalism. Together, they form the grey pillar, which we could think of as "Conflict among the Nations."

Part III tells how different levels of tolerance toward inequality affect people's placement on the political spectrum, and part IV finds the roots of egalitarian and hierarchical personality traits in the biology of parent-offspring conflict. Together, these two parts form the black pillar, which arises through "Conflict within the Family."

Finally, part V explains how differing perceptions of human nature impact people's left-right political orientations, and then part VI reveals the roots of our true human nature by describing the biology of altruism. These last two rectangles form the white pillar, whose theme is "Conflict inside the Conscience."

This roadmap is a bit abstract. But some concrete examples will make much more sense. So let's dive into the details of our political nature.

PART I.
TRIBALISM ON THE POLITICAL SPECTRUM

Part I explains how the tribalism cluster of personality traits influences left-right orientation. In relation to the political spectrum, tribalism breaks down into three components: (1) ethnocentricity, (2) religiosity, and (3) sexual (in)tolerance. High measures of ethnocentricity, religiosity, and sexual intolerance are commonly associated with one another. Individuals with this cluster of traits tend to have political views on the right. On the other end of the spectrum, attraction to out-groups (xenophilia), secularism, and higher sexual tolerance are well correlated with one another and with political views on the left.

What, specifically, would be the objects of sexual intolerance? Any form of sexuality deemed less likely to contribute to normatively sanctioned reproduction, economic stability, and social harmony of a given in-group. Examples include nonreproductive sexuality (such as homosexuality), premarital sexual choices made by young people, and extramarital affairs.

What is the logic between these three components of tribalism? The more ethnocentric, religious, and sexually intolerant people are, the more likely they are to reproduce with a mate from their own in-group. Moreover, conservatives are more likely to emphasize group values, such as prioritizing the reproduction and defense of their ethnic group, over other possible competing interests (e.g., personal pleasure, and education or career choices made at the expense of family).

On the other hand, more xenophilic, secular, and sexually tolerant people are more likely to see equal (or even greater) value in out-groups, and to reproduce with them. Thus, liberals place relatively greater importance on individualism and less on in-group values.

In short, the tribalism cluster of personality traits relates to the contrasting kinship and reproductive patterns found in all groups—marrying in (endogamy) and marrying out (exogamy).

The three chapters in part I are dedicated to each of the three components of tribalism. We'll look at the specific RWA questions related to ethnocentrism (in chapter 6), religiosity (in chapter 7), and sexuality (in chapter 8) to see how these parts of Altemeyer's scale predict left-right voting. These chapters present engaging news events that illustrate each concept, and they also reveal public-opinion statistics and fascinating political-science discoveries.

6.
ETHNOCENTRISM VS. XENOPHILIA

Glenn Beck has been called a leader of the ultraconservative opposition to President Obama, and a voice of the far-right Tea Party movement. During the depth of the Great Recession, in 2009, Beck was quickly rising to a position of national prominence, despite—and because of—his extremely controversial statements. In July 2009, for example, Beck declared: "Everything that is getting pushed through Congress" is "driven by President Obama's thinking on . . . reparations [for slavery]," and is conceived of as a way to "settle old racial scores."[1] Beck also asserted that a "deep-seated hatred for white people" drives President Obama.[2]

Examining Beck's political psyche is a fruitful exercise because his bigger-than-life personality magnifies the traits that interest us on his end of the spectrum. But do Beck's statements characterize a typical position of the Tea Party or of the far right in general?

In 2009, 2.8 million viewers tuned in to Beck's television program. Moreover, Beck's live shows and merchandise business have earned $35 million in a single year.[3] Considering that Beck is essentially selling political ideas, it's fair to say that his ideological orientation carries substantial currency among a particular segment of the US spectrum.

As interesting and as popular as Beck's personality may be, he is still only one person in one country. What we really want to know is whether his ideas on race are meaningful and typical. Is it possible to objectively determine how much people's views about race affect their left-right orientation? And are ethnocentrism and xenophilia[a] associated with the same segments of the political spectrum in all countries around the world?

Let's start by looking at the RWA scale, our cross-cultural political litmus test. Unlike Glenn Beck, the statements on the RWA scale do not reference any particular ethnic group—or even interethnic relations. This "omission" is surprising, since most conservatism scales overtly mention these topics. Why the absence?

The mother of the RWA scale, you may remember, was the "pre-fascism" F-scale devised by the authors of *The Authoritarian Personality*. But the original F-scale also had a couple of sister tests, including the Anti-Semitism scale and the Ethnocentrism scale. The Berkeley professors kept these scales separate because they wanted to demonstrate the strong correlation between ethnocentrism (including Anti-Semitism) and "authoritarianism," as measured by the F-scale. Because of this original separation, Altemeyer's RWA test inherited only oblique references to ethnocentrism.

So what specific content of the RWA items *does* figure 10 code in the "ethnocentrism" category? Below are a couple of the statements that qualified:

- "*Some of the worst people in our country nowadays are those who do not respect our flag . . . and the normal way things are supposed to be done.*"
- "*Students in high school and university must be encouraged to . . . criticize the customs and traditions of our society.*"* [The asterisk shows that this second statement is worded in the opposite direction of the first one, to prevent an acquiescence bias. Answering "yes" to the first question increases an RWA score, but answering "yes" to this second question *decreases* RWA.]

A couple of RWA items also appeal to a sense of nationalism, by mentioning, for instance, "*the duty of every patriotic citizen.*"

The RWA scale's performance would likely be even stronger if it contained an item or two that overtly elicited attitudes toward race relations. On the other hand, the test's lack of questions about specific racial groups has a couple of advantages: first, this absence ensures that the test reliably produces similar results in extremely different countries. Second, avoiding racially controversial language may reduce the self-censorship of people who would otherwise feel socially inhibited from expressing explicitly discriminatory attitudes.

In any case, people around the world who score highly (i.e., conservatively) on RWA are significantly more likely to score highly on explicit ethnocentrism scales, even when adapted to their geographic region. Consider a few sample statements from the "Manitoba Prejudice Scale," which is designed to measure ethnocentrism in this Canadian province:

- "*There are entirely too many Chinese students being allowed to attend University in Canada.*"
- "*The main reason certain groups like our native Indians end up in slums is because of prejudice on the part of white people.*"*

- *"Certain races of people clearly do not have the natural intelligence and 'get up and go' of the white race."*
- *"The Pakistanis and East Indians who've recently come to Canada have mainly brought disease, ignorance, and crime with them."*
- *"There is nothing wrong with intermarriage among the races."* *

High scores on this Prejudice Scale moderately correlate with Manitobans' RWA scores.[4] Other social scientists have adapted this twenty-item prejudice scale to measure hatred toward Jews and other ethnic minorities in the Soviet Union, and toward blacks among white South Africans. They have found strong correlations with RWA.[5] Among political elites in various US state legislatures, the Americanized version of the Prejudice Scale correlated very strongly with conservatism, as measured by RWA.[6]

ETHNOCENTRISM AND THE RIGHT

In every country, the right is more ethnocentric than the left. Conservatives have more positive feelings toward members of their in-group and higher levels of patriotism toward their country. In the United States they are more likely than liberals to have a flag in their bedroom.[7] On the other side of the coin, the right is more xenophobic toward other countries and members of out-groups.

In America, the modern political era began when the Civil Rights Movement pushed the race issue squarely into the politics of the 1960s. During this decade, opposition to civil rights was a cornerstone of American conservatism.[8] The Republican candidate in the 1964 presidential election, Barry Goldwater, made the fateful decision to fight against the Civil Rights Act. This historic piece of legislation sought to outlaw racial segregation, including—most prominently—in schools. In his 1960 book *The Conscience of a Conservative*, Goldwater wrote:

> I am firmly convinced—not only that integrated schools are not required—but that the Constitution does not permit any interference whatsoever by the federal government in the field of education. It may be just or wise or expedient for Negro children to attend the same schools as white children, but they do not have a civil right to do so which is protected by the federal constitution, or which is enforceable by the federal government.[9]

Although many Americans have been taught to see this debate in the frame of "states' rights" versus "federal powers," the universal issue underlying the political controversy revolves around interethnic relations—just as a substantial portion of politics do anywhere else in the world. This is why there is no "states' rights" personality cluster in figure 11. Rather, the pertinent rectangle is labeled "tribalism"—and for good reason; the specific laws that states have disagreed on with the federal government have concerned racial issues, beginning with slavery. A more recent point of contention pertained to "anti-miscegenation laws." Until the US Supreme Court banned these laws in 1967, some sixteen states in the American Southeast prohibited interracial marriages.

Forty-six years after the era of Goldwater, the Civil Rights controversy resurfaced again—during the racially charged midterm elections of the nation's first black presidency. This time, however, politicians reframed the Civil Rights Act as a "free-market" issue (as opposed to a racial one). In 2010, the Tea Party–backed senatorial candidate from Kentucky was Rand Paul. When asked if he would have voted for the Civil Rights Act of 1964, Paul responded:

> I don't like the idea of telling private business owners [what to do]. I abhor racism. I think it's a bad business decision to ever exclude anyone from your restaurant—but at the same time I do believe in private ownership. And I do think there should be absolutely no discrimination in anything that gets public funding. And that's mostly what the Civil Rights Act was about.[10]

With this statement, Rand Paul asserted his belief that private interests have the right to racially discriminate.

In commenting on the controversy, reporter John Stossel explicitly said: "It's time now to repeal that part of [the Civil Rights Act that concerns the private sector]. Because private businesses ought to get to discriminate . . . it should be their right to discriminate."[11] Rand Paul won his election in Kentucky.

Interethnic controversies seem to assume endless "nonracial" frames. Perhaps these frames serve as moral camouflage in partisan wars of ideas to capture swing voters, who tend to be more moderate and averse to conflict. Because of this camouflaging and the passionate politicization of racial political controversies, perhaps we still need stronger evidence to definitively determine whether ethnocentrism is truly a core component of conservatism.

Conservatism Predicts Ethnocentrism Best

Hard statistical studies repeatedly find that there is no better way to predict ethnocentrism than to measure political conservatism. Stanford University's Paul Sniderman is a world expert on the political psychology of race. He once asked 659 nonblack individuals how many negative stereotypes of black people they would accept. Self-placed political conservatism predicted the acceptance of the negative stereotypes more than three and a half times better than family income did.[12]

In another survey, Sniderman asked his respondents whether or not they supported government guarantees of equal opportunity for blacks. Their answers are quite revealing: 75 percent of college-educated liberals supported this legislation favoring African-Americans, compared with only 38 percent of college-educated conservatives.[13]

With such a pronounced partisan contrast on race attitudes, it's no wonder that the 113th US Congress (2013–2015) has forty black Democrats and only one black Republican.[14]

Beyond the United States, is there such a strong relationship between ethnocentrism and the political spectrums of countries that have totally different minorities, histories, and cultures? In 1988, researchers asked four thousand Europeans their opinions about their local ethnic out-groups. In this 30th Eurobarometer Survey, the French expressed attitudes about North Africans, Vietnamese, and Cambodians; the Dutch about Surinamers and Turks; the British about West Indians and South Asians; and the West Germans about Turks. In addition to their views on ethnic minorities, the surveyors collected sixty possible indicators of ethnic prejudice, including every imaginable demographic variable—from income to education to region. Out of all these sixty potential factors, self-placed political conservatism was *the major predictor* of prejudice across the four thousand Europeans.[15]

In many countries, the political spectrum itself offers clues about ethnocentrism. In Israel, for instance, political parties on the far right draw supporters from ethnically homogeneous populations. The Shas party represents very religious Jews of Sephardic and Mizrahi ancestry (controlling 9 percent of the Knesset), while the United Torah Judaism coalition represents very religious Jews of Ashkenazi ancestry (with 6 percent of legislative seats). On the far left, in contrast, Hadash (The Democratic Front for Peace and Equality, which controls 3 percent of the legislature) has formed interethnic partnerships between Arabs and Jews of diverse ethnicities.

Beyond Conscious Attitudes

The measurements that we've considered so far have come from volunteered attitudes. But how do we know if other people are as ethnocentric or xenophilic as they consciously claim to be? Some people might argue that they oppose racially charged issues like the Civil Rights Act because of "states' rights" and "free markets," while claiming to completely oppose all kinds of racism. What is the skeptic to believe?

A group of psychologists from Harvard, the University of Virginia, and the University of Washington developed an ingenious way to bypass conscious inhibitions against expressing ethnocentric attitudes. They call it the "Implicit Associations Test" because it asks people to pair a word with a picture on a computer screen. The pictures include images of white faces and black faces, and the words have positive and negative meanings. The idea behind the test is that a more ethnocentric white person, for example, hesitates relatively longer in pairing a positive word, rather than a negative word, with a black face.

When researchers gave this implicit racism test to 130,000 white Americans, a substantial majority of them, including both liberals and conservatives, tested positive for some amount of antiblack prejudice. However, when the psychologists correlated these results with voting patterns in the 2004 presidential election, they discovered a significant difference: of the 435 congressional districts covered, those with higher levels of "implicit" ethnocentrism were more likely to vote for the conservative candidate, George W. Bush.[16]

Racial Threat and Distorted Demographic Perceptions

Just as a majority of white Americans show implicit levels of antiblack ethnocentrism, over half of *all* Americans greatly overestimate the proportions of blacks in the population.[17] They also have similarly distorted perceptions of other minority groups. In 1990, the average American believed that the country was 32 percent black, 21 percent Hispanic, and 18 percent Jewish (even though the actual figures then were 12 percent, 10 percent, and 2 percent).[18]

White Americans who give even higher estimates for the proportion of black America also tend to (1) mistakenly believe that the average African American has an equally good or better income, job, housing, healthcare, and education than the average white American, and (2) take conservative positions against affirmative action and against food stamps for the poor in general.[19] These right-wing perceptions and policy preferences reflect a view that minority groups constitute a demographic threat and a competitive drain on the in-group's resources.

It's not surprising, then, that conservatives who over-overestimate the proportion of minorities in their society also tend to oppose immigration more than liberals. In 1920s America, conservatives lobbied for lower immigration quotas for Irish, Italians, Jews, and Poles, whom they called "inferior races."[20] Ninety years later, the immigrants were different and the politicians shied away from overt prejudice, but conservatives were still the group most opposed to immigration. During the 2010 midterm elections, for example, Louisiana Republican senator David Vitter ran a campaign advertisement featuring a group of suspicious men of Mexican appearance sneaking across the border. On the American side, a large sign read "Charlie Melancon [Vitter's Democratic opponent] welcomes you to the USA!" White Americans handed the immigrants a giant check payable to "illegals." Vitter won the election.

The United States certainly has no monopoly on ethnocentrism or anti-immigration sentiment. In other countries, again, the conservative end of the spectrum opposes immigration most.

In Switzerland, some 65 percent of the incarcerated population lack Swiss citizenship. In November 2010, Swiss voters debated a referendum proposing the automatic expulsion of non-Swiss citizens for crimes such as murder, rape, breaking and entering, drug dealing, and social-security fraud. The measure was proposed by the right-wing Swiss People's Party (SVP). Opponents of the legislation on the left of the Swiss political spectrum claimed that the referendum was motivated by xenophobia, and they opposed deporting the children of criminal immigrants, who did not automatically receive citizenship. The right-wing SVP's campaign probably would not have made many international headlines, however, were it not for a hugely controversial poster. This ad depicted a group of white sheep literally kicking a black sheep out of Switzerland. In the end, the deportation referendum won 53 percent of the vote.[21]

Equal-Opportunity Hatred

In 2010, the Tea Party movement backed far-right-wing businessman Carl Paladino's candidacy for the governorship of New York. In the course of the race, Paladino's opponents accused him not only of racism, but also of bearing more than one type of hatred. One of multiple gag e-mails sent by Paladino featured a photograph of an African tribal ritual, titled "Obama Inauguration Rehearsal." The press also quoted a remark that Paladino made about Sheldon Silver, who had been speaker of the New York State Assembly for sixteen years, and who was Jewish. Paladino said: "If I could ever describe a person who would fit the bill of an Antichrist or a Hitler, this guy is it."[22]

These accusations of racism against Carl Paladino seem to address two quite different forms of prejudice. If one does *not* give Paladino the benefit of the doubt, and assumes that these incidents reflect genuine xenophobic attitudes toward blacks and Jews, then how unusual are Paladino's double prejudices?

Not unusual at all, according to political psychologists. In fact, the most interesting thing about prejudice is that, ironically, it's equal opportunity for all out-groups. Consider the following survey, in which Paul Sniderman asked 676 people whether they agreed with these six negative stereotypes about blacks:

(1) "*Blacks on welfare could get a job.*"
(2) "*Blacks need to try harder.*"
(3) "*Black neighborhoods are run down.*"
(4) "*Blacks have a chip on their shoulder.*"
(5) "*Blacks are more violent than whites.*"
(6) "*Blacks are born with less ability.*"

Of those who agreed with these six statements, 84 percent also agreed with at least two out of these three anti-Semitic statements:

(1) "*Jews engage in shady practices.*"
(2) "*Jews don't care about non-Jews.*"
(3) "*Jews are pushy.*"[23]

In the 1950s, the authors of *The Authoritarian Personality* came to the same conclusion as Sniderman. Those who scored highly on the Anti-Semitism scale also tended to be antagonistic toward "Negroes," "Japs," "Okies"[b] and "foreigners in general."[24]

The same principle holds true in Italy. According to Sniderman: "If you know how many problems Italians blame on [immigrant] Eastern Europeans, you can tell just as well how many unfavorable characteristics [Italians] will ascribe to African immigrants, as if you know how many problems they blame on Africans themselves."[25]

Israeli social scientists have found similar trends in their country. If you know the correlation between Jewish Israeli RWA scores and xenophobic attitudes toward foreign workers (from Thailand or the Philippines), nearly the same correlation defines attitudes toward Palestinians, and also toward immigrants from the former Soviet Union. This finding is quite surprising, considering the enormously different relationships between Jewish Israeli society and these three out-groups—especially since the study was published in 2003, amid the Second Intifada (the Palestinian rebellion that took over 6,500 lives).[26]

Why is this same pattern repeating everywhere? This quandary is precisely the point where social scientists run into a wall. Paul Sniderman, for example, is befuddled by this equal-opportunity hatred of such diverse out-groups. He muses: "That [Americans'] dislike of Jews is tied so closely to dislike of blacks—two groups whose history and circumstances are so different in so many ways—ought to make palpable the irrationality of intolerance, whether racial or religious."[27] But the fact remains that plenty of people dislike both blacks and Jews. In the eyes of most social scientists, this fact just proves human irrationality. End of story. But to an extraterrestrial biologist, this equal-opportunity hatred would make perfect sense.

Suppose that a Martian biologist were observing the American president Richard Nixon. During his presidency, the Supreme Court passed *Roe v. Wade*, the landmark case that legalized abortion. Nixon, who was a conservative, never made a public statement at the time. In 2009, however, the Nixon Library publicized previously secret tapes in which Nixon expressed concern that greater access to abortions would promote "permissiveness." Nevertheless, Nixon told an aide: "There are times when an abortion is necessary. I know that. When you have a black and a white . . . or a rape."[28]

Why was Nixon willing to make an exception to his anti-abortion stance in the case of mixed-race fetuses? Was it only because Nixon didn't like African Americans? Perhaps. But plenty of anti-Semitic statements came to light in the Nixon tapes as well.

Based on his numerous prejudiced statements, Nixon would have likely scored highly on ethnocentrism. By creating a distaste for blacks, Jews, and other minority groups all at the same time, Nixon's ethnocentrism loaded the biological dice in two ways: (1) it decreased the chance that he would marry or reproduce with members of out-groups, and (2) it increased the probability that he would marry and reproduce with someone much closer to his own group. Indeed, Nixon married a woman who was a Methodist, like his father.

Nixon's hating and mating behavior represents a larger trend. According to the Pew Research Center, today's Republicans are nearly twice as likely as Democrats to agree that "people of different races marrying each other" is "a bad thing for society."[29] Thanks to their different levels of tribalism, liberals and conservatives clearly have different attitudes toward the transmission of DNA. This evolutionarily important fact would not be lost on the Martian biologist—nor on an impartial human one, for that matter.

It was fairly easy for Nixon to tell that his wife wasn't black or Jewish. Throughout human history, however, populations native to adjacent territories have often looked a great deal more similar to one another than blacks, whites, and Semites. And yet they have gone to great length to differentiate themselves from their slightly less related neighbors. Biologists have compared markings of ethnic separateness—including hairstyles, dress, ornamentation, languages, and laws—to a symbolic "pseudo-speciation."[30]

These cultural differences likely function to exaggerate differences between the in-group and its out-groups in order to encourage marriage within the in-group (i.e., endogamy).[c] It makes sense, then, that more conservative members of a given group are more likely to wear—and to encourage or even enforce the wearing of—these ethnic markers of pseudo-speciation. By doing so, they promote endogamy. The public debate over the Islamic headscarf in Europe partially relates to this issue.

If we look at entire populations, people in the most conservative countries tend to reproduce with more closely related individuals. Dutch political psychologist Jos Meloen has discovered a strong correlation (0.74) between the cultural preference for endogamy and countries with right-wing authoritarian governments. Populations in these countries have weaker incest taboos that permit or favor cousin marriage, and customs whereby parents tend to arrange their children's marriages. In more liberal countries, on the other hand, cultures are more likely to have stricter incest taboos, individuals have greater freedom to select their own marriage partners, and reproduction tends to occur between more distantly related individuals (i.e., these countries are more exogamous).[31]

THE EXTREME RIGHT

When we shift our focus from the political right to members of the *extreme* right, the trait that most quickly becomes apparent is a conscious and constant concern with the reproduction of genetically different populations. Raphael Ezekiel, an ethnographer of white hate groups across the United States, has studied a neo-Nazi group in Detroit that he pseudonymously called the "Death's-Head Strike Group." In reference to these young men, Ezekiel noted: "In all our discussions, the group members spontaneously, with no prompting, spoke about race mixing: Cross-racial sex, cross-racial childbearing. The images upset them, and I strained to grasp how real the emotional disturbance was." For example, a group member whom Ezekiel calls "Paul" lamented: "from the white viewpoint, where the whites stay with the whites and blacks with the blacks, seeing this white girl who was bad enough to mix with Arabs, now she's got a baby by one, that's what we call the ultimate sin." Paul called the mixed-race baby "a symbol of everything that's wrong and evil."

Members of American hate groups not only loathe interracial reproduction; they also have a deep and explicit fear of whites becoming a minority—or worse. Paul commented: "Soon we [whites] won't even be the majority, because the blacks and Hispanics are breeding like flies and they're going to outnumber us." Members of the "Death's-Head Strike Group" mistakenly believed that black people composed 60 to 70 percent of the American population; this perception doubled the average American overestimation of the black population, and it multiplied the actual proportion of African Americans (12 percent) by at least five. Tom Metzger, the founder of White Aryan Resistance (WAR), has echoed the same fear: "The whites must face the issue of extinction. All groups must concern themselves with their own extinction. And so whatever method you have to do to resist the extinction, has to be done."

What, then, do extremists believe should be done? The answer surely varies, but the response that a member of the Southern White Knights of the Ku Klux Klan gave at a rally in Stone Mountain, Georgia, deserves attention: "What I'd like to do . . . whites ever want to get into power, that would be the thing to do: Castrate the males, Jewish and nigger males, castrate them and sterilize the females, chain them in the yard like dogs." In other words, Ezekiel's interviewee did not want to simply eliminate the out-groups that he hated; instead, his fantasy specifically involved destroying their reproductive capacities while keeping them alive like "lower" animals.

In addition to dehumanizing members of out-groups, many members of the Ku Klux Klan overemphasize the genetic relatedness between the members of their in-group. They accomplish this by calling each other "kinsmen."[32]

Hitler himself, of course, also made eliminationist ethnocentrism a pillar of Nazism; Hitler and his followers believed that Jews, Romani, and other ethnic minorities constituted a specifically biological threat to the existence of Germanic peoples, via sexual reproduction. In his autobiographical manifesto, *Mein Kampf*, Hitler wrote: "Blood mixture and the resultant drop in racial level is the sole cause of the dying out of old cultures; for men do not perish as a result of lost wars, but by the loss of that force of resistance which is contained only in pure blood." Elsewhere he accuses: "With satanic joy in his face, the black-haired Jewish youth lurks in waiting for the unsuspecting girl whom he defiles with his blood, thus stealing her from her people."[33]

Although these words may seem outdated or irrelevant to some readers, Hitler's ideas remain present in the political spectrum of many countries. In 2005, for instance, *Mein Kampf* became a bestseller in Turkey.[34] The fascination—and even infatuation—with Nazism has also popped up closer to home, on the American far right. In October 2010 Rich Iott, the Tea Party–backed candidate for Ohio's 9th District in the House of Representatives, fell to a scandal. Photographs surfaced of Iott dressed up as a Nazi. Not just as any Nazi, but as an SS officer named Reinhard Pferdmann, who was the would-be congressman's alter ego in a World War II reenactment group.[35] Iott lost the election.

WHAT ABOUT THE LEFT?

Is the political left truly less ethnocentric than the right? Consider the situation in La Paz, Bolivia, in October 2010: the president of this land-locked Andean country was strongly backing a "Law against Racism and All Forms of Discrimination." President Evo Morales's far-left-wing "Movement for Socialism" party argued that the law was necessary to rectify centuries of oppression against Bolivia's indigenous majority. During the General Assembly of the United Nations in New York the previous month, President Morales, who is an Aymara Indian, had defended the bill by recalling discrimination against his own mother.

What made the proposed law controversial? The legislation would permit the government to close media outlets and jail journalists for disseminating "racist material." The Bolivian media protested and accused Morales of attempting to censor the freedom of the press.[36] Does Morales's "Law against Racism" represent a genuine will to stamp out all forms of racism, or is it left-wing authoritarianism? In other words, has Bolivia experienced a one-way ethnic backlash against the country's white-mestizo population? Or is the left really less ethnocentric than the right?

Around the world, the *moderate* left has a reputation for xenophilia. Consider a political-humor e-mail circulated among conservative Israelis, titled "What Does It Mean to Be a Leftist?" According to this piece of political jokelore, to be a leftist in Israel is:

- "To be ready to make painful concessions [of land to the Palestinians], but to sue your [in-group] neighbor over the stray branch of a tree that poked into your yard," and
- "To prefer the hummus [a chickpea-based spread] from Damascus [Syria] over the hummus from the Golan Heights [in Israel]."[37]

The underlying message insinuates, if derisively, that liberal Israelis love outsiders more than insiders.

The far left sometimes seems xenophilic as well. During Hugo Chavez's "21st-century socialist" government, Venezuela admitted masses of immigrants from around the world, including some fifteen thousand Haitian refugees from the 2010 earthquake, four million Colombians, fifty thousand Chinese, and thousands of immigrants from Lebanon, Syria, Jordan, Pakistan, and India. On the other hand, thousands of Venezuelan Jews have emigrated because of rising anti-Semitism, encouraged by Chavez's antagonistic relationship with Israel.[38]

What do the hard data say about the left's apparent xenophilia? Let's jump back to postwar Germany. At the turn of the 1950s, psychiatrist David Levy was tasked with the critical responsibility of producing a "de-Nazification study." At the time, the victorious Allies wanted to do everything possible to prevent a feared resurgence of Nazism. To do so, the Allies needed to have some way of screening individuals for their political orientation. Thus, a fascinating research question was born: Who could have easily joined the Nazi party before the end of the War, but instead became an anti-Nazi dissident? People with their characteristics, the Allies believed, would logically be the most resistant to fascist ideology.

To answer this question, Dr. Levy needed a group of anti-Nazis, as well as a control group of Nazis. The latter were easy to find among imprisoned SS officers and a group of "run-of-the-mill" Nazis from Miesbach, Bavaria. What about the dissidents? The Allies had needed German civilians to serve as "vetters" for the military government. To clear German civilians for these positions, Americans had conducted thorough background investigations on these potential anti-Nazis. The background checks included questionnaires, interviews of neighbors and acquaintances, searches for increased income under the Nazi regime, and reviews of police, civil-service, military, and party-membership records.

The de-Nazification study found some very interesting differences between the two groups. Compared with the Nazis, anti-Nazis were more likely to read books written in non-German (i.e., foreign) languages. By the same logic, having traveled and lived abroad for at least six months decreased the chances that a subject was a Nazi.[39] Among the best ways of predicting left-wing dissidence against the fascist regime, then, were these signs of xenophilia.

"Bedroom cues" from contemporary Americans tell the same story. New York University psychologist John Jost has found significant correlations between left-wing identifications on the political spectrum and having each of the following items in one's bedroom: books about travel, cultural memorabilia (such as trinkets from vacations), travel tickets, and books about "ethnic matters." The presence of international maps showing countries other than the United States, as well as world-music CDs, also had very significant correlations with left-of-center self-placements.[40] On the flip side of the coin, Jost has also discovered that conservatives have a greater distaste for unfamiliar music."[41]

Geographical regions within a country can also be xenophilic—both literally and politically. In the United States, the greater the proportion of a given state born outside America, the more likely the state is to support the Democratic Party. The "foreign born" percentage even predicts a higher acceptance of interracial dating. Republicanism, on the other hand, is strongest in regions with higher concentrations of Anglo-Americans and fewer immigrants.[42]

"Liberals!"

THE ANTI-ETHNOCENTRISM OF THE FAR LEFT

The xenophilia of moderate liberals sometimes transforms into something very different as one moves to the far left of the political spectrum—into a phenomenon that could be called "anti-ethnocentrism." Communist ideologies, such as that propounded by Lenin, demanded the suppression of ethnic and religious allegiances; in short, the destruction of the "vestiges" of tribalism.

Soviet Communism was so intent on forging a supernational identity and a "new Soviet Man" that the government even promoted disloyalty to "reactionary" family members. For instance, a semilegendary thirteen-year-old boy named Pavel Morozov supposedly denounced his father as a counterrevolutionary to the government. Pavel's family killed him in retaliation, and the Communist regime made Pavel into a national hero and a celebrated martyr. "Pavlik," as he was known by the affectionate diminutive, became the subject of compulsory readings, propaganda films, and songs.[43]

The Soviet leaders, who had to govern a diverse country spanning nine time zones, may have had selfish motives for jealously repressing all competing allegiances other than loyalty to their own ideology. But the self-interested purposes of a tiny proportion of elites cannot explain the anti-ethnocentric attitudes prevalent in the mass public opinion of the far left. What is the origin of this desire to destroy ethnic identity but not the individual bearers of culture themselves? How does this anti-ethnocentrism relate to xenophilia?

Both of these concepts lie on the left half of the spectrum, and both may serve a similar ultimate function. A liberal xenophile on the moderate left highly values foreign cultures and is attracted to their bearers. This xenophilic attraction increases the chance that the xenophile's chromosomes may recombine with the more distantly related DNA of an admired out-group member. The anti-ethnocentrist on the far left may also yearn for a universal equality between all people, but believe that this dream can only be achieved by (forcefully) eliminating the tribal identities that prevent brotherhood between peoples—and that prevent exogamy.

A current example of this anti-ethnocentrism is the "New Anti-Semitism." This ideology is nearly synonymous with anti-Zionism, which in the Western world comes from a position on the extreme left that rejects the right of the Jews to have a state of their own like any other nation. By opposing the Jewish state, anti-Zionists also oppose one of Israel's functions, which is to prevent assimilation and allow the continued existence of Jewish ethnic groups as such. The other fundamental function of Israel is to protect Jews from the old anti-Semitism; this ideology, which lies on the extreme right, has sought to destroy both Jewish ethnicity as well as Jewish lives.

To generalize from this example, the anti-ethnocentrism of the extreme left pushes for assimilation and exogamy, even at the cost of ethnocide; the most extreme right's plan, in contrast, ends with genocide, which impedes exogamous mixing entirely.

The political narrows that have squeezed Jewish existence from both ends is by no means unique to the Jewish people—contrary to beliefs prevalent on both the extreme right and left. Indigenous peoples without two-thousand-year diasporas have also felt the same pinch. On the one hand, colonial societies have tried to suppress indigenous people's ethnocentrism by interfering with the transmission of native languages and cultures. These anti-ethnocentric tactics have historically formed part of an ethnocidal agenda to assimilate indigenous genes into a larger national gene pool. On the other hand, indigenous peoples have faced displacements and genocides; these mass killings have destroyed indigenous genes, thus preventing them from mixing with foreign populations of colonialists.

In recent South American history, the association between the extreme left and anti-ethnocentrism is quite clear. In Peru, the Maoist Shining Path rebels grew frustrated that the tribal allegiances of indigenous peoples made many of them difficult to bring into the fold of a Communist agenda. A narrative emerged within factions of the radical left that Indians couldn't be assimilated. Some leftist militants even rationalized this problem within their own worldview, by writing off Indians as members of a "petty bourgeoisie" that was too tied to its land to join the movement. Ironically, the traditional land-tenure systems of many indigenous communities, where they remained intact, were about as communalized as humanly possible.

In the historic experiments where Communist governments *did* succeed in implementing extreme, anti-ethnocentric policies for several generations, these measures eventually failed notoriously in their goal of erasing the ethnic markers of tribalism. Jerrold Post, director of the Political Psychology program at the George Washington University, has perceptively commented on the long-term futility of these policies. Within only a few years of the fall of the Communist empire, the world witnessed "the revival of age-old hatreds in exaggerated form, as Serbs slaughtered Croats, Slovaks asserted their autonomy and split from Czech lands, and hatred of minorities was given free rein, often becoming a major theme in political campaigns."[44] In Poland, where 99.97 percent of the pre-War Jewish population had been eliminated for two generations, the end of Communism in 1989 brought a surge in anti-Semitism: "The largely nonexistent Jewish population was blamed for Poland's economic distress, with an invocation of the international Zionist conspiracy," points out Post.[45]

In this chapter, we've learned the following:

- Levels of ethnocentrism and xenophilia vary greatly within all groups.
- These dispositions are among the most important factors that influence political orientation. Ethnocentrism is associated with conservatism, a greater degree of xenophilia is associated with liberalism, and anti-ethnocentrism is associated with the far left.

In the following chapter, we'll meet ethnicity's sister in the political family of tribalism: religion.

7.

RELIGIOSITY VS. SECULARISM

Those who say that religion has nothing to do with politics do not know what religion means.
—Mohandas K. Gandhi[1]

Religion and politics are necessarily related.
—Ronald Reagan[2]

In June 2009 a Christian woman named Aasiya Bibi was working as a farmhand in a rural Punjabi village in the Islamic Republic of Pakistan. Aasiya was asked to fetch water, but the local Muslim women with whom she was picking berries refused to drink it. They said that the touch of a Christian had polluted the water.

A few days later, one of the Muslim women claimed that Aasiya had insulted the Prophet Muhammad during the water incident. Aasiya, for her part, maintained that the dispute had begun much earlier, when her goat escaped into her accuser's yard and damaged the property.

Whatever engendered the original quarrel, a lynch mob supported by the local clerics soon formed, and the local police had to rescue Aasiya. The authorities locked Aasiya in jail, presumably for her safety, and charged her for infringing Pakistan's blasphemy laws. But Aasiya sat in prison for seventeen months, until a court finally sentenced her to execution by hanging. In the meantime, her terrified husband and five children went into hiding. Pope Benedict XVI called for Aasiya's release, while mullahs offered $6,000 for her death.[3]

The governor of Punjab at the time, Salman Taseer, happened to be one of Pakistan's more liberal politicians; his political organization, the Pakistan Peoples Party, is affiliated with the Socialist International. Taseer openly opposed the blasphemy laws and appealed for a pardon. The governor's bodyguard, however, disagreed. With twenty-seven shots in the back, he assassinated Taseer. At the bodyguard's trial in January 2011, he proclaimed: "I haven't killed anyone unlawfully. I have taught a lesson to apostate Salman Taseer in the light of the teachings of the Koran and the Tradition of the Prophet."

Masses of supporters outside hailed the bodyguard as a hero, showered him with rose petals, and even brought him Valentine's Day cards. A conservative religious party, Jamaat Ahle Sunnat, released a statement warning that "no Muslim should attend the funeral [of the murdered governor] . . . or even express any kind of regret or sympathy over the incident," lest they meet the same fate.[4]

Next, Pakistan's Minorities Minister Shahbaz Bhatti, who was also a Christian, spoke out against the blasphemy law. Bhatti clearly understood the risk of doing so, since he had predicted his own death in a video four months earlier, even as he vowed to defend his community against this law. Sure enough, Minister Bhatti was assassinated shortly thereafter. The Punjabi Taliban proudly claimed credit; at the scene of the crime, they even left their pamphlets.

An aide to President Zardari told the international press that the country faced "a concerted campaign to slaughter every liberal, progressive and humanist voice in Pakistan." Yet Zardari's government sensed so much popular opposition to the reform of the blasphemy law that it dared not act. The female parliamentarian who had proposed the reform bill, Sherry Rehman, was receiving death threats at the rate of two per hour.[5]

Nine months after the assassination of the governor of Punjab, a judge on the Lahore High Court finally sentenced the bodyguard to death. But after delivering this verdict, the judge received so many death threats himself that he had to flee to Saudi Arabia.[6]

Though extreme and tragic, the Aasiya Bibi affair highlights the strong connection between religion and politics. According to the Turkish political scientist Yilmaz Esmer, "worldly concerns have yet to enjoy a complete reign over voting decisions anywhere on the globe."[7] Just how significant a role *does* religion play in politics?

Surely, the answer to this question depends on where and when. At one extreme is governance by divine guidance: theocracy. Other polities have official religions, explicitly religious political parties, and various degrees of religious control over the judiciary. Many Western liberal democracies, in contrast, are governed by constitutions that deliberately separate church from state; and yet religion still exerts a notable force on our electoral politics.

According to the great historian Bernard Lewis, the notion that religion *should* be separated from politics dates back only to the Judeo-Christian Enlightenment of eighteenth-century Europe. But Lewis traces the idea that religion and politics *could* be different back to Jesus' instruction to "Render . . . unto Caesar the things which are Caesar's, and unto God the things that are God's" (Matthew 22:21).[8]

Before this Christian enjoinder, however, separate religious and political authorities *did* technically exist. Positions of power that combined king, god incarnate, and high priest into a single role—such as the institution of the Dalai Lama—were rare and, arguably, unstable. The Kingdom of Hawaii exemplified a more typical configuration, in that it had a governing caste (the *ali'i*) that was distinct from the priestly caste (the *kahuna*). Likewise, Hindu society has traditionally had a separate ruling elite (the *Kshatriya*) from the *Brahmin* priests. Yet in spite of these distinct castes, rulers and priests were once commonly part of the same system—or part of the same body, according to Hindu scripture, in which the priests originated from the mouth of the primordial man, and the warrior rulers, from his arms (*Rig Veda* 10:90).

In the contrasting case of early Christianity, however, the religion was born in a conquered province, and so was foreign to the Roman Empire's political elite. Thus, Jesus' kingdom of heaven remained a world away from the control of the temporal Roman authority—at least until Christianity became the sole authorized religion and the state church of the Roman Empire in the fourth century. But so began the long path toward the secularization of Western states.

Despite the modern separation between church and state in many liberal countries, individual citizens still very much bring religion with them into the voting booth. This chapter sheds light on how religiosity impacts political orientation; we'll also consider the ways in which religious politics affect reproductive patterns and their evolutionary implications.

RELIGIOSITY AND SECULARISM ON THE POLITICAL SPECTRUM

In the story of Aasiya Bibi, the most conservative religious leaders and political parties sought to enforce the strictest Islamic punishment against her alleged blasphemy; the liberal Muslim politicians and non-Muslim minority groups, in contrast, opposed the Islamic punishment. Are the links between conservatism and religiosity, and between liberalism and secularity, universal?

RWA Scores

People who score higher (more conservatively) on the RWA scale usually agree with test items such as:

- *"It is always better to trust the judgment of the proper authorities in government and religion than listen to the noisy rabble-rousers in our society who are trying to create doubt in people's minds."*

And high scorers tend to *disagree* with the following statements:

- "*Atheists and others who have rebelled against the established religions are no doubt every bit as good and virtuous as those who attend church regularly.*"*
- "*People should pay less attention to the Bible and the other old traditional forms of religious guidance and instead develop their own personal standards of what is moral and immoral.*"*

According to Altemeyer, those who score highly on the entire RWA test (not only the questions related to religion) attend religious services, pray, read scripture, discuss religious moral codes, observe religious holidays, and participate in religious youth groups and education at higher rates than low scorers. Conversely, the average score of people who were not raised in a religious home is lower than those raised in any religion.[9]

Altemeyer once distributed to about five hundred of his Canadian students a fictitious law that proposed the following policy: to teach public school students, starting in kindergarten, to believe in God, pray together in school, memorize the Ten Commandments, and learn about Christian morality, and to eventually encourage them to accept Jesus Christ as their personal savior. Only 5 percent of those in the lowest RWA quartile would have voted for the law; in contrast, 48 percent of the highest quartile agreed with the proposal.[10]

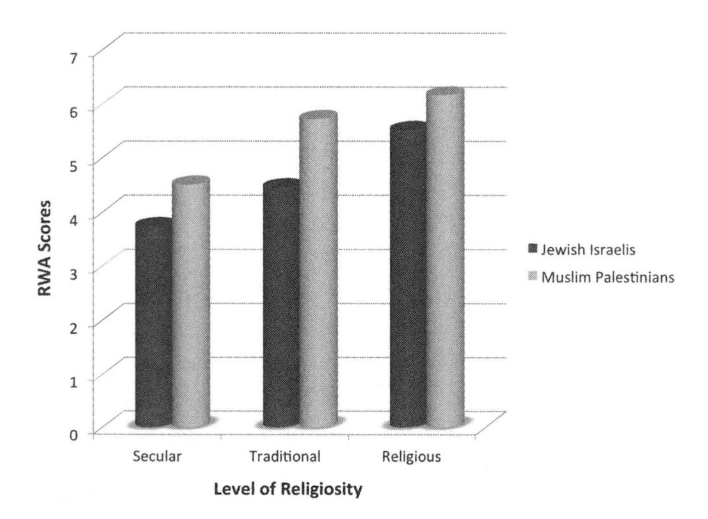

Figure 13. RWA Scores of Jewish Israelis and Muslim Palestinians, by Religiosity (1996).

RWA measures religiosity very effectively in non-Christian regions as well—at least among Abrahamic religions. Figure 13 shows how RWA scores increase among Jewish Israeli undergraduate students and their Muslim Palestinian counterparts as one moves from the group defined as secular to the groups defined as traditional and religious. The graph also suggests that the Palestinian Muslims were more politically conservative than the Israeli Jews within each of the three categories of religiosity. Figure 13 does not, however, show another interesting finding—that 69 percent of the Jewish sample belonged to the secular group compared with only 23 percent of the Muslim sample.[11]

Religious Behavior and Partisanship

Glenn Beck strongly identifies with his own religious tradition. One of his live shows, held at the Kennedy Center for the Performing Arts, was called "Divine Destiny." The mega-event featured religious music, speeches, and appearances by well-known priests and pastors. However, Glenn Beck isn't very fond of religions outside of the Judeo-Christian tradition—especially Islam. In November 2006, Beck appeared in a Headline News interview with Keith Ellison, the Democratic congressman from Minnesota who had become the first Muslim to sit in the House of Representatives. Beck said to Ellison:

> I have to tell you, I have been nervous about this interview with you, because what I feel like saying is, "Sir, prove to me that you are not working with our enemies." And I know you're not. I'm not accusing you of being an enemy, but that's the way I feel.[12]

On another occasion, Beck asked the builders of the so-called Ground Zero Mosque to build it elsewhere. Beck's antipathy to building a mosque near Ground Zero is commonplace among conservatives, and it distinguishes them from most liberals, at least in the Empire State; according to a poll of New York State voters, 90 percent of Republicans thought that it was wrong to build this mosque, compared with only 34 percent of Democrats.[13]

In general, religiosity predicts the partisanship of Beck's compatriots very well: Americans who attend church once a month vote Republican 66 percent of the time, while attending church on a weekly basis raises Republican voting up to 75 percent.[14] Religious opinions also vary by partisanship: 60 percent of Republicans believe that God created humans in their present form ten thousand years ago, as compared with only 38 percent of Democrats.[15]

By comparison with the United States, Europe is generally a more secular region, where most political campaigns lack explicitly religious issues. Nevertheless, the European Social Survey, which covers seventeen countries, has discovered that higher rates of church attendance and prayer significantly increased the chance that a European will vote for the political right (including Christian Democratic parties). The small minority of cases where this relationship did not hold true were all in extremely secular and northern populations, such as in Estonia and Scandinavian countries. These recent findings are consistent with a range of studies from the 1960s through the 1990s, which have shown that religiosity far surpasses social class as a determinant of party choice in Europe.[16]

This same link between European religion and politics goes back further, to at least the 1950s and 1940s. A de-Nazification study (mentioned in chapter 6) discovered that Germans who married outside of their own religion, as well as those who had no religious affiliation whatsoever, had a greater chance of being anti-Nazi dissidents. That is, their political orientations likely lay to the left—both religiously and politically—of their average peers.[17]

Whenever possible, it's vital to cross-compare findings from extremely different populations to see whether these trends represent Western cultural idiosyncrasies, or whether we are observing something of truly universal import. Japan serves as an excellent control case because the roots of its religious culture lie in the East. Moreover, Japan is among the ten most secular countries in the world (as defined by the proportion of people who disagree that religion is an important part of their daily lives). So one might not expect religion to play a major role in the Japanese political spectrum. Nonetheless, a clear political pattern emerged in 2007 when researchers asked a sample of Japanese citizens whether they agreed with the following statement: "*I draw comfort and strength from religion*." On the far left, only 27 percent of members of the Communist Party (Kyosanto) agreed with this statement on religion. On the center-left of the spectrum stood the Democratic Party of Japan (Minshuto); 35 percent of its supporters drew "*comfort and strength from religion*." Occupying the center-right niche was the Liberal Democratic Party (Jiminto), for which religious comfort rose to 47 percent. Finally, on the socially conservative right stood the Clean Government Party (New Komeito), which is based on the Sōka Gakkai sect of Buddhism. The Komeito platform included reducing government bureaucracy and bolstering the private sector. Fully 87 percent of Komeito voters drew "*comfort and strength from religion*."[18]

Of course, not all religious people vote for the right. After all, we even speak of some denominations as being more "egalitarian" or less "conservative" than others. Some people describe themselves as "spiritual." Spirituality usually entails a more xenophilic attraction to religious traditions of out-groups, with a simultaneous rejection of the limitations of one's own religion. In other words, "spirituality," from a political point of view, may have very different—if not the opposite—properties of religiosity, which measures the frequency of organized religious behaviors. In fact, semantic analyses of these terms in their linguistic contexts suggest that they are extremely different concepts; "spiritualism" clusters together with transcendentalism and existentialism.[19]

The bottom line is this: many political liberals practice religion too. But an objective quantification of religious behavior shows that the people who dedicate the most time to organized religious activities are more politically conservative. And this relationship is presumably universal.

RELIGION AS A NATURAL PHENOMENON AND A PRODUCT OF EVOLUTION

Indeed, religion itself is universal as well. In 2009, Gallup conducted a survey in 114 countries. They found that 84 percent of adults in the world consider religion to be an important part of their daily lives. This figure had remained stable over the previous four years they'd measured it.[20] Other studies show that 77 percent of the population in the average country believes in God; in no country does this figure drop below 36 percent.[21]

Numerous books in recent years have pondered the evolutionary origins of religion. Nicholas Wade, who has reported for *Nature*, *Science*, and the *New York Times*, has summarized their consensus view:

> Religious behavior has occurred in societies at every stage of development and in every region of the world. Religion has the hallmarks of an evolved behavior, meaning that it exists because it was favored by natural selection. It is universal because it was wired into our neural circuitry before the ancestral human population dispersed from its African homeland.[22]

Although evolutionary theories of religion are still in their infancy, often speculative, and vulnerable to numerous legitimate critiques, they are nonetheless fascinating and will improve with time.

For now, however, we'll try to stick to the hard facts that concern political orientation. One noteworthy point is that religiosity shows a very similar heritability to that of political attitudes (discussed in chapter 2). Researchers rated 275 of the Minnesota twins on a comprehensive scale to determine their religiosity. The measurements entailed all of the religious activities mentioned above, as well as having friends with similar beliefs, discussing religious teachings with family, and basing moral decisions on religious teachings. The religiosity scores of the identical twins correlated strongly (.62), while those of fraternal twins had only a moderate correlation (.42).[23]

RELIGION AND STRESS

Like most other social phenomena, religiosity's heritability and probable evolutionary origins do not shield it from environmental influences. Two such factors seem to have the greatest impact on religiosity: (1) poverty, and (2) the fear of death. As either of these two stress factors increases, so does religiosity and political conservatism. This finding does not mean that the poor are necessarily more likely to vote for the right; but this relationship between poverty and religion does help to explain why the poor do not overwhelmingly vote for leftist agendas to redistribute more wealth to themselves more quickly.

Poverty and Religiosity

Most people think of poverty in highly human terms, as a social condition. Some see poverty as a consequence of greed and injustice. Others see it as a reflection of ability. Or bad luck. Or God's unknowable will. Karma. An artifact of bad governance. These perspectives, of course, are colored by particular political orientations.

There is another way to think about poverty as well, though, which will be increasingly useful as we switch to a more evolutionary perspective later on: poverty is also an environment that some human beings inhabit, just as populations of nonhuman animals sometimes live in environments of lesser abundance. Economics, then, which is the science of scarcity, is not detached from evolution; to the contrary, economics is inseparably connected.

When fewer resources exist in an environment, or when a larger population competes for them, there is poverty. This environment of scarcity changes the social behavior of both human and nonhuman animals. Political behavior and religious belief are not exceptions to this rule. Just because the environment of poverty may influence these social phenomena, however, does not mean that religion or political orientation lack an evolutionary origin.

Figure 14 shows that people living in poorer environments are more likely to agree that religion is an important part of their daily lives. As wealth increases, religion loses importance at increasingly higher rates, which gives the graph a downward curve. Of the 114 countries covered in the graph, the most religious ones were Bangladesh and Niger (with over 99 percent agreement that "*religion is an important part of my daily life*"), followed by Yemen, Indonesia,

Malawi, and Sri Lanka, at 99 percent, and Somaliland, Djibouti, Mauritania, and Burundi at 98 percent. Each of these extremely religious countries has a per capita income below $5,000. Seven of the ten are Islamic.

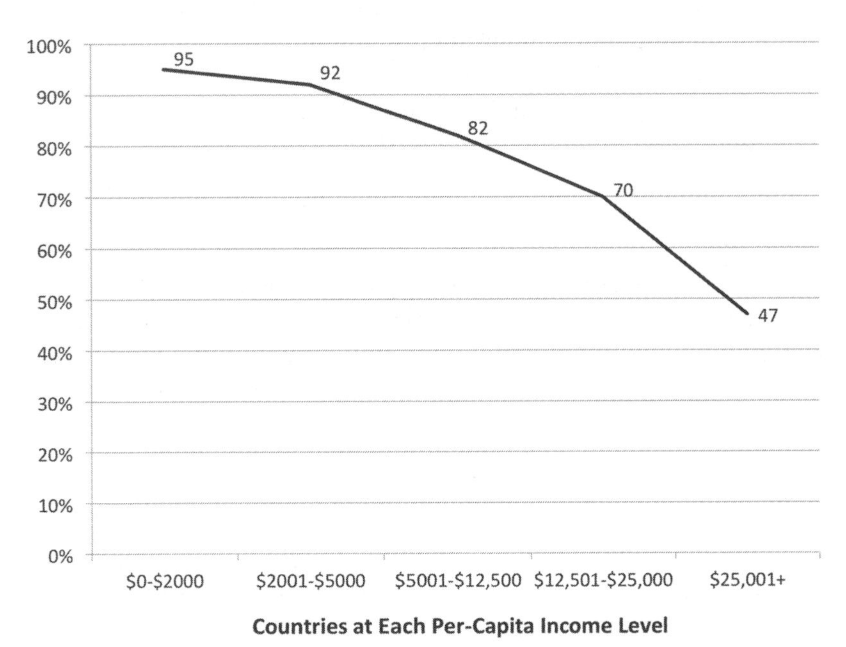

Figure 14. Proportion of People Who Agree Religion Is an Important Part of Their Daily Life (median responses by countries' income level; 114 countries surveyed in 2009).

The least religious countries were generally richer and at higher latitudes (or, in the case of Vietnam, Communist): Estonia, Sweden, and Denmark were in the high teens, below 20 percent; Japan, Hong Kong, and the United Kingdom were in the mid-20s; while Vietnam, France, Russia, and Belarus were in the low 30s (China was not surveyed).[24]

In addition to income, other aspects of poverty also affect religiosity. Specifically, having lower levels of education (as measured by the UNDP's Human Development Index) strongly correlates with placing greater importance on God. Lower life expectancies are also moderately associated with increased importance placed on God.[25]

Having less education usually means that people are able to extract fewer resources from the environment, sometimes through even harder labor. Perhaps the greater hardships associated with lower education and poverty push people to think of the next life more than this one. Perhaps poverty causes a person to value his or her group a bit more than individual resources. Or to rely somewhat more on faith in divine help than in the harsher forces at play in the environment. For whatever reasons, the environment of poverty increases religiosity.

A fascinating poll from the wealthy United States offers some support to the "divine help" hypothesis. This study compared the relatively high-stress economic period of 1978–1982 with the low-stress period of 1983–1987.[a] During the harder times, 29 percent of Americans believed in astrology; in the good times, however, only 12 percent looked to the stars for guidance.[26]

Fear of Death and Religiosity

The notion that the uniquely human, existential fear of death may relate to religion dates back to the ancients. In 50 BCE, the Roman philosopher Lucretius declared that "fear begets gods."[27] Bronisław Malinowski, one of the fathers of twentieth-century anthropology, articulated this conviction even more explicitly: "Death," he wrote, "which of all human events is the most upsetting and disorganizing to man's calculations, is perhaps the main source of religious belief."[28]

Is there any proof that fear of death increases religiosity? Indeed, dozens of studies conclude that death anxiety is associated with religiosity. In the 1982 Lebanon War, secular Israeli soldiers reported a greater fear of death than their religious counterparts.[29] So even if there are some atheists in the proverbial foxhole, religion would better soothe them in this harsh psychological environment.

A meta-analysis of thirty-six studies found that in two thirds of them the religious groups feared death less than the control groups.[30] This is not to say that the other third of the studies found no relationship. Sometimes, depending on the definition of these terms, the opposite can be shown. For instance, among a group of gay men suffering from AIDS, higher death anxiety correlated with greater church attendance.[31] Either way, the relationship between the fear of death and religiosity remains significant.

RELIGION'S IMPACT ON TRIBALISM AND REPRODUCTION

The time has come to ask, "What is religion?" Some of the greatest minds that have most recently addressed this question in the name of science have done so with a fervent agenda to argue *against* the phenomenon that they've set out to explain. It's not that these arguments are necessarily illegitimate. Many of the points they make are valid and interesting. But at the end of the day, they serve a political objective and miss the scientific point, which is to dispassionately understand reality, as it exists. And religion is not the outer layer of an onion that can be peeled away by disabusing people of "mistaken beliefs" so as to expose a shinier, smoother self. In this line of thought, political psychologist Jonathan Haidt incisively points out:

> Many scientists today think of religion primarily as a set of *beliefs* about God, the world, and the origins of humankind. Because many of these beliefs are demonstrably false, religion is then dismissed as a foolish and virulent delusion (Dawkins, 2006; Harris, 2006). But many of the great sociologists, most of whom were atheists or agnostics, thought that religious beliefs were the surface manifestations of something deeper.[32]

What, then, is this deeper something that religion reflects?

People universally fear death. The greater the fear of death, and the greater the stress of life, the more religious people become (above whatever their genetic baseline of religiosity may be). Religion alleviates this anxiety by promising the continuation of life, in some form, after death. Religion also confronts the problem of selfishness that arises from the finite amount of resources in the world; religions teach that control over one's own selfishness is linked to the outcome of the life to come, and even suffering in this lifetime. In short, religion fights death with life, and selfishness (extreme forms of which are considered evil) with altruism.

Religion accomplishes these feats by diverting the individual's attention from his own suffering and focusing it instead on the meaning of belonging to the group. Regardless of the reproductive success of the individual, he or she can celebrate the reproduction of the tribe. However much selfishness one confronts in the world, at least altruism is higher among coreligionists. And despite fears of death, one can take solace in the eternal continuity of the tribe. In many traditions, the next life promises a reuniting with the entire tribe in the next world. Some conceptions of heaven exclude members of out-groups.

Thus, the founder of sociology, Émile Durkheim, fundamentally understood religion when he wrote: "God and society are one of the same . . . the god of the clan . . . can be none other than the clan itself."[33] The more that people worship a god, then, the more they care about their clan—about its defense, reproduction, and integrity. And the more they fear threats from other groups.

An incident that occurred in December 2010 exemplifies Durkheim's principle. Iran sentenced a young man named Habibollah Latifi to death for "waging war against God." How did Latifi manage to commit such a momentous crime? The twenty-nine-year-old Kurd had supported a banned rebel group fighting for greater autonomy for his "clan." The Iranian government saw this rebel group (the Party for a Free Life in Kurdistan) as a threat to the divinely guided

Iranian theocracy, and therefore to the Iranian people. By Durkheimian logic, then, the young Kurdish political activist threatened God (that is, the God of ethnic Persians). It didn't matter that most Kurds are Muslims, too. Although common sense tells us that Latifi and his accusers share the same God, both the Durkheimian definition and the Iranian government disagree.[34]

Although Islam has transcended ethnic boundaries with various degrees of success, this doesn't change the fact that the earliest religions began with ancestor worship and local gods. At some level, God and the tribe remain one. Religion is a far more complicated phenomenon than this, of course. Yet this is the principle most useful for understanding religion's influence on politics as well as its evolutionary implications.

The Tribal Relationship among Coreligionists

Outside of the great faiths, almost all traditional religions have a tribal or geographic basis. The Incas would literally kidnap the local gods of the tribes they conquered, holding stone idles hostage in the imperial capital to extort good behavior from the tribes. Within the great faiths, the oldest ones in both East and West also retain a more tribal identity. This observation is true of Hinduism and Judaism. Native Chinese religions, including folk traditions, Taoism, and Confucianism, prominently feature ancestor veneration. In contrast, the major daughter religions—Buddhism, Christianity, and Islam—have developed tribe-transcendent ideologies. Thus, these relatively newer religions have spread around the world to believers of very distantly related populations.

In the case of both the older *and* the newer religions, the key issues at stake, where politics intersects with religion, revolve around the relatedness of "the tribe"—that is, how successfully an ethnic group is reproducing, how much it mixes with other groups, and which resources accrue to which gene pools.

Traditional Religions

Some of the most salient laws of traditional religions function to prevent the tribe from reproducing with out-groups (i.e., these laws enforce endogamy). Consider the Polynesian taboo system. In ancient Hawaii, a hierarchical caste system prevailed. Strict religious rules mandated what could enter one's body, depending on which caste one belonged to. Eating with people of lower social groups was prohibited. And sexual intercourse with them was "polluting."[35] The very highest noble caste was even supposed to marry incestuously, between siblings.[36]

To ancient Polynesians who lived in the more hierarchical societies, castes were by no means arbitrary, "socially constructed" units; rather, members of higher castes had greater *mana*, which meant a closer genealogical closeness to an ancestor god. From this inherited connection, the higher-born castes were more in touch with the god's productive power over nature. At the top of the social pyramid, the chiefs or kings were believed to have an almost supernatural power over natural resources. This power justified their station as well as the sacred taboos that protected it.

An important Marquesan chief, for example, attained his status by descent from the deities associated with the island's staple food, the bread-fruit.[37] The kings of Tonga traced their lineage back to the original ruler, the Tu'i Tonga, believed to be the offspring of the god Tangaroa and an earth-woman whom he impregnated.[38] The Hawaiian nobles had as many as ten levels of genealogical seniority, based on proximity to deity ancestors.[39]

The Polynesians' perceived level of kinship with deities might have overlapped with terrestrial genetic differences as well. So caste-based endogamy over many years could have conserved genetic differences between social groups. Regardless, religion in caste-based Polynesian societies controlled the flow of genes.

In traditional Hindu society, the castes are also endogamous . . . most of the time. When individuals try to cross the line, tragedy can ensue. In June 2010, two teenage lovers on the outskirts of Delhi were beaten, stabbed, and electrocuted to death by the girl's parents, uncle, aunt, and cousin. Asha had made the fatal mistake of loving Yogesh, who belonged to the lower, untouchable Yadav caste. When asked why it happened, one of Asha's uncles later said the killing was to save the family's honor and to avoid "a bad precedent for the other children in the family."[40]

The uncle's justifications are cultural and psychological. We can consider his explanations to be his "proximate" motivations, or those that he is conscious of. From an ultimate, evolutionary perspective, could there have been any objective difference at all between Yogesh and a parentally approved mate choice from Asha's own, "higher" caste?

When the Indo-Europeans arrived in India from Central Asia some five thousand years ago, they mixed with native populations. Most of the newcomers must have been men, because the maternally inherited gene pool of contemporary Indians (their mitochondrial DNA) is more uniform. But studies of paternally inherited, Y-chromosome DNA have found that the higher castes have Y-chromosomes that more closely resemble those of Europeans; the Y-chromosomes of the lower castes are more similar to those of Asians.

The international team of geneticists that carried out the research concluded that the Indo-European colonists had established themselves as upper castes upon arriving in the Indian sub-continent. The data also suggested that a caste-like class structure had already developed among the native South Asian tribes, *previous* to the influx of Indo-European colonists.[41] The indigenous caste system would likely have also segregated reproductive groups by some genealogical criteria—as in complex Polynesian societies. In any case, Asha likely had fewer genes in common with Yogesh than she would have shared with a parentally desired mate choice from her own caste.

Needless to say, these findings do not justify honor killings! However, they do show that, in the Indian caste system, religion is linked to genetics and ancestry. And politics is linked to both. Indeed, one of the various meanings of the word *varṇa*, which refers to the high-level caste divisions in Hindu society, is "color."[42] Many have disputed or embraced this connotation of varṇa, based on their political views.

One of the mechanisms that have prevented exogamy in traditional Indian society has been an elaborate system of food taboos. Where they exist, these taboos vary by region and group. But one of their overarching principles is that members of other castes and religions pollute food.[b] That's why Aasiya Bibi was accused of polluting the water.

Like the South Asians and the Hawaiians, the Israelites also had their own dietary laws. As in the previous cases, preventing certain foods from entering the body symbolically relates to preventing marriage to out-groups. Indeed, the eating instructions given in Deuteronomy (the fifth book of the Old Testament) come hand in hand with laws prohibiting exogamy with various local tribes, such as the Canaanites. The text explains the disastrous Babylonian conquest and the exile of Judean elites as the result of the Israelites' failing to honor ethno-religious marital exclusivity.[43]

Even today, there seems to be a distinct tribal logic to political controversies involving Judaism. In the following two recent cases, perceptions of ancestral relatedness trump religion. In the 1980s and 90s, Israel organized covert airlifts of Ethiopian Jews to Israel to rescue them from civil war and famine. Operations Moses and Solomon transferred over twenty thousand members of this population to Israel, where they and their children now make up over 2 percent of the Israeli population. These Ethiopian Jews legally qualified to immigrate to Israel under the state's Law of Return because they had Jewish ancestry stretching back to Menelik I, the son of the biblical King Solomon and the Queen of Sheba.

A controversy arose over another population of Ethiopians, however, knows as the Falash Mura. This group of Ethiopians once belonged to the Jewish community, but their ancestors converted to Christianity in the nineteenth century—

sometimes under force and sometimes voluntarily. In addition, it's not clear how many of the Falash Mura truly have Jewish ancestors and how many are simply eager to emigrate from Ethiopia to Israel.

In either case, the Falash Mura do not legally qualify to enter Israel under the Law of Return because they had long lost their religious cultural identity. Nevertheless, the perception of genetic relatedness, combined with compassion for cousins in difficult circumstances, triumphed over religious and national laws in the end. Upon the cabinet's approval of a plan to take in eight thousand Falash Mura, Prime Minister Netanyahu said: "These are the seeds of Israel—men, women, and children who currently find themselves in the worst living conditions." Thus, Netanyahu appealed to the moral duty of Israelis to intervene in a "complex humanitarian crisis."[44]

In Israeli politics, perceptions of shared ancestry also lead to less compassionate ways of overriding religion, even in the name of religion. In June 2010, eighty Ashkenazi parents sat in jail. Thousands of ultra-Orthodox Jews, also of Ashkenazi descent, were protesting in support of the parents, who belonged to a strict Hasidic sect called the Slonim. The Slonim had just been sentenced for contempt of court. The legal proceedings had involved the parents' effort to segregate their schoolchildren in separate classrooms from other Jewish children of Sephardic and Middle Eastern origin.

Yakov Litzman, a legislator from the far-right Ashkenazi party United Torah Judaism, explained the Ashkenazi parents' objection to keeping desegregated schools:

> There is a set of rules [in the Ashkenazi ultra-Orthodox community]. We don't want televisions in the home, there are rules of modesty, we are against the Internet. . . . I don't want my daughter to be educated with a girl who has a TV at home.[45]

In other words, the most conservative of the conservative Jews who had spent two thousand years in Europe believed that the Jews who had remained in their Middle Eastern neighborhood weren't religious enough to sit in the same classroom. Of course, the ultimate reason that the Slonim wanted to segregate their children from the Middle Eastern Jews was to prevent them from marrying each other; instead, they prefer more closely related mate choices for their children.

Extremely religious Ashkenazim already have slightly incompatible forms of dietary laws from Sephardim and Mizrahim. But the Slonim parents must have feared that their dietary reproductive barrier would no longer suffice to keep their children safely away from other communities of coreligionists—especially in a quickly melding modern Israel.

Transcendental Religions

Where do the newer, transcendental faiths fit in? First, let's put the big picture in perspective. Figure 15 shows the percentage of the major religious groups in the world, as estimated by the CIA[46] in 2010.[c]

The pieces of the pie chart that are pulled out from the center are the newer religions that could be considered globalized, because they have successfully proselytized, or easily accept converts from all peoples. The two largest religions in the world come from this category: Christianity and Islam. Together, their believers make up more than half of the people living on earth. Have religions such as these truly managed to transcend the forces of tribalism underlying older religions? Sometimes they have; in other instances, not quite so well.

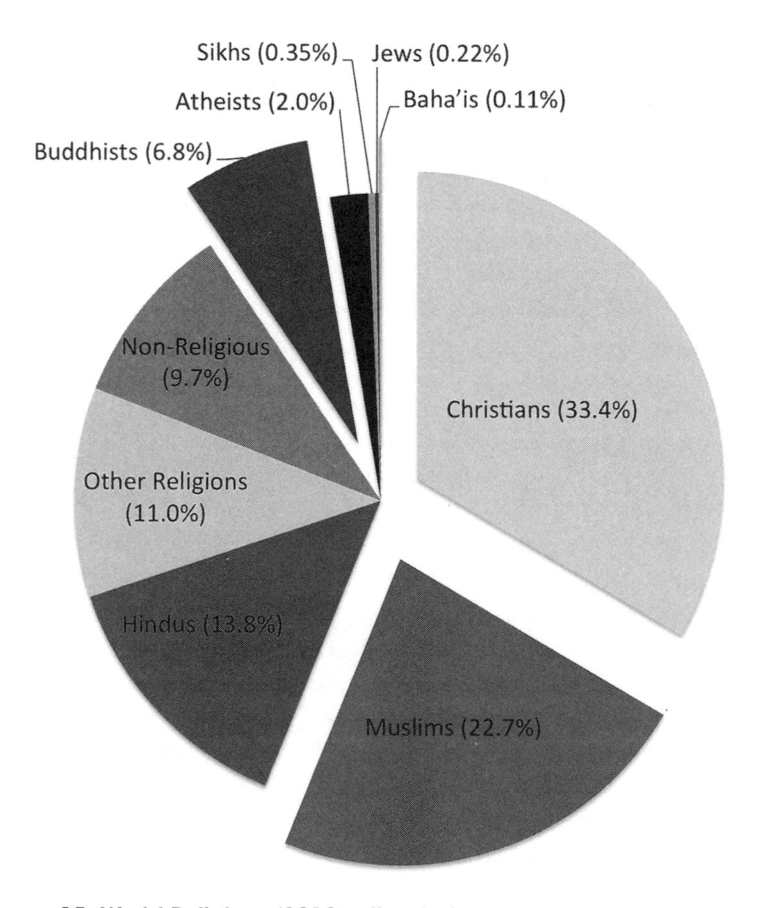

Figure 15. World Religions (2010 estimates).

Christianity has a universal message that has resonated with diverse populations around the world. Still, the religion often uses kinship terms: one can speak of "brothers and sisters in Christ"; people address nuns as "sisters," priests as "fathers," and the Pope (which means "father") as "the Holy Father." God has a son, and a father, and a mother.[47]

Certainly there are countless Christians who have been transformed by the religion's transcendent message. Still, there is a paradox—one that the RWA test makes apparent. When people score higher and higher on RWA, there is a greater chance that they are more religious as well. At the same time, higher RWA scores translate into higher levels of ethnocentrism. Yet as a universal religion with no barriers against ethnicity, Christianity, in theory, is incompatible with ethnocentrism. The more conservative and religious a Christian becomes, however, the force of tribalism often grows as well, just as it would within any other human being.

As one moves further and further to the right of the political spectrum, it becomes easier to find Christians who discriminate against their coreligionists of different ethnic backgrounds. At Glenn Beck's "Restoring Honor" rally, for instance, he rejected President Obama's Christianity, calling it "a perversion of the Gospel of Jesus Christ as most Christians know it." Instead, Beck peddled the conspiracy theory that Obama is a closet Muslim, citing the fact that Obama held a Ramadan dinner in the White House.[48] Beck didn't mention that George W. Bush—a white, conservative coreligionist—held eight such dinners during his terms in office. Beck is obviously biased against Obama. But it's difficult to *prove* that the cause is racial. Perhaps Beck just hates liberals.

Moving from the religious opinions of the far right to those of the *extremely* extreme far right helps to clarify the race issue. Members of churches aligned with the Christian Identity movement have resolved the ethnocentrism paradox by radically changing the rules of Christianity. According to their worldview, God is white. The Israelites of the Old Testament were not Jews, but rather whites, and Moses was a great Aryan leader. The various tribes of Israel founded Germany and the Nordic countries. White people are the only children of God. Other races do not have souls. They call people of color "mud races," believing that they were engendered by bestiality. Jews, they imagine, come from a separate lineage, going back to Satan's seduction of Eve, in the form the serpent. According to their religious beliefs, only the members of their own Aryan tribes come from the seed-line of God.[49]

Needless to say, the great majority of Christians would reject these views as historically inaccurate and theologically heretical. Yet the case is of interest to political psychology.

Christian Identity adherents are not the only revisionist group to change the ethnic identity of biblical characters to their convenience. Some far-right, ethnocentric Rastafarians believe that they (people of West African heritage) are the real Jews, while the Jewish people are impostors. Zion is not really Israel, but rather Ethiopia. This worldview differs from that of most Ethiopian Jews, who have yearned for their Zion in Israel.

Have Muslims been able to transcend ethnocentrism better than Christians? Have they succeeded in making the *ummah* (the global community of the Muslim faithful) their primary tribe? In April 2010, Iran was scheduled to host the Islamic Solidarity Games, an international sporting competition for the fifty-seven member states of the Islamic Conference. This organization's mission is to represent the "collective voice of the Muslim world."

The Solidarity Games ran into trouble when Iran wanted to call a certain body of water the Persian Gulf, while the people on the other side wanted to call it the Arabian Gulf. The disagreement sunk efforts to design logos and medals. In the end, the Islamic Solidarity Games—ironically—had to be canceled.[50]

Of course, the Iranians and their coreligionists on the Arabian Peninsula did not cancel the games over petty differences. They are divided not only by the controversially named body of water, but also by sectarian differences. Most Iranians are Shia Muslims, while most Arabs across the gulf are Sunni Muslims. This sectarian schism in Islam, in turn, corresponds not only to differences of opinion over the rightful successors to Muhammad, or over which collections of Hadith (sayings and deeds ascribed to the Prophet Muhammad) are valid. Although the peoples of Iran are diverse, the dominant ethnic group is Indo-European. The word "Iranian" is related to the word "Aryan." The Arabs across the water, on the other hand, are a Semitic people.

These differences may play some distant role in the cancelation of the Solidarity Games. On the more immediate, political level, a long history of enmity divides the two regions. This geopolitical tension has most recently been exacerbated by Iran's nuclear program and support of Shia organizations (including the terrorist group Hezbollah) and other armed groups in the Arab world.

A tragic case of tribalism overpowering Islamic solidarity has befallen the Sudan. Arab Sunni Muslims have been accused of establishing apartheid for, and committing genocide against, black Sunni Muslims. The United Nations estimates that, between 2003 and 2010, Arab Sudanese have killed at least three hundred thousand black Sudanese. Although natural resources are part of the equation, Western journalists have described the state-sponsored "Janjaweed" militias as Arab supremacists, who espouse a blatant antiblack racism.

Tribal Conflicts between Religions

Conflicts between different religious groups often entail which genes flow where, and which genes reproduce at which rates. Examples of this principle come from both personal disputes and international political controversies.

In the series of movies adapted from J. K. Rowling's Harry Potter books, there's a girl who resides in the Ravenclaw House at the Hogwarts School of Witchcraft and Wizardry named Padma Patil. In *Harry Potter and the Goblet of Fire*, Padma's sister convinces her to go on a date with Ron Weasly, a redheaded European boy. Padma is visibly displeased with Ron and his appearance. In any case, Ron has no interest in Padma; rather, Ron is unbearably attracted to a white girl named Hermione.

In real life, the name of the actress who plays Padma is Afshan Azad. The twenty-two-year-old Afshan grew up in the United Kingdom, raised by a Muslim Bangladeshi family. In a story quite different from the movie, the real Afshan fell in love, quite *against* the will of her family, to a young man from a different ethno-religious group. One day Miss Azad's brother heard her talking on her cell phone in her bedroom, and he assumed that she was speaking with her Hindu boyfriend. Afshan's brother dragged her away by her hair, punched her repeatedly, and branded her a "slag." He shouted: "Marry a Muslim or you die!" Afshan's mother called her a "prostitute." Her father threatened to send her back to Bangladesh for an arranged marriage, where she would surely marry someone more closely related to her.[51]

Many religious disputes at the international level also concern how particular genes are replicating. The bestselling work of political nonfiction produced by a German author in the past decade is called *Germany Does away with Itself*. The author is former member of the Executive Board of the Deutsche Bundesbank Thilo Sarrazin. In his book, Sarrazin laments that Turkish and Moroccan immigrants are "constantly producing new little headscarf-girls," while "living by welfare [and] denying the legitimacy of the very state that provides that welfare." Sarrazin goes on to claim that "entire clans have a long history of inbreeding and a correspondingly high rate of disability," such that "genetic factors could be partially responsible for the failure of parts of the Turkish populations in the German school system," and that immigrants are making Germany "dumber."[52]

Sarrazin worries that the demographic growth of Muslims in Germany may overwhelm the ethnic German population within generations to the point where Germans would become "strangers in their own land." The controversial author claims that "in no other religion is the transition to violence, dictatorship and terrorism so fluid." Polls suggested that nearly half of the German population agrees with Sarrazin's views and that 18 percent would vote for his party if he were to start one.[53]

Translated into evolutionary terms, many of Sarrazin's followers apparently fear the following propositions: (1) their gene pool is reproducing at a much slower rate than the immigrant gene pool; (2) the immigrants do not reproduce frequently enough with other populations, suggesting a lack of good will or alliances with other groups; (3) the foreign, endogamous gene pool is extracting too many resources from the country's economic environment; (4) the demographic imbalance could eventually enable the foreign population to take political control over the state's institutions of power; (5) once in charge, this population's political culture may subvert the democratic safeguards protecting the rights of minorities to their share of resources and freedoms; and (6) if steps one through five transpire, ethnic Germans would suffer reduced fitness within their native habitat.

Sarrazin did not come from the far right. Rather, he belongs to Germany's Social Democratic Party (which is affiliated with the Socialist International), even though the party leadership considered revoking his membership because of his inflammatory statements. Coursing through his veins runs blood from English, Italian, German, French, and probably Arab or Berber ancestors. "Sarrazin" comes from the term used by medieval Europeans to refer to Muslims.

Sarrazin's attitude toward Turks and Arabs likely stems more from anti-ethnocentrism than from extreme ethnocentrism. In other words, he is angry with these populations because he believes they will not sufficiently mix and identify with their adoptive nation. Translated into much more politically correct lingo, Sarrazin essentially argues—as has German chancellor Angela Merkel—that Germany's post-War multiculturalist policies have failed.

Why do many Muslim immigrants in Germany not assimilate? Perhaps because, *on average*, the group is more religious than average contemporary Germans. And underlying greater religiosity is greater conservatism, which entails greater endogamy.

Conservatives, according to cognitive linguist George Lakoff, think in terms of a "morality is wholeness" metaphor, accompanied by the corollary belief that "immorality is degeneration." Even if Lakoff may not have thought of this conservative moral frame in reproductive terms, it corresponds well with endogamy (keeping an ethnic group "whole"). Islamic law, in fact, forbids women from marrying outside of their religion. Liberals, on the other hand, more frequently believe that many aspects of morality change with time and circumstances, and even that morality should improve.[54] This liberal metaphor is compatible with exogamy (the ex*change* of genes between ethnic groups).

Economic Prosperity, Secularism, and Exogamy

While Germany's Muslims may be relatively more conservative, religious, and endogamous, and less affluent than the average citizen, the population of Germany as a whole has trended in the opposite directions. In fact, Europeans in general have grown richer, more secular, and more exogamous in recent decades.

Between 1981 and 1999, real GDP per capita, adjusted by purchasing power, increased in Belgium, Denmark, France, (West) Germany, Great Britain, Iceland, Ireland, Italy, the Netherlands, Spain, and Sweden. Put simply, this means that Europeans could buy more of the same goods and services for their money, making them feel richer as a result. Over these same eighteen years, every one of these countries also grew less religious; the proportion of people who claimed to belong to no religious denomination rose across the board.

When the European Values Survey asked Europeans during these years whether "shared religious belief is very important for marriage," the proportion of people who agreed in each of these eleven countries fell.[55] That is, not only were Europeans becoming less religious, but their attitudes also suggested that they were becoming less culturally endogamous. More exogamous attitudes translate into higher probabilities that different ethnicities will reproduce with one another. Europeans were becoming more and more different from their Muslim immigrants. And a large fertility differential was contributing to the imbalance.

Today Turkish immigrants in Germany have an average birthrate of 2.4, compared with the average ethnic German birthrate of 1.38. At this rate, the population of ethnic Germans will be halved in about six generations, while the number of Muslims in Germany will more than quadruple.[56]

Religiosity and Fertility

According to a right-wing Israeli joke, to be an Israeli leftist is "to make fun of a [religious] settler with five kids, and then to go out for a stroll in the National Plaza with five poodles."[57] Is there actually any link between how religious people are and how many children they have?

Psychologist Frank Sulloway, a recipient of the McArthur Genius Award, set out to determine the possible evolutionary consequences of religious belief. While working at UC Berkeley's Institute of Personality and Social Research, Sulloway gathered data from eighty-six societies from the World Values Survey. The sample totals over two hundred fifty thousand people interviewed between 1981 and 2004, and it represents more than 85 percent of the world's population.

Even after controlling for numerous demographic differences that could be confounded with fertility (such as socio-economic class), Sulloway found that both religiosity and religious beliefs independently predicted higher fertility. This finding means that the frequency of attending religious services, praying, and so forth (religiosity) is associated with having more children. And religious beliefs—such as the belief in God, a soul, and life after death—also predict more children, with or without the religiosity behaviors mentioned above. The people with the strongest religious convictions had 1.48 times as many children as the most secular people.

In many industrialized countries, fertility has greatly declined within the past couple hundred years (chapter 12 explains this phenomenon and its political repercussions). In these modernized countries that are reproducing at or below replacement rate, does religion still predict high fertility? The answer, according to Sulloway's data, is an unequivocal "yes."[58] Even in low-fertility developed countries, religious people have more children.

A Brief Word on Abortion and Religion

Sulloway's data show that an anti-abortion stance weakly correlates with religious beliefs, such as the belief in heaven. Moreover, pro-lifers had 1.45 times as many children as those who had the most pro-choice attitude.[59] Global religious beliefs about abortion therefore clearly have evolutionary implications; having an anti-abortion attitude is associated with greater fitness.

Religions have two reasons to oppose abortion: (1) discouraging or prohibiting abortion multiplies the faith's believers more quickly, and (2) if a principle function of religions is to counter death anxiety, religion accomplishes this through a pro-life stance on abortion (the thought of killing unborn fetuses simply doesn't assuage death anxiety).

These global trends, however, belie some fascinating differences *between* religions; and these cultural idiosyncrasies are worth considering, since they explain why abortion is a prominent political issue in some countries more than in others.

For Buddhism, the question of exactly when life begins is not quite as important a point as it is in the Western, Abrahamic religions. Samsaric existence implies a cycle of countless lives. Nevertheless, ancient and modern authorities agree that a given life starts at conception. Since the Pali Cannon takes an anti-abortion stance, the more conservative Theravadin Buddhist countries, such as Sri Lanka and Thailand, have enforceable laws prohibiting the practice, with the exception of rape and endangerment of the mother's life.

Among a sample of Thai medical staff interviewed in 1998, 36 percent expressed the concern that performing abortions would incur bad karma. Fifty-five percent of the staff, however, favored a liberalization of Thai abortion laws. Thai monks oppose abortion if asked, but do not make public statements about the issue, much less get involved in anti-abortion politics. In fact, the Thais have a religious mechanism to atone for the guilt they feel over the three hundred thousand abortions carried out in their country each year. In Bangkok, people leave offerings of milk cartons and children's toys at a shrine in a Buddhist sanctuary. In November 2010, authorities found 2,002 aborted fetuses in the temple morgue complex, believed to have arrived from illegal clinics.[60]

In Japan, where Buddhism is not the state religion (as it is in Thailand), abortion is legal. The Japanese alleviate their conscience through the Mizuko Kuyo memorial service for "water children."[61]

As for Tibetan Buddhists, their highest religious authority is the Dalai Lama. The current and fourteenth Dalai Lama believes that abortion is permissible if a baby will be born retarded or will cause serious problems for the parent.

In regard to the Judeo-Christian tradition, the Bible does not specifically mention abortion. The overwhelming majority of liberal and Reform rabbis support a woman's right to choose. Even Orthodox rabbis believe that a fetus does not have equal rights to a born person, which is reflected by their granting women permission to abort a fetus if it endangers her life or health. Jews generally believe that the decision to abort a fetus is a personal matter that should not be decided by government. Official statements taking this position have been expressed by the Central Conference of American Rabbis, the United Synagogue of Conservative Judaism, the National Council of Jewish Women, and Hadassah, among others.

In Israel, abortion is not a major political issue, yet it does cause controversy. Predictably, the far left parties (such as Meretz and Shinui) advocate for even fewer restrictions on abortions, while the far-right religious parties would like to see abortion become illegal.

In Islamic belief, the human soul is *not* created upon conception. According to the Sahih al-Bukhari collection of Hadith, which Sunni Muslims consider the most authentic:

> Allah's Apostle, the true and truly inspired said, "(as regards your creation), every one of you is collected in the womb of his mother for the first forty days, and then he becomes a clot for another forty days, and then a piece of flesh for another forty days. Then Allah sends an angel to write four words: He writes his deeds, time of his death, means of his livelihood, and whether he will be wretched or blessed (in religion). Then the soul is breathed into his body." (vol. 4, book 55, no. 549)

Although most Muslim scholars believe that abortion is wrong, many will permit it under certain conditions. The dispute usually revolves around how late the practice may be carried out. When the Serbian army raped thousands of Muslim Bosnian women, a fatwa was issued allowing the women to abort, but it encouraged them to do so before 120 days (when the soul is believed to enter the fetus).

The religion that is most vehemently opposed to abortion is Christianity. Every one of the six states in the world that outlaw abortion under *all* circumstances are Christian: the Vatican, Malta, Chile, Nicaragua, El Salvador, and the Dominican Republic. In the wider Christian world, abortion often constitutes a more controversial political issue than elsewhere.

The religion's opposition to abortion stems from the central Christian narrative, which begins with the Virgin Birth. For Jesus to be born, God had to take control of Mary's body. If this divine act of reproduction had been interrupted by human choice, human salvation would not have been possible.

According to this logic, branches of Christianity that place greater emphasis on the worship of Mary, mother of Jesus, should oppose abortion even more than branches that place relatively less emphasis on her. Indeed, the six countries with the strictest abortion laws are all Catholic. Among Christians in the United States, white Catholics have traditionally opposed abortion by about seven percentage points more than white Protestants have.[62]

These variations in abortion beliefs highlight how culture can impact reproductive success, and therefore evolutionary fitness.

Where Are We Headed from Here?

Over the last couple of centuries, the wealth of more and more industrialized societies has increased, and so has secularism. At the same time, these populations' birth rates have fallen. One might think, then, that overall the world is becoming a more secular place. Not true. The *proportion* of religious people in the world is increasing. And the rate of increase is faster than ever before, because the fertility differential between religious and secular populations has never been greater.[63] Even in a wealthy country like the United States, the congregations of conservative religious denominations have grown the fastest because of higher fertility rates and earlier childbearing ages.[64]

All else being equal, a world with proportionally more religious people may shift further to the right on the political spectrum. On the other hand, increases in the wealth of developing countries could also lower their citizens' religiosity. For now, we can only speculate.

What *is* abundantly clear from this chapter are the following principles:

- Religiosity (i.e., the frequency of religious behaviors) goes hand in hand with political conservatism;
- Religiosity is a facet of tribalism, one of the universal underlying determinants of political orientation; and
- Religious beliefs influence how populations reproduce with one another and how many children they have; thus, religion has evolutionary consequences (in addition to its probable evolutionary origins).

The following chapter is all about the last facet of tribalism: sexual attitudes. We'll see why opinions about masturbation, pornography, oral sex, and gay sex—among other acts—divide public opinion in countries around the world.

8.

ATTITUDES TOWARD SEXUALITY, HOMOSEXUALITY, AND GENDER ROLES

In an address on state television in 2003, Iran's supreme leader Ayatollah Ali Khamenei declared that he feared unrestrained sexuality more than the sword:

> More than Iran's enemies need artillery, guns and so forth, they need to spread cultural values that lead to moral corruption. They have said this many times. I recently read in the news that a senior official in an important American political center, said: "Instead of bombs, send them miniskirts." He is right. If they arouse sexual desires in a given country, if they spread unrestrained mixing of men and women, and if they lead youth to behavior to which they are naturally inclined by instincts, there will no longer be any need for artillery and guns against that nation.[1]

Why is sexual morality invariably such a central and contentious issue in the politics of every people? And why is it so in some countries more than in others?

TOLERANCE OF NONREPRODUCTIVE AND NONMARITAL SEXUALITY

The last two chapters have shown that ethno-religious conservatives prefer more endogamous mate choices, and that they have a greater number of children than the relatively more secular people to the left of them on the political spectrum. To achieve this reproductive outcome, conservative cultures around the world are less tolerant toward all forms of sexuality *other* than sexual reproduction within their group's sanctioned form of marriage.

To satisfy their instinctual sexual drives, then, young people in conservative cultures must get married, which they do at a younger age. To marry in many of these societies, young people need the consent of the older generation; and the elders generally choose mates who are more closely related to their family than the premarital sexual partners whom young people would otherwise choose on their own, if given greater freedom.

The political left, on the other hand, is not as concerned about reproducing the in-group as quickly as possible. Instead, liberals favor relatively more sexual freedom. Their greater tolerance toward individual choice is compatible with exogamy. This tendency of the left is *not* merely a historical artifact of modern times. Before the sexual revolution of the 1960s sought to break down sexual norms throughout the Western world, the early Soviet Union liberalized Czarist prohibitions on divorce and homosexuality (which remained in place until the Stalinist dictatorship). And long before the invention of Communism, an early Christian sect called the Adamites rejected marriage and practiced nudism in second through fourth-century North Africa.[2]

RWA and Sexual Tolerance

What grounds are there for assuming that the Adamites were leftists, like the Bolsheviks and the 1960s sexual revolutionaries? Low scorers on the RWA test tend to agree with the following statements:

- "*There is absolutely nothing wrong with nudist camps.*"*
- "*There is nothing wrong with premarital sexual intercourse.*"*

Conservatives, on the other hand, usually agree with RWA items like this one:

- "*The facts on . . . sexual morality . . . all show we have to crack down harder on deviant groups and troublemakers if we're going to save our moral standards.*"

Over the many years that Altemeyer gave his Canadian undergraduates the RWA test, he also discovered striking differences in the sexual behavior of these students, most of whom were unmarried. Those who scored in the top quartile (the most conservative students) were more than twice as likely as the group that scored in the lowest quartile to report being virgins.[3] Ideological orientation had an extremely significant influence on sexual behavior, even in a relatively liberal society.

Around the world, it's easy to see how conservative cultures, political parties, and governments actively discourage all outlets that do not channel sexual drives toward marital reproduction. As we'll see in the sections below, they do so by suppressing public displays of sexuality, banning sexual education for fear that it might arouse premarital sexual desires, prohibiting masturbation and pornography, punishing premarital and extramarital sex, and stigmatizing nonreproductive sexual acts—even between married heterosexual partners.

Arousing Sexual Desire in Public

In the ultraconservative Islamic Republic of Iran, immodest dress angers God enough to kill tens of thousands of people, according to Tehran's Friday prayer leader Hojjat ol-eslam Kazem Sediqi. In a 2010 televised sermon at Tehran University's campus mosque, the imam warned: "Many women who do not dress modestly lead young men astray and spread adultery in society, which increases earthquakes." The youth's "prevalence of degeneracy," he admonished, required a "general repentance."[4] If Iran's youth would dress more *conservatively*, he argued, then God would be less likely to cause another catastrophic earthquake.

The United Arab Emirates, on the other side of the controversially named gulf, also enforces very conservative laws regarding the public display of sexual affection. In 2010, a British couple was enjoying a meal at the restaurant of a beach resort in Dubai. A local woman reported to the authorities that the two foreigners had passionately kissed each other on the lips in public. She hadn't actually seen the kiss herself, but rather claimed to have heard about it from her two-year-old daughter. Although the British couple protested that they had only kissed each other on the cheek, the court sentenced them to a month in jail.[5]

How do these incidents contrast with a political segment and a country that lie on the opposite end of the sexual-tolerance spectrum? In August 2010, Venezuelan politician Gustavo Rojas was running for a seat in the National Assembly, with the center-left Justice First Party. To finance his campaign, Rojas decided to sell six-dollar raffle tickets for a breast-enlargement operation. In a country obsessed with physical beauty, where some thirty thousand Venezuelan women buy breast implants every year, Rojas succeeded in raising funds, increasing his popularity, and winning his election.[6]

Sexual Education

The issue of sexual education commonly divides the political left from the right. Invariably, conservatives are more opposed to teaching young people about sexuality; they believe that doing so will encourage youth to engage in premarital sexual relations (and possibly with a partner whom the parents may not approve of). Conservatives are also more opposed to contraception, which separates reproductive consequences from sexual relations.

In the 2008 US presidential election, Republican candidate John McCain supported his right-wing party's position on sexual education: that schools should teach children that abstinence until marriage is the only appropriate way to avoid pregnancy and disease. On August 29, McCain announced the choice of Sarah Palin as his running mate. Political commentators suggested that the young, relatively inexperienced governor of Alaska would strengthen McCain's credibility among religious conservatives, thanks to Palin's prominent identification with her faith and her record of taking socially conservative stands on controversial issues. For example, during Palin's 2006 gubernatorial campaign, she had prominently opposed "explicit sex-education programs, school-based clinics and the distribution of contraceptives in schools";[7] instead, she supported only abstinence-until-marriage programs.

Then, just three days after Palin's unexpected introduction to the American electorate, and two months before the election, Palin announced that her unmarried, seventeen-year-old daughter was five months pregnant. Liberals ridiculed her. For Palin's socially conservative base, however, one of the biggest problems of her personal crisis was that, for the first time, Palin began to describe herself as "pro-contraception." She even recommended that sex-education programs discuss condoms in addition to abstinence.

The issue of sex education continues to divide Americans. In the liberal state of California, the state education code *prohibits* abstinence-only programs in public schools. Ninety-six percent of California's schools offer sex education.[8]

The outcome of this partisan debate in the United States has serious ramifications for the rest of the world. Moisés Naím, the internationally syndicated political commentator and former editor in chief of *Foreign Policy*, has pointed out that the United States is the main source of finance for sexual-education programs, family planning, reproductive health, and contraceptives for countries in the developing world.[9] Political decisions made in Washington can affect the fertility and reproductive health of millions of people around the world.

Developing countries, of course, have their own political spectrums—and their own debates on sex education. In Pakistan, a liberal doctor named Mobin Akhtar has published a book titled *Sex Education for Muslims*. Actually, in the Urdu-language version, the book has a more modest title: *Special Problems for Young People*.

Dr. Akhtar's mission is to teach young, unmarried people about sex in a way that he considers compatible with the Islamic faith. He believes that not doing so has harmful consequences. According to Dr. Akhtar, adolescent boys in Pakistan reach puberty and believe that the changes in their bodies are caused by some kind of disease: "They start masturbating," he explains, "and they are told that is very dangerous to health, and that this is sinful, very sinful." This predicament can cause depression and even suicide, he says.

Dr. Akhtar has faced substantial resistance from many forces in his conservative country. No public schools teach sexual education, and few vendors will sell his book. Teachers, parents, and even many doctors avoid broaching the topic of sexuality, because many people believe that doing so would encourage the youth to act in an "un-Islamic" way. A conservative provincial politician has accused Dr. Akhtar of promoting pornography. Others have sent him threats.[10]

Masturbation and Pornography

Masturbation and the consumption of pornography are two sexual outlets that serve no direct reproductive function. They may even reduce some sexual tension that would otherwise compel young people more quickly along the path toward marriage and reproduction. Consequently, conservatives in every political spectrum around the world oppose these sexual outlets more than liberals.

Although masturbation is an embarrassing topic for many, it hasn't stopped at least one politician on the far right of the US spectrum from crusading against the solo sex act. In the 2010 Senate race in Delaware, the Tea Party backed Christine O'Donnell, a candidate with over a decade's worth of experience as a staunch opponent of masturbation. In the 1990s, O'Donnell led a conservative Christian lobby called the "Savior's Alliance for Lifting the Truth." While representing her organization on television, an attractive young O'Donnell preached: "The Bible says that lust in your heart is committing adultery. So you can't masturbate without lust."[11]

Not everyone agreed with her. In 2010, the substantial pro-masturbation majority of Delaware voted against O'Donnell by a seventeen-point margin.

Before O'Donnell became a born-again evangelical Christian in college (where she joined the College Republicans), her Italian mother and Irish father raised her as a Catholic. Catholics, of course, have their own religious left-right spectrum. Not all Catholics agree with all aspects of Catholic doctrine. As with other religions, the more conservative a Catholic, the more religious rules they tend to agree with.

One such rule concerns masturbation. The *Catechism of the Catholic Church* takes a firm position against this sexual practice: "Both the Magisterium of the Church . . . and the moral sense of the faithful have been in no doubt and have firmly maintained that masturbation is an intrinsically and gravely disordered action."

The Catechism explains that "the deliberate use of the sexual faculty, for whatever reason, outside of marriage is essentially contrary to its purpose," because the prohibited act of "sexual pleasure is sought outside of . . . human procreation."[12]

The archenemy of anti-masturbation campaigners is pornography. High RWA scorers typically agree with the statement: "*It would be best for everyone if the proper authorities censored magazines and movies to keep trashy material away from the youth.*" Not surprisingly, then, conservative politicians around the world oppose pornography, while their liberal opponents defend it. This generalization holds true for quite conservative, religious countries like Indonesia, as well as for much more secular, liberal countries like Japan.

Pornography has been an extremely controversial political issue in Indonesia over the past two decades—even though hard pornography has long been illegal. In 2005, the Indonesian legislature revived a strict bill from the 1990s called Rancangan Undang-Undang Anti Pornografi dan Pornoaksi. The key word in the bill was *Pornoaksi*, which would translate into English as "pornoaction." This term's definition, however, is problematically vague. The law, proposed by the conservative Islamist Prosperous Justice Party, would have criminalized the following acts as "pornoaction": kissing in public (even for spouses), the exposure of female body parts (such as legs, shoulders, or belly-buttons), or nudity in artwork. These and other pornoactions would have been punishable by a $220,000 fine or up to twelve years in prison. Liberal opponents of the bill criticized it as part of a covert Sharia agenda to keep women indoors.[13]

By 2008, Indonesia passed a more moderate version of the bill, which protected traditional artwork from the Hindu island of Bali. In 2010, however, an Indonesian rock star named Ariel Peterpan appeared on an Internet sex video with two female celebrities. Ariel's teenage fans broke down in tears of grief at his trial when the judge sentenced the singer to "only" three and a half years in prison (as opposed to twelve). Hardline Muslim political organizations, on the other side, launched a jihad against pornography, which they blamed for the country's "moral decline." In response, the Indonesian government told the Blackberry telephone provider to filter out pornographic content from their network, lest they lose their distribution license in the archipelago nation.[14]

On another island some 3,600 miles away, the political establishment of Japan had become embroiled in a passionate argument over anime films and *manga* comics. The later contain romance stories and classic dramas, but sometimes depict rape, incest, and other sex crimes. The controversy involved a piece of legislation proposed in the Tokyo Metropolitan Assembly that asked the cartoon pornography industry to tone down its content. The bill also imposed a $3,800 fine on retailers who sell such material to minors under the age of eighteen.

The person who proposed the bill was Tokyo's governor Shintaro Ishihara, whom the Japanese consider one their most far-right politicians. Ishihara appealed to the conscience of his people, admonishing them: "You cannot possibly show such things to your own children."[15]

The definitions and legality of pornography differ enormously between Indonesia and Japan. In Indonesia, showing a female shoulder could constitute pornoaction. Japanese pornography, on the other hand, features such a unique range of graphic themes that the industry translates cartoons into foreign languages for export. Despite this immense cultural variation, the nature of the political spectrum remains constant: conservatives in both countries seek to decrease sexual tolerance of pornography, while liberals seek to increase or protect tolerance.

Premarital Sex

In the United States, technology has contributed to a dramatic change in people's attitudes toward premarital sex. In 1969, in the first decade of the birth control pill's availability, 68 percent of Americans believed that sex before marriage was wrong; by 2009, however, only 32 percent opposed premarital sex.

Despite this unusually large shift in public opinion, partisanship still defines the remaining divergence on this controversial issue. In 2010, more than twice as many Republicans as Democrats disapproved of unmarried couples raising children (61 percent versus 30 percent). Republicans were also more likely than Democrats to be married with children (57 percent versus 38 percent). Almost half of Republicans believed that alternative family arrangements are bad for society, as compared with only 17 percent of Democrats.[16]

In more conservative cultures, such as in the Arab world, attitudes toward premarital sex sometimes turn into a matter of life and death. Young women have been honor killed by relatives or have taken their own lives over doubts concerning their premarital virginity. Nevertheless, here, too, technology is changing the picture. To avoid ostracism or violence, every week two or three Parisian Arab women turn to Dr. Abecassis, who performs an operation called "hymenoplasty." As the name suggests, the procedure involves reconnecting the torn tissues of the hymen. For $3,000, women can conceal the broken taboo.[17]

Girls in Egypt who lose their virginity before marriage typically resort to a cheaper, more discreet technology that costs between $15 and $35. It's called "the Chinese hymen" because they mail order the item from China. The insertable piece of plastic contains fake blood, but requires a bit more of a theatrical performance than the hymenoplasty.

Members of the conservative Muslim Brotherhood party have demanded the punishment of anyone vending the virginity-faking gadget. One of the Brotherhood's parliamentarians even called the sale of the plastic hymen kit "a blot on the [pre-Revolution Mubarak] government."[18]

Extramarital Sex

Extramarital affairs jeopardize the stability of families and siphon away emotional and financial resources that would otherwise be invested in children. Therefore, conservative societies strictly punish adultery.

In the northern Afghan province of Kunduz, a twenty-three-year-old engaged woman and a twenty-eight-year-old married man made the tragic decision to elope together. Unfortunately for them, the crime of passion allegedly took place within the de facto jurisdiction of the Taliban. When the Taliban caught them, the captors used a mosque's loudspeakers to summon the townspeople to the local bazaar. The Taliban threw the first stones, and then asked the crowd to join in. According to witnesses, the man did not die quickly. But the Taliban returned to shoot him dead, and to warn the community that any un-Islamic behavior would incur the same fate.[19]

Afghanistan's interpretation of Islamic punishment for adulterers falls on the strict end of the Sharia spectrum; however, Iran, Saudi Arabia, and Nigeria also permit execution by stoning (*rajm*) for this crime. Malaysia, in contrast, punishes adulterers less strictly, and according to a more Qur'anic prescription (verse 24:2 of the Qur'an stipulates one hundred lashes). In February 2010, the state caned three women convicted of having extramarital sex. These three adulteresses were surely of Malay ethnicity, since Malaysia's large Chinese and Indian minority groups are not subject to the country's Islamic laws.[20]

In the United Arab Emirates, modern technology has intersected quite unexpectedly with an ancient taboo. During a divorce lawsuit in 2009, the estranged husband of an Emirates Airlines stewardess brought a number of flirtatious text messages to the attention of the court. The SMS notes suggested, according to the presiding judges, that the stewardess and her male cabin-crew colleague had planned to "commit sin." Although extramarital affairs are illegal in the UAE, the court could not prove that the two Indian nationals had actually broken the law. Consequently, the "sexy texts" resulted in "only" a three-month prison term.[21]

In contrast, more liberal societies (including those in Europe and in most US states), have effectively decriminalized adultery.

Nonreproductive Heterosexuality

Religious conservatives often express antipathy toward nonreproductive sexual acts, even between heterosexual married couples. They believe that sexual energy is sacred and that it should be channeled, ideally or exclusively, into acts that could result in the multiplication of their coreligionists.

This generalization easily transcends the Abrahamic religions. The Buddhist traditions consider marriage a largely secular affair, and they play much less of a role in this rite of passage than other religions. Nevertheless, the fourteenth Dalai Lama, the maximum authority of Tibetan Buddhism and a celibate monk, goes beyond a pro-marriage position; in addition, His Holiness has taken a pro-reproductive position—by expressing his aversion to oral sex.[22]

In comparison, the Catholic Church is much more famous for its condemnation of nonreproductive sex acts. In particular, the Church has opposed the condom, as a latex piece of technology that sinfully separates the reproductive potential from the procreant act (only to be consummated within the sacrament of marriage).

According to comments made by Pope Benedict XVI, the Catholic Church's problem with the condom is not actually the condom itself, but rather the condom's contraceptive property. Remarks that the Pope published in his book *Light of the World* (2010) elucidate this distinction: receptive male prostitutes may use condoms, since in this case prophylactics can reduce the risk of HIV infection *without contracepting pregnancy*; however, "to use condoms to avoid an unwanted pregnancy is . . . in no way justified," the Vatican later clarified, in response to confusion over this subtle difference.[23]

In other cases, the authorities of the Catholic Church have opposed condom use because they believe it encourages premarital sex—or even far worse behavior, according to Juan Luis Cipriani, the archbishop of Lima and a member of the ultra-conservative Opus Dei organization. In November 2010, the Peruvian Minister of Health was busily distributing some eighteen million condoms to impoverished regions around the country, courtesy of the United States Agency for International Development. Minister Ugarte explained that the mission of Plan Condom was to ensure that low-income Peruvians could avoid unwanted pregnancies and STDs. The extremely conservative Cardinal Cipriani was outraged. During a fiery sermon, he declared:

> Instead of looking out for people's health, the Minister of Health is promoting a campaign of licentiousness . . . of more sexual abuse. What could possibly be going on in his head for him to think that a public authority can simply go about his duty to prevent illnesses and deaths—he's a combination between the Marquis de Sade, Juan Casanova [*sic*], and an Internet-café pedophile.[24]

Ironically, the dispute occurred a day after UNESCO's International Day of Tolerance.

On the opposite end of the political spectrum, leftists are far more likely to celebrate sexuality for the sake of pleasure, irrespective of reproduction. During the 2010 regional elections in Catalonia, Spain, the Young Socialists ran a campaign ad that created quite a stir. In the video, an attractive young woman appears to experience sexual pleasure as she inserts her vote into the ballot box, withdraws it, then slips it in and out several times—finally to cast her vote for the Socialist Party as she appears to experience an orgasm. The spot concludes with the phrase, "Voting is a pleasure."

The conservative opposition, the Popular Party of Catalonia, didn't think the video was very funny. Alicia Sanchez-Camacho, the leader of the right-wing party, called the ad "an attack on the dignity of women." The left wingers, on the other hand, chuckled under their breath. The Socialist Equality Minister, Bibiana Aido, responded: "If it was true [that voting causes orgasms], electoral participation would go up greatly, but I think we are dealing with a misleading advert." The leader of Catalonia's Green Coalition joked that it would be "very difficult to reach orgasm voting for any of the candidates, [himself] included."[25]

HOMOSEXUALITY

Conservative individuals and governments virtually everywhere oppose homosexuality more than liberals. The more conservative regions of the world have legislated homosexuality to be a criminal offense. The most liberal countries, in contrast, have legalized gay marriage.

Uganda lies on the conservative end of this spectrum. Ugandans convicted of committing homosexual acts face fourteen years in prison. As anywhere else, however, the issue causes substantial political controversy. But in this East African state, some people do not think the punishment is strict enough. In 2009, Congressman David Bahati sought to introduce a new crime called "aggravated homosexuality," punishable by death. In addition to targeting people infected by HIV, the bill would have permitted the extradition of Ugandans accused of violating homosexuality laws abroad. Uganda's minister of ethics and integrity, James Nsaba Buturo, allied himself to the cause, declaring: "Homosexuals can forget about human rights."[26]

Due to a barrage of international criticism, the Ugandan legislature dropped the anti-homosexuality bill. One of the Ugandan gay rights activists who campaigned against the proposed law was David Kato. In October 2010, the country's *Rolling Stone* newspaper published Kato's photograph, along with those of ninety-nine other people accused of being gay. Their images appeared with headlines that read: "100 PICTURES OF UGANDA'S TOP HOMOS," and "HANG THEM; THEY ARE AFTER OUR KIDS!!" Although Mr. Kato won a lawsuit against the newspaper at the beginning of 2011, three weeks later someone walked into his home and brutally beat him to death with a hammer.

The editor of the *Rolling Stone* newspaper responded: "We want the government to hang people who promote homosexuality, not for the public to attack them." In the United States, Secretary of State Hillary Clinton condemned Kato's death and reaffirmed that "human rights apply to everyone, no exceptions, and that the human rights of [LGBT] individuals cannot be separated from the human rights of all persons."[27]

Under pressure from international donors threatening to cut off development aid, the Ugandan antigay legislators declared in November 2012 that the controversial bill would punish certain forms of "aggravated homosexuality" with a life sentence instead of the death penalty. At the same time, Speaker of Parliament Rebecca Kadaga affirmed that the bill would soon be passed as a "Christmas gift" to Mr. Bahati and his supporters.[28]

Is homosexuality and the politics surrounding it a Western import, as some Africans claim? Scientifically, we know that homosexuality exists in all human populations, and among hundreds of species of birds, mammals, and even a few other creatures. The important question, then, is whether homosexuality is universally a politically emotive issue, and whether it polarizes the left-right spectrum in a predictable way—as have other forms of nonreproductive sexuality.

RWA and Homosexuality

Only one of the thirty RWA statements specifically references homosexuality. Liberals usually agree with the item, "*There is nothing immoral or sick in somebody's being a homosexual.*"*

Nevertheless, the entire thirty-question RWA test has succeeded in predicting antigay attitudes in the following experiment: Altemeyer distributed a story about a fictitious trial to 545 students. In one version, a certain Mr. Langley, the head of "Canadians for Gay Rights," was accused of leading a protest in which his supporters attacked counterdemonstrators. In another version, Mr. Langley was the founder and president of "Canadians against Perversion," and had been accused of inciting violence against counterdemonstrators carrying signs reading "Gay Power" and "Rights for Gays." When asked to sentence each Mr. Langley, the high RWA scorers (conservatives) sentenced the pro-gay Langley to a significantly longer jail term than the anti-gay Langley. Low RWAs (liberals) did the opposite; they gave the anti-gay Langley a harsher sentence than the pro-gay Langley (although the difference was not significant in this case).[29]

Attitudes toward Homosexuality and the US Political Spectrum

One of the areas in which US public opinion has drastically shifted over recent decades concerns the rights of homosexuals. In 1950, the original question on homosexuality from the *Authoritarian Personality*'s F-scale, which inspired Altemeyer's question above, was: "*Homosexuals are hardly better than criminals and ought to be severely punished.*"[30] As late as 1977, only 56 percent of the respondents to a Gallup poll believed that homosexuals should have equal job opportunities. By 1999, however, support for homosexual rights had increased to 83 percent, in spite of the emergence of HIV and the stigmatization of gays during these decades.[31]

Even with this rise in tolerance toward homosexuals, partisan differences have *not* changed. During the 2010 mid-term elections, many members of the ultra-conservative Tea Party made antigay remarks. Christine O'Donnell, for example, called homosexuality an "identity disorder." Jim DeMint, the South Carolina senator whom the media have called the Tea Party's "kingmaker," suggested a ban on openly gay schoolteachers.[32] And Glenn Beck cracked jokes about how to name an imaginary gay bar for Muslims (possibilities included "Turban Cowboy" and "You Mecca Me Hot").[33]

According to the Pew Research Center's 2010 "Social and Demographic Trends" survey, more than twice as many Republicans as Democrats believed that it's bad for society for more gay and lesbian couples to raise children (65 percent versus 30 percent).[34] In 2004, *before* the emergence of the far-right Tea Party, the American National Election Study found that 52 percent of Democrats, but only 17 percent of Republicans, supported gay marriage.

Perhaps Judeo-Christian religiosity accounts for some of the homophobia on the right side of the US political spectrum. After all, the Bible takes an anti-gay stance. Genesis implies that the sin that leads to God's destruction of Sodom and Gomorrah is homosexual relations (19:4–8). Leviticus explicitly commands: "Do not have sexual relations with a man as one does with a woman; that is detestable" (18:22). Two chapters later, Leviticus declares male homosexuality to be a capital offense (20:13). The New Testament condemns homosexuality in several places, one of which proclaims that homosexuals "will not inherit the kingdom of God" (1 Corinthians 6:9–10).

These traditional Judeo-Christian positions on homosexuality came into play in the 2010 gubernatorial race for New York. Tea Party–backed candidate Carl Paladino, who had been accused of anti-Semitism, nonetheless needed the support of the Jewish electorate. His campaign strategy entailed reaching out to a very conservative minority of this population (Orthodox Jews) in the name of what Paladino, a conservative Catholic, believed would be a winning, common-denominator issue: antigay sentiment. Paladino shopped around until he found an ally in Yehuda Levin, who headed a small congregation of two dozen Orthodox Jews in Brooklyn. Rabbi Levin seemed like an appropriate candidate; several years earlier, he had spoken out against a gay-pride festival in Jerusalem, declaring: "This is not the homo land, this is the Holy Land!"[35]

At a speech to a group of Orthodox Jewish leaders, Paladino appealed to a pro-reproductive attitude strong among religious conservatives: "I just think my children and your children would be much better off and much more successful getting married and raising a family, and I don't want them brainwashed into thinking that homosexuality is an equally valid and successful option—it isn't."[36] During the campaign, Paladino criticized his Democratic rival, Andrew Cuomo, for taking his daughters to a "disgusting" gay-pride parade, where men in "little Speedos . . . grind against each other."[37] In the end, Paladino lost the race, in part because of his anti-gay comments' poor reception in the liberal state of New York.

In the same election cycle, Tea Party candidate Ken Buck was challenging Colorado Senator Michael Bennet, a Democrat, for his seat. Buck had enjoyed a modest lead during the final few months leading up to November—until he compared homosexuality to alcoholism. Buck conjectured: "I think birth has an influence over [sexual orientation], like [it does over] alcoholism and some other things, but I think that basically you have a choice." As a result of this comment, the gap between Buck and the Democratic incumbent grew narrow, jeopardizing Buck's chances. Suddenly, it seemed as though control over the nation's Senate potentially hung on the outcome of Colorado's poll. Over $25 million poured into the state's senatorial race, as funders hoped to sway the results one way or the other. In the end, Buck lost the race, and the Democrats kept control over the Senate.[38]

The further to the right one travels on the US spectrum, the more anti-gay sentiment increases. Dave Holland, the "Grand Dragon" of the Southern White Knights of the Ku Klux Klan, represents the extreme endpoint both of the spectrum and of homophobia. Ethnographer Raphael Ezekiel recorded Holland's views about homosexuals: "A queer, as you well know, is not, I don't care what he says, he can't live a Christian life. How could he? . . . But I think that if they're going to have a perverted life, let them die. Because they're living contrary to the laws of God."[39]

On the opposite end of the American political spectrum, the left generally supports the rights of homosexuals to marry, to adopt children, and to serve openly in the military.

Virtually all the political-science research on homosexuality asks what the public at large thinks about gays. But what do gay people think about politics? According to Gary Gates, an expert on lesbian, gay, and bisexual (LGB) voting patterns at UCLA's Williams Institute, the LGB vote "has been absolutely consistent since 1992: a three-to-one split" in favor of the left-leaning Democratic party.[40]

According to exit-poll data from the National Election Pool (the consortium of media outlets that keeps the public informed on the night of the election), Democratic presidential candidates can count on at least 70 percent of the LGB vote. This trend held true for Obama in 2012 and 2008, John Kerry in 2004, Al Gore in 2000, and Bill Clinton in 1996. Since bisexuals vote more like heterosexuals, Dr. Gates believes that an even higher proportion of the homosexual population would be shown to vote for the liberal party if bisexuals were excluded from these samples.[41]

Homosexuality in Global Politics

In December 2010, the General Assembly of the United Nations voted to restore the category of "sexual orientation" to a resolution against the killing of minority groups. A number of Arab and African member states had previously lobbied for the removal of the clause, but the United States succeeded in reinstating the protection of homosexuals. In total, ninety-three countries voted in favor of homosexuals' rights *not* to be unjustifiably killed; on the other side, fifty-five countries voted against gay rights to life, and twenty-seven countries abstained. Zimbabwe's ambassador to the UN, Chitsaka Chipaziwa, protested: "We cannot accept this, especially if it entails accepting such practices as bestiality, pedophilia and those other practices many societies would find abhorrent in their value systems."[42]

Homosexuality clearly incites political controversy at the global level. But do attitudes toward gay rights polarize the left-right spectrums within countries around the world in the same way that they do in the United States?

In October 2010, the city of Belgrade had its first Gay Pride parade in nearly a decade. No one had dared organize the event since 2001, when far-right extremists injured one hundred police officers, threw petrol bombs, and lit the headquarters of the center-left Democratic Party on fire. The Serbian Orthodox Church condemned the gay parade, and rioters yelled "Death to homosexuals!"

The 2010 Gay Pride Parade in Belgrade was even more violent. Fourteen protesters received jail sentences for their involvement in riots that injured 147 police officers. Among the protesters was Mladen Obradović, the leader of the far-right Obraz movement, which the Ministry of the Interior has labeled "clero-fascist."[43] In Serbia, then, the fiercest anti-gay attitudes come from the far right.

The Tibetan political spectrum is a world away from the Serbian one. Nevertheless, the highest religious authority, the Dalai Lama, has expressed an aversion to gay sex, according to his co-autobiographer, Alexander Norman. Norman does emphasize, though, that the Dalai Lama conveys his opinion in the most tolerant way possible: "He will say it's your choice, it's up to people's own conscience. He is very conscious of not giving people offence."[44] Despite a huge difference in tolerance between the Serbian and the Tibetan religious right, both oppose homosexuality more than leftists and secularists on their respective spectrums.

At the beginning of the 2011 general elections in Peru, it appeared as though gay rights would be a nonissue. Then, unexpectedly, it became a campaign issue: the campaign chief of the center-left Peru Posible Party, which was leading in the polls, told the press that his candidate would legalize gay and lesbian civil unions if elected president. Not to be outdone, the leftist Fuerza Social Party announced its intention to support gay marriages. The right-wing archbishop of Lima called the proposals an "error," and reminded Peruvians that "the Church teaches something else."[45] In short, the Peruvian political spectrum responded predictably to homosexuality—according to the same principles that apply in Serbia and Tibet.

In each part of the world, the ethno-religious right opposes gay rights more than the left. Whatever the proximate explanations given by conservatives themselves, the ultimate evolutionary reasoning underlying this antipathy is that openly gay individuals usually do not reproduce. The more homophobia that people transmit in a conservative society to their gay children and siblings, the higher the chance that these relatives will remain closeted, marry, and reproduce the family genes; in the opposite scenario, the more tolerance and rights that homosexuals gain to marry each other in liberal societies, the less they will reproduce.

GENDER ROLES AND EQUALITY

The political battle over the reproduction of nations is fought over the bodies of women. The political right typically seeks greater control over women, since it sees them as the reproductive vehicles of the in-group's future generations. The left, on the other hand, which is less concerned about reproducing the in-group, supports more equality and behavioral freedom between the sexes.

Why is it that the right generally favors greater female subservience and male dominance? One reason is that tighter control over women lowers the chance that women will be able to reproduce with members of out-groups. But men are also able to reproduce exogamously, so why do conservative cultures put the burden mostly on women?

When Altemeyer discovered that the high quartile of RWA scorers were more than twice as likely to be virgins than the low quartile, he was referring to a mixed sample of both men and women. However, when he separated his survey by sex, he discovered two notable asymmetries: (1) in the case of males, roughly the same proportion on each end of the spectrum had lost their virginity by the beginning of college, with no significant difference between them; in the case of the most liberal quartile of females, this same proportion (as the men) had also lost their virginity, but a *significantly larger* proportion of the conservative women remained virgins; (2) for the entire sample, the males of Altemeyer's Canadian college freshmen anonymously reported having had twice as many sexual partners as the females.[46]

Altemeyer's findings suggest that political orientation succeeds in controlling the virginity of women more easily than it does the virginity of males—even in the liberal country of Canada. If female sexuality is relatively more flexible than male sexuality, then maybe this is why conservatives focus more of their energy on restricting women's sexuality. Perhaps this is easier than trying to curtail the indomitable libido of males, who reported twice as many sexual partners.[a]

In any case, the important point is that the left favors greater equality between the sexes, while the right prefers greater protectionism of female sexuality, and greater male dominance.

RWA and Gender Roles

According to Altemeyer, high scorers on the RWA scale believe that women should "keep to their traditional roles in society" and be "subservient to their husbands."[47] Conservatives also endorse traditional sex roles. They are more likely to agree with the following statements, from an "Attitudes toward Women" scale:

- "*Women should worry less about their rights and more about becoming good wives and mothers.*"
- "*A woman should not expect to go to exactly the same places or to have quite the same freedom of action as a man.*"
- "*Feminism contributes to the breakdown of family values in this country.*"[48]

The gender-role questions on the actual RWA test itself include the following, more subtle items:

- *"It may be considered old-fashioned by some, but having a decent, respectable appearance is still a mark of a gentleman and, especially, a lady."*
- *"A lot of our rules regarding modesty and sexual behavior are just customs which are not necessarily any better or holier than those which other people follow."**

The second question is phrased in the opposite direction, of course, to avoid an acquiescence bias.

Gender Roles and the US Political Spectrum

During the Cold War, the United States polarized to the right, away from its Communist adversaries—including on gender issues. A current of American culture even associated the far-left threat with female domination over males. The late Berkeley political scientist Michael Paul Rogin has written about this phenomenon. One of his examples pertains to the demonization of Julius and Ethel Rosenberg, the American Communist couple executed for passing nuclear secrets to the Soviets. A psychological report on the Rosenbergs sent to President Eisenhower claimed that "Julius is the slave and his wife, Ethel, the master." In reality, Ethel's involvement in the espionage was far more dubious than that of her husband.

Another of Rogin's examples comes from a 1940s military manual, which argued that Americans were not taking the Communist threat seriously because they were being softened by their mothers.

Rogin's work has even revealed how some Cold War–era Hollywood films linked Communism to female supremacy. In the movie *Them!* (1954), atomic radiation creates gigantic, mutant ants. A queen rules over the colony, controlling a collectivist, matriarchal society. Males fertilize the queen, only to die. A scientist warns that ants are "chronic aggressors that make slave laborers out of their captives." He emphasizes the "industry, social organization, and savagery" of the ants. "Unless the queens are destroyed, man as the dominant species on this planet will probably be destroyed," the narration forewarns.[49]

Even today, popular culture implicitly feminizes the left half of the American political spectrum. Humorist Dave Barry has poked fun of the stereotype that blue-state residents are "Volvo-driving . . . left-wing communist latte-sucking . . . perverts."[50] The Volvo, of course, comes from a socialist country and has a female-sounding name. The phrase "latte-sucking" evokes milk and breast-feeding, implying that liberals are overly reliant on females, like a nursing infant. Stereotypes aside, political scientist John Jost did find that there's a significantly higher chance of finding a book about a feminist topic in the bedroom of a liberal than in the sleeping place of a conservative.[51]

The great majority of US liberals, in fact, favor government efforts to equalize opportunity for women, while most conservatives oppose government intervention on behalf of women's equality (see figure 16). This graph shows the polarizing effect of a college education *in both directions* of the spectrum. That is, having a higher education makes liberals more liberal and conservatives more conservative on this issue. In the survey graphed in figure 16, men and women did *not* differ significantly in their opinion on this matter.[52]

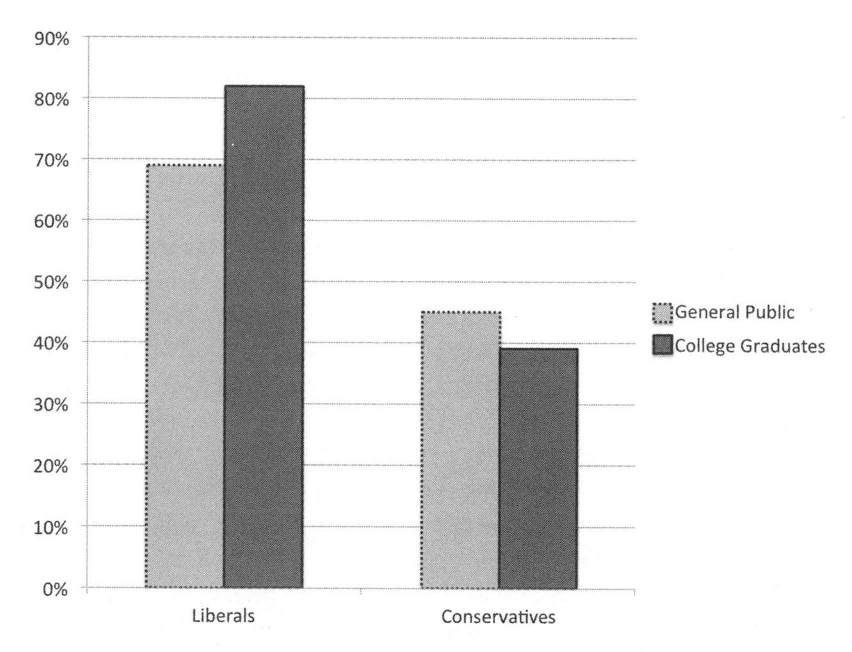

Figure 16: Support for Government Guarantees of Equal Opportunity for Women, by Ideology and Education.

The further to the right of the spectrum one looks, the more politicians are opposed to women's control over their reproductive tract. On the far right, this attitude can turn into a general derision of women. For example, Tea Party mouthpiece Glenn Beck once called his rival's wife on the telephone while broadcasting his radio show. He then proceeded to ridicule her for her recent miscarriage.[53]

Ezekiel has observed that on the most extreme right of the American political spectrum, "highly traditional ideas of sex roles, and fears of losing male dominance, fill the conversation and the speeches" of white hate groups.[54]

On the far left, in contrast, some people believe that there is little inherent difference between the sexes—that people are and should be treated equally regardless of gender. Since their radical view holds that gender is socially constructed, they make an effort to overcome what they believe to be the patriarchal influence of society by emphasizing equality between the genders. *New Yorker* cartoonist Edward Koren pokes fun at this mentality in one of his drawings: a very liberal-looking woman kneels down on the sidewalk to shake the hand of a baby in a stroller, while asking the smiling mother, "How is it gendered?"

Gender in Global Politics

Dutch political scientist Jos Meloen discovered a striking relationship when he studied a sample of eighty-four states: the world's more authoritarian countries tended to support male dominance, to prepare men for military service, and to emphasize women's role in raising children away from public life. He even found a strong correlation between a country's level of authoritarianism and an independent ranking of gender inequality.[55]

People in the world's more conservative countries enforce three forms of gender inequality. These cultures reduce women's freedom in: (1) sexually attracting male strangers in public, (2) choosing a spouse, and (3) divorcing. The underlying logic of these cultural patterns is to control women's reproductive behavior; this control ensures a conservative outcome: early marriage, endogamy, and a high birth rate. These results increase the fitness of the tribe but come at a high price to women.

How do some of the more conservative countries reduce women's ability to attract men? Through a culture that permits sexual harassment of women who tempt the male libido. In Bangladesh, for example, the Education Ministry declared June 13, 2010, to be "Eve Teasing Protection Day." This euphemism refers to the sexual harassment of girls and women. Some government officials had become alarmed that fourteen girls and women had committed suicide in only four months to escape gender-based bullying. But not all Bangladeshis agreed that there was a problem; three men who publicly protested against "Eve Teasing" were killed by their more conservative countrymen.

The highest-profile victim of sexual harassment was a thirteen-year-old named Nashfia Akhand Pinky. She aspired to be a model, but unfortunately her beauty and independence brought trouble. On her way to school, a group of young men would make lewd jokes, catcalls, and slap her body. In her suicide note, Pinky criticized her society's acceptance of abusive behavior toward unaccompanied women in public: "When [my tormentor] pulled my scarf and harassed me physically in front of the house, onlookers at the scene laughed. Nobody protested. None of my family members are responsible for my suicide."[56]

In Egypt, one poll has found that a majority of men admitted to harassing women, which is considered "culturally acceptable" under certain circumstances. This behavior intimidates women into not walking around alone, and into doing everything possible to avoid the provocation of the male sexual appetite. Covering with a headscarf helps, but it does not prevent harassment for an unaccompanied woman in public.

In 2008, the chair of the Egyptian Center for Women's Rights surveyed one thousand women from all over her country. The poll discovered that about eight out of ten Egyptian women, and 98 percent of foreign women, reported having experienced sexual harassment. Nearly two-thirds of the men she interviewed admitted to harassing women. The issue came to the world's attention when CBS correspondent Lara Logan, a blonde woman reporting on the Egyptian Revolution, was sexually assaulted and beaten by a mob of men while covering the celebrations in Tahrir Square.[57]

Cultures with the greatest male dominance over women restrict female sexuality through this type of intimidation. People may explain away harassment as "simply" male chauvinism. Nevertheless, the ultimate *function* of the conservative cultures that permit this behavior (whether conscious or otherwise) is to control reproduction.

Extremely conservative cultures also have marriage customs designed to control the in-group's reproduction. Again, these customs achieve their protribal goals through male dominance over women. Few places could illustrate the links between male-dominated harassment, marriage customs, and endogamy better than Chechnya.

In 2007, Chechen president Ramzan Kadyrov banned women and girls without a headscarf from educational facilities and other public spaces—in violation of Russian law. To enforce his headscarf edict, plain-clothed men drove around the streets of Grozny to shoot paintballs at women with uncovered heads. The men left leaflets warning of the "more persuasive measures" that they vowed to take against "immodestly dressed" women in the future. In an address on state television, President Kadyrov claimed not to know anything about who were coercing women to comply with his gender-targeted law; nevertheless, Kadyrov promised: "When I find them, I will express my gratitude."[58]

To avoid being shot, Chechen women must cover their heads and avoid walking about without a male companion. Without male protection, in fact, even a Chechen woman's own family may not be able to choose a mate for her. About 20 percent of Chechen women fall victim to the ancient tradition of bride stealing. In the movie *Borat*, the "Kazakhstani" protagonist makes fun of this Central Asian custom[b] by pretending to capture *Baywatch* star and *Playboy* playmate Pamela Anderson. In real life, though, the actual tradition isn't so funny for the women, as it involves kidnapping and forcing them into marriage.

The experience, which can involve gunfire and screaming, often traumatizes Chechen women. Many of them suffer mental breakdowns in their forced marriages. As a result, their mothers-in-law take them to the Center for Islamic Medicine in central Grozny, which conducts exorcisms. One of the mullahs who specialize in this procedure is Mairbek Yusupov. During the exorcism, Mullah Mairbek beats the young woman with a stick while shouting verses from the Qur'an. He justified the seemingly cruel thrashing of a woman to a visiting BBC reporter, explaining: "She feels no pain. . . . We beat the genie and not the patient."

The evil spirit tormenting the kidnapped young bride, according to the exorcist, was a 340-year-old genie that had fallen in love with the possessed girl. The genie, unfortunately for his victim, had made an exogamous love choice; he was a non-Muslim Russian named Andrei.[59] In the logic of Chechen culture, then, resisting male domination may be related to resisting endogamy.

To make sure that their women don't leave them without permission, the male authorities of the ultra-conservative Kingdom of Saudi Arabia have developed a more high-tech solution. When a girl or woman tries to leave the country, her male guardian receives a text message.[60] This SMS warning system, which went from optional to automatic in November 2012, helps enforce two Saudi laws controlling women: it is illegal for them to travel without permission from a male guardian, and they may not mix with unrelated men. Electronically guarding a political border against female flight, then, could theoretically prevent exogamy and prevent wives from escaping undesired marriages.

Other extremely conservative cultures have more conventional and low-tech methods for ensuring that their women marry the "correct," endogamous choice. In Yemen, for instance, over a quarter of girls are married off by their families before the age of fifteen. At this age, girls are a decade away from mental maturity (chapter 21 explains why mental maturity occurs around age twenty-five). Marriage at such a young age eliminates the chance that a Yemeni girl's own preferences may defy the will of her parents.

Very young brides, however, incur other types of risks. In April 2010, a thirteen-year-old Yemeni girl died from internal bleeding three days after her marriage. The cause of death, according to the medical report, had been a tear to her genitals inflicted by intercourse. The year before, the Yemeni legislature had tried to raise the minimum age of marriage to seventeen, but conservative lawmakers repealed the change as "un-Islamic."[61]

The medical term for the girl's fatal injury is "obstetric fistula." The condition, whose causes are usually early childbirth and limited medical attention, is also prevalent in Bangladesh, Pakistan, Afghanistan, and Nepal. As a result, many young girls become incontinent, leaking urine and feces. Their husbands abandon them, leaving the women unmarriageable and often unable to pay the several hundred dollars needed to repair their bodies.

Extremely conservative cultures control marriages not only within the desired in-group, but also for economic and political purposes. In these contexts, women are considered property. The male elders in such societies typically trade women for material resources or alliances. On occasion, however, trades take place under less happy circumstances. One Afghan family had to give up a beautiful fourteen-year-old girl named Aisha as a "blood debt" to a Taliban fighter (many Pashtun peoples pay this kind of debt, which is called a *baad*, as a traditional tribal form of resolving disputes). After suffering abuse at the hands of her husband's family for four years, Aisha ran away to her biological family. But Aisha's father returned her to her in-laws, who punished her by cutting off her nose and ears. Aisha came to the world's attention when *Time* magazine featured her on its cover. The title read: "What Happens If We Leave Afghanistan?" The controversial intention of the cover was to emotionally manipulate liberal Americans (who opposed the war but supported gender equality) to support keeping US troops in the country instead of turning it over to the Taliban.[62]

Some cultures with extreme male domination also keep women subordinate by making it extremely easy for men to divorce them. In India, for instance, Muslim men in the Hanafi tradition can divorce their wives in only minutes. They do so by simply repeating the word *talaq* ("divorce") three times. If done in the presence of two male witnesses, the divorce comes into force immediately. The divorced woman is only entitled to three months' alimony. Some Sunni traditions do not even require witnesses for the triple-*talaq* to take effect. A wife, however, cannot divorce her husband in the same instantaneous, extrajudicial fashion. If governed by Sharia law, she must turn to a religious court, run only by men, to request a divorce. Sometimes she requires the consent of her husband to do so.[63]

Liberal societies, in contrast, may not have completely eliminated male dominance; however, women have enforceable legal protection against harassment, much greater freedom to make their own mate choice, and can typically seek a divorce without even finding a husband "at fault." Compared with the conservative societies described above, these liberal societies with greater gender equality also have later marriage ages, higher rates of exogamy, and lower fertility rates.

These facts require a deeper explanation—one that goes beyond the social sciences. Part II, "The Biology of Tribalism," pursues the evolutionary roots of the cross-cultural phenomena described in part I.

PART II.
THE BIOLOGY OF TRIBALISM

The three chapters of part I have shown how every nation has a political spectrum running through it. This spectrum exerts the centripetal pulls of tribalism and also the centrifugal pushes of xenophilia. These opposing psychological forces influence each population's levels of endogamous and exogamous reproduction. To understand why these forces exist, we now turn to a parallel conflict in biology—the conflict between inbreeding and outbreeding: there are benefits to reproducing with an individual who is closely related, but there are risks if the genetic relationship is too close; likewise, there are both benefits and tradeoffs to mating with an individual who is more distantly related.

The chapters of part II explore this set of biological conflicts, which have shaped the structure of human political psychology. Chapters 9 and 10 focus on phenomena called "outbreeding enhancement" and "inbreeding enhancement"; they explain how numerous factors impact survival and fertility, depending on the genetic distance or relatedness between mates. Chapter 11 reveals how nature encourages people (and other sexually reproducing creatures) to engage in "optimal outbreeding"; that is, it shows how sexual attraction often guides individuals to make mate choices that maximize fitness, while avoiding excessive inbreeding or outbreeding.

In addition to outbreeding and inbreeding, other forces related to tribalism also impact the reproductive rates of groups. Chapter 12 explains why variations in gender inequality across human history have dramatically affected the number of children that women bear in their lifetime. Finally, chapter 13 discusses the natural history of war and genocide.

Each of these chapters takes on a topic that has had crucial bearing on the survival and fertility of human populations over the ages. Seeing the markings that these forces have left on our political personalities has been fairly straightforward in part I. The coming chapters of part II, however, require us to attune our perception to the immense time scale of evolutionary change. This adjustment is a bit less intuitive because our mortality limits our normal perspective.

Imagine, however, if you could observe vast eras of human history from on high. Decades pass by in seconds, and millennia in minutes.[a] You would be able to view enormous geographical areas along with their inhabitants, the ethnolinguistic identities of populations, different gene pools, and even pathogens. As time unfolds below, you would see environmental change, wars, periods of stability, migrations, epidemic and pandemic diseases, displaced populations, alliances, ethnocide, assimilation, bottlenecks, founder effects, fertility booms, colonization, commerce patterns, and genocide. Each event would affect the mortality rates and reproductive success of different demographics in different ways. Over time periods impossible to perceive through human eyes, these events would interact with genes and behavior, shaping the distribution of ethnocentrism and xenophilia accordingly.

The next several chapters detail some of the myriad factors that have likely contributed to the pulls of inbreeding and the pushes of outbreeding.

9.

WHEN OUTBREEDING IS FIT AND INBREEDING ISN'T

Interest in the effects of inbreeding and of outbreeding is not confined to the professional biologist. Historically, these are old, old problems, practical problems of considerable significance bound up with man's gravest affairs.
—Edward M. East, Harvard geneticist and president of the American Society of Naturalists, 1919[1]

If you're reading this sentence, you're probably a modern human—or a *Homo sapiens*. But there's also a chance that your ancient great-great-grandparents outbred with other species of premodern humans. This prehistoric comingling could mean that you also have some Neanderthal in you.

Sometime between 270,000 and 440,000 years ago, modern humans and Neanderthals split paths from their common ancestor. Geneticists have calculated the time period of this divergence by comparing our genome with DNA extracted from the femur bones of our paleo-cousins. They have also found that we share 99.84 percent of our genetic material with *Homo neanderthalensis*.[a]

The Neanderthals occupied Europe and Western and Central Asia long before modern humans left Africa. When humans *did* migrate into the Middle East some one hundred thousand years ago, and into Europe around forty-five thousand years ago, they would have encountered this slightly shorter, much stronger being, which had red hair and pale skin. For many years, paleoanthropologists have hotly debated whether modern humans interbred with these Neanderthal populations, or whether a Neanderthal displacement and genocide took place at the hands of *Homo sapiens*.

Then, in 2010, Swedish geneticist Svante Pääbo and his colleagues at Germany's Max Planck Institute for Evolutionary Anthropology announced the results of their Neanderthal genome project: modern Europeans and Asians share between 1 and 4 percent of their DNA sequence with the Neanderthals that once occupied their regions. Modern Africans, on the other hand, do not have Neanderthal DNA. No Neanderthal remains have been found to date in Africa.[2]

When early humans left their African homeland and expanded into Asia, Neanderthals weren't the only other hominins on the scene. The Neanderthals also had cousins to the East called Denisovans. Russian archeologists named them after a Siberian cave where they discovered a forty-thousand-year-old finger bone and tooth. DNA tests carried out by Pääbo's lab in Germany revealed that Denisovans shared a common ancestor with Neanderthals sixty-four thousand years ago. And this common ancestor shared an even earlier common ancestor with *Homo sapiens* one million years ago, before the Denisovans' forebearers left Africa to spread across Eurasia (see figure 17).

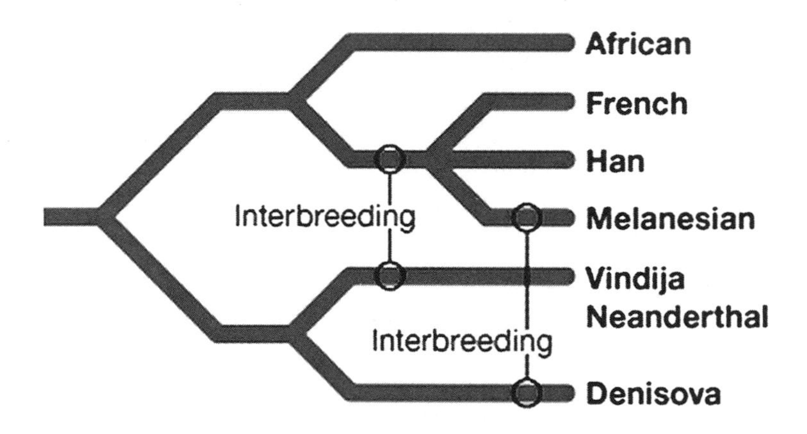

Figure 17: The Human Family Tree (not drawn in proportion to time periods).

Pääbo compared the Denisovan DNA with numerous populations of contemporary humans and found that 4 to 6 percent of the DNA of modern Melanesians contains Denisovan genes. Yet no other population shared Denisovan DNA. What happened between Siberia and Papua New Guinea—and why Asian populations lack Denisovan DNA—remains a mystery. Professor Chris Stringer of London's Natural History Museum reasons that only fifty successful matings between Denisovans and an ancestral population of one thousand Melanesians could account for the proportion of Denisovan DNA in modern-day Melanesians.[3]

Why did early modern humans outbreed with Neanderthals and Denisovans? What were their relationships like? Did "conservative" Denisovans disapprove? Were there "liberal" Neanderthals who were attracted to the exotic humans as they migrated into their territory?

No Neanderthal genes have turned up in the mitochondrial DNA of living humans, which we inherit exclusively from our mothers.[b] But since we have Neanderthal DNA in parts of our nuclear genome, some of us have inherited genes from Neanderthal males.

Whatever the nature of these prehistoric inter-subspecies or interspecific relationships, they did produce viable offspring that have passed their genes down for thousands of generations to our contemporaries. Since outbreeding has been a successful reproductive strategy (even in this extreme case), xenophilic personalities have evolved to ensure that it occurs. This chapter explains when and why outbreeding is fit, and how it has shaped human political orientations.

OUTBREEDING ENHANCEMENT AND INBREEDING DEPRESSION

What exactly is outbreeding? Does it refer to matings between our early modern ancestors and other hominins, whose ancestors were separated by hundreds of thousands of years? Or does it mean reproducing with a different race of humans, separated by only tens of thousands of years? A definition is surely in order.

"Outbreeding" is a somewhat relative concept. It entails mating between individuals that have a greater genetic distance between them than one would expect at random, given the size of the population. "Inbreeding," on the other hand, occurs when kin mate with one another more often than expected by chance.

It's important to note that outbreeding forms a continuum with inbreeding.[4] Figure 18, for instance, plots out the distribution of a wild Great Tit population in the United Kingdom along an inbreeding-outbreeding continuum. The horizontal x-axis shows the distances that female birds traveled from their natal site to the place of their first breeding event. The vertical y-axis indicates how many birds dispersed a given distance. This large sample represents over two thousand five hundred dispersions, measured within the same population over forty-four years.

The geographic distance flown by the Great Tits corresponds, on average, to the genetic distance between mating pairs. Most birds disperse over half a kilometer from their birthplace; this prevents them from landing on the inbreeding side of the curve. The birds that dispersed less than two hundred meters were almost three and a half times more likely to mate with a sibling or parent, compared with the average bird. This close inbreeding reduces the fitness of the Great Tits; the ones that reproduce with siblings or parents have 55 percent fewer viable grandchildren than their outbreeding counterparts.[5]

One could think of the outbreeding continuum in figure 18 as an ethnocentrism-xenophilia political spectrum for Great Tits! Some conservative birds like to stay closer to home; other, more liberal ones fly off to reproduce with foreigners.

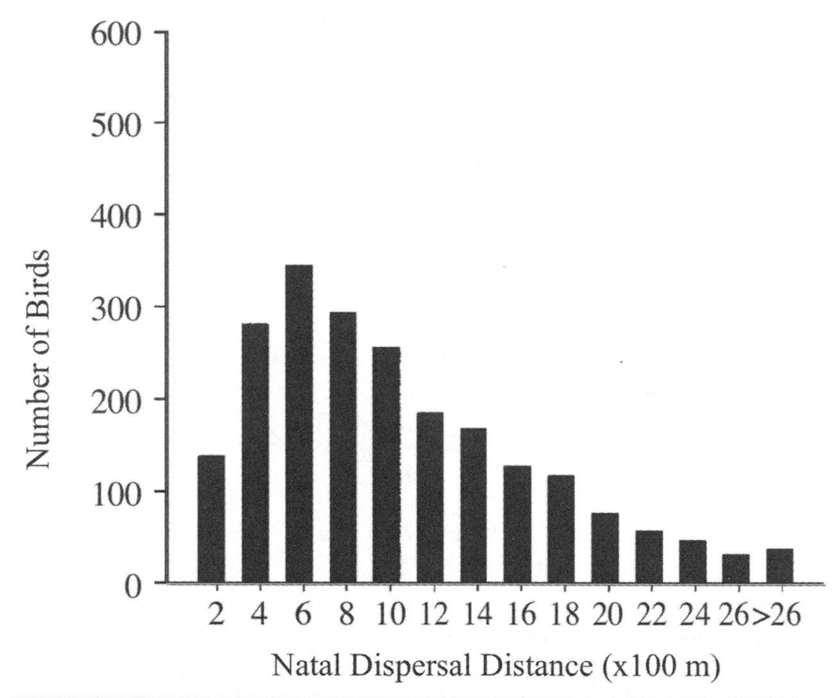

Figure 18: Dispersal Distribution for Female Great Tits (*Parus major*), Wytham Woods, United Kingdom (1964–2007). Photo: Luc Viatour.

This chapter explores what pushes some creatures away from the inbreeding end of the curve and pulls them toward the outbreeding end. We call the pulling aspect of this force *outbreeding enhancement* because attraction to relatively less related mates can enhance the fitness of the resulting offspring. The pushing part of this force is called *inbreeding depression*. This doesn't mean that some birds feel depressed if they mate with siblings; rather, the word "depression" refers to the lowering of evolutionary fitness (i.e., having fewer viable offspring) that can happen when closely related individuals reproduce. What are the specific biological problems that arise through too much inbreeding, and which select for some degree of outbreeding behavior and the xenophilic personalities that encourage it?

The blueprints for all known living organisms are written in DNA. Most cells in our body contain DNA, which is a molecule made by a double-helix backbone and a "spiral staircase" of complementary nucleotides. The four types of nucleotides in DNA are like four different letters that spell out enormous strings of instructions. These instructions are arranged into chromosomes. In a human egg or sperm cell, three billion of these "letters" form twenty-three different chromosomes, which contain some twenty-three thousand protein-coding genes.

During sexual reproduction, the new human organism receives one set of twenty-three chromosomes from its mother and a homologous set of chromosomes from its father. Whether specific genes from a parent come from the parent's mother or father is a matter of chance. Thus, sexual reproduction creates enormous diversity.

Individuals have two forms of each gene, which are called alleles: one allele comes from the maternally inherited gene, and one comes from the paternally inherited gene. The two alleles may have the same "spelling" of nucleotides, or they may code for different forms of the gene.

To make all of this more intuitive, consider hair color. If Ralph inherited the same brown-hair-color alleles from both parents, then the gene instructs Ralph's body to make his hair brown. If Ralph's friend Tony inherited one "brown" hair-color allele from his father, and a "red" one from his mother, then Tony will be a "carrier" for red hair. But the brown form of the hair-color gene is "dominant," so Tony's hair will be brown. The "red" allele is "recessive," which means that an individual needs two red alleles to grow red hair. If Tony marries Jessica, who is also a brown-haired carrier of the red-hair allele, and they reproduce, some of their children may inherit *two* copies of the recessive red allele. These kids will be redheads.

Sometimes mutations occur that change the sequence (or spelling) of nucleotides in a gene. Mutations can arise from radiation, viruses, or spontaneous errors made by our cells. Whatever their cause, mutations can create new types of alleles. For example, human chromosome 7 has a gene called *CFTR*, which is about 189,000 base pairs long. *CFTR*'s code tells the body how to make the CFTR protein, which comprises 1,480 amino acids. A common mutation on the *CFTR* gene involves the deletion of only three of the nearly two hundred thousand nucleotides. The consequences are devastating.

Because of the three deleted "letters," the 508th amino acid in the CFTR protein goes missing. Without this particular amino acid (phenylalanine), the ever-so-truncated protein does not fold properly, and the body quickly gets rid of the malformed molecule. Therefore, the damaged CFTR protein cannot successfully accomplish its function, which is to regulate the components of sweat, digestive juices, and mucus.

People with this type of damaged CFTR protein suffer from diarrhea, poor growth, a buildup of thick mucus in the pathways of the lungs, and the numerous other symptoms of cystic fibrosis. Half of cystic fibrosis patients in the United States today survive into their mid-thirties. However, throughout most of human history the disease killed young children in the first years of their lives—long before they reached reproductive maturity.

How, then, has this mutated *CFTR* allele survived? Why hasn't natural selection weeded it out of the gene pool completely? How could it be that one in twenty-five people of European descent carries a copy of the mutated allele, and that one in two thousand five hundred babies in the EU is affected by cystic fibrosis?

The mutated allele is recessive, which means that having a normal copy of the *CFTR* allele on one chromosome prevents cystic fibrosis. It takes *two* of the recessive, mutated *CFTR* alleles to cause the devastating disease. Scientists believe that the shorter *CFTR* allele has managed to survive over the past fifty-two thousand years since the mutation originated because carriers who have only one copy have benefited from greater tolerance to a host of deadly infectious diseases. These might include typhoid fever, cholera, diarrhea, and tuberculosis.[6]

A major problem with too much inbreeding, then, is that offspring have a greater chance of inheriting two deleterious recessive alleles (like those who suffer from cystic fibrosis). Outbreeding can reduce this risk by "masking" these deleterious genes with normal copies of dominant alleles.[7]

Another disadvantage of inbreeding is that it reduces genetic variation within a population. A genetically homogeneous group could be decimated if the environment changes drastically and most of the overly similar organisms cannot adapt.[8] This is precisely the situation that's currently endangering bees and the humans who depend on them.

Although bees tend to specialize in particular species of flowers, cross-pollinating bees can wander between different subspecies of sexually reproducing plants. Thus, it's especially ironic that certain bumblebees—the great promoters of floral diversity—have succumbed to the ills of their own inbreeding.

Off the west coast of Scotland, biologist Penelope Whitehorn has studied populations of moss carder bumblebees on nine of the Hebridean islands. Each island creates various degrees of isolation and, therefore, inbreeding. Whitehorn discovered that the most inbred bumblebees, with the lowest level of genetic diversity, also suffered from the highest prevalence of the intestinal parasite *Crithidia bombi*.[9]

Why worry about protozoa floating about in the guts of island insects? As humans increasingly modify their environment, habitat fragmentation creates more isolated populations of bees. Mainland bees also begin to inbreed more, as their island counterparts do, which reduces their genetic diversity and increases their susceptibility to parasites. The Bumblebee Conservation Trust worries, as these insects become endangered, that humans will lose one of the key pollinators of commercial crops, which we rely on for our food supply.[10]

Now, imagine a group of overly inbred human beings. In addition to the possible risks from diseases mentioned above, this hypothetical tribe might have less diverse personalities. Suppose that the group's average measure of "Openness" indicates that its members are extremely closed to new experiences. This personality trait, fixed at an unusually high frequency through generations of close inbreeding, may have served the tribe well for some time.

Suppose, however, that environmental degradation in a distant river valley displaces a foreign tribe into the first tribe's territory. The leaders would have to decide whether to welcome the new group of migrants, keep a safe distance, or even drive them away. A wider, normal distribution of Openness (as found in reality) would foster arguments between the more open and more closed leaders, allowing the tribe to better contend with both potential allies *and* enemies. Too much homogeneity, in contrast, could bias the group toward *always* expecting enemies (or allies).

Even if a population were isolated in perpetuity, too much inbreeding could produce offspring that are too similar to one another in other problematic ways. This greater likeness would cause brothers, sisters, and cousins to compete more intensely for the same type of resources within the same environment, to everyone's disadvantage.[11] If greater genetic variation adapted some individuals to eating one type of food, and other individuals to extracting a slightly different food source, all would benefit.

But these concepts are quite speculative and theoretical. Let's take a look at what, concretely, goes wrong with excessive inbreeding in reality, and what benefits greater outbreeding bestows. Below, we begin by looking at nonhuman animals. Then the chapter explores inbreeding depression and its political consequences among humans.

NONHUMAN ANIMALS

Inbreeding depression in animals has a long history of documentation. In 1868, Charles Darwin wrote extensively on the topic in his two-volume book *The Variation of Animals and Plants under Domestication.*[12] More recently, biologists in the 1990s documented the effects of inbreeding depression in plants, fishes, amphibians, reptiles, zoo animals, and primates.[13] Primatologists, for instance, have shown that inbreeding results in higher infant mortality rates than noninbred matings in over a dozen species of primates.[14]

A moderate degree of outbreeding can prevent these undesirable effects of inbreeding depression; but what enhancement can come from more extreme outbreeding? One clear example of outbreeding enhancement comes from mice. Modern European house mice live a dangerous existence. Human beings don't like rodents scurrying about their kitchen floors, so they have manufactured numerous pesticides to kill them. One pesticide contains a drug called warfarin, which is also used in small quantities in medicine as an anticoagulant. But too much warfarin not only prevents blood clots; it also induces fatal bleeding.

Exposure to warfarin-based pesticides has exerted a strong survival pressure on mice in many countries. In response, some mice populations have evolved resistance to this strong poison. The development of this resistance is slow, costly, and limited to particular regions. But there's a quicker and easier way for mice to gain resistance to this pesticide—by outbreeding.

Biologist Michael Kohn of Rice University has found that German and Spanish mice have gained this valuable resistance to warfarin by breeding with Algerian mice. But because the European and Algerian mice belong to different species separated by 1.5 to 3 million years, most of their hybrid offspring are sterile. Nonetheless, a few of the hybrid females are fertile, which has introduced drug-resistant "super-mice" into Europe. This hybrid fertility has opened "a small window," Kohn explains, "for genes to be moved from one species to the other" through outbreeding. Now, the great majority of Spanish mice, as well as increasing numbers of German mice, are better adapted to survive in their warfarin-fraught environments.[15]

Hybridization through extreme outbreeding occurs more often than many people realize. Evolutionary geneticist James Mallet, at University College London, estimates that up to 10 percent of animal species and a quarter of plant species breed, on occasion, with organisms from other species.[16] Many hybrid offspring are infertile, however, when their parent species have different numbers of chromosomes. Such is the case with mules, which come from the mating of a donkey with a horse. The same applies to zebroids, which are crosses between zebras and horses or donkeys. Even viable hybrids cannot always compete with their parent species, which are often better adapted to their environments.

What will happen to the habitats of brown bears and polar bears, however, is an open question. The giant white bears originally split off from their common ancestor about six hundred thousand years ago.[17] But the rapid global warming of recent decades has endangered polar bears' existence in their current environment; rising temperatures melt sea ice, making it difficult for the Arctic Ursidae to hunt seals. At the same time, higher temperatures allow brown bears (grizzlies) to expand their territory to the north—well into polar-bear country.

In 2006, an Idaho man hunting in Canada's Northwest Territories shot a bear with brown patches on white fur. The animal turned out to be a grizzly-polar hybrid. Then, in 2010, a Western Canadian Inuit hunter shot what he believed was a polar bear. DNA testing found that the strange, creamy brown bear's mother had been a grizzly-polar hybrid, and its father had been a grizzly bear.[18] The hybrid was fertile!

If drastic environmental changes persist, there's a chance that this viable new hybrid would prove the most adapted to its new climate. Theoretically, a new species could even be born. If pure polar bears ever went extinct, their genes may survive only in pozzly bears (or grizzlars?)—just as Neanderthal outbreeding helped save a part of their species. In these scenarios, extreme outbreeding increases the fitness of the outbreeders.

HUMAN BEINGS

Excessive inbreeding in human beings undoubtedly produces ill effects. Consanguineous unions (that is, where mates are descended from the same recent ancestor) elevate rates of infant mortality, birth defects, and genetic diseases like cystic fibrosis.

To see why, let's begin at the logical extreme. The closest genetic relationship between two human beings who can reproduce together is that between full-blooded siblings, or between a parent and child. In both cases, these individuals share about 50 percent of their genes.[c]

Czech geneticist Eva Seemanova has studied cases of children born to such unions between 1933 and 1970. The children of sibling and parent-child unions had a 9.3 percent mortality rate by age one, as compared with a control group's 5.3 percent mortality rate. Of the incestuously begotten children, 40.8 percent developed a congenital disease or severe intellectual handicap, compared with only 4.5 percent of the more outbred control group.[19]

It makes biological sense, then, that inbreeding within the nuclear family is universally considered incestuous and forbidden. Only a few royal families have ostensibly broken this strictest of taboos. The desire to consolidate political power and economic resources has motivated their mate choices. Even within royal families, though, marriages between full-blooded siblings have been rare throughout history. Unions with slightly less related close relatives, however, have occurred quite commonly among royals. In the case of the Spanish Habsburg dynasty (1516–1700), the consequences of this reproductive pattern led to a calamitous fortune.

Within this royal family, uncle-niece and first-cousin marriages occurred very frequently over a long period of time. In 2009, a team of geneticists and physicians in Spain used computers to calculate the precise level of inbreeding that accumulated in the Spanish Habsburgs' tangled family tree. Their analysis spanned sixteen generations and included more than three thousand individuals. By the time Charles II was born in 1661, so much ancestral inbreeding had occurred between cousins, uncles, and nieces that the heir to the throne was slightly *more* inbred than a child of full-blooded siblings would be—even though his own parents were "only" uncle and niece. Charles II's genome was such an unmitigated disaster that his moniker was "the bewitched." In an attempt to cure himself, the king even resorted to exorcism, but to no avail.

Charles II (see figure 19) suffered from grave physical, intellectual, emotional, and sexual problems. To begin with, he had inherited the famous "Habsburg jaw"—a mandibular prognathism. In Charles II's case, however, his lower jaw stuck out *so* far that he was unable to chew his food. His tongue was so big that it made his speech almost unintelligible. He drooled profusely. As a child, Charles II drank breast milk until the age of four, but needed fourteen wet-nurses; they had to constantly rotate, because he kept biting their nipples (due to his jaw condition).

Charles II didn't learn to walk until the age of six. By his tenth year, he still had difficulty perambulating. Charles II's mild mental retardation was debilitating enough that he never learned to read or write, despite having access to the best tutors.

When it came time to marry, Charles II took as his bride his second niece, the French princess Maria Luisa of Orleans. The French ambassador at the time wrote back to his court: "The Catholic King [of Spain] is so ugly as to cause fear, and he looks ill."[20]

Figure 19: Charles II of Spain (1661–1700).

Under Charles II's rule, Spain's body politic also suffered from grave symptoms. The crown's authority grew weak in many provinces, so neighboring states exerted their influence over the Spanish court. The economy stagnated, nurturing political extremism. During these years, the Spanish Inquisition flared up with some of its largest-ever *autos da fé* (these "acts of faith" entailed burning alleged heretics at the stake).

In review of the historical record and autopsy, a contemporary group of Spanish urologists have concluded that Charles II likely had monorchism and an atrophic testicle (in plain English, a single, shrunken ball) and premature ejaculation. In addition, he was likely an XX male hermaphrodite (with intersex genitals), probably in association with fragile X syndrome. Because of the king's impotence and infertility, neither of his two wives could give him an heir.[21]

To make matters worse, geneticists believe that Charles II suffered from a grave genetic disorder likely caused by the unmasking of recessive alleles: renal tubular acidosis.[22] Consequently, he died of chronic kidney failure at the age of thirty-eight, effectively bringing his dynastic line to extinction. A power vacuum opened at the heart of the Spanish Empire, sparking the eleven-year, bloody, global War of Spanish Succession. By the end of the war, Spain had lost land in modern-day Belgium, Luxembourg, France, Italy, Sicily, and Sardinia. Over half a million people perished.

The Spanish Habsburgs' mate choices do not represent a typical pattern for large populations of humans. There are, however, numerous regions where consanguineous marriages are the norm. In fact, over one billion people live in cultures where 20 to 50 percent of marriages are between partners descended from a recent ancestor. The most common union of this type is between first cousins, which accounts for approximately half of all marriages in Pakistan.[23]

Cousin marriages in Pakistani communities in the United Kingdom have become a contentious political issue over recent years. Some groups have called for the banning of this marital practice, since consanguineous unions engender higher rates of genetic diseases, which weigh down on the national health system.

In May 2008, the United Kingdom's Royal Society of Medicine held a meeting to evaluate the issue.[24] In England and Wales, the general infant mortality rate is 5.2 deaths per thousand births. But a four-year study of over forty thousand children discovered that having a Pakistani mother more than doubles the chance of infant mortality—raising the rate to 10.5 deaths per thousand births. Low socioeconomic class could not be responsible, since more affluent Pakistani mothers had even higher infant mortality rates than their poorer counterparts. Therefore, the researchers concluded that inbreeding was to blame.[25]

Birmingham, Britain's second most populous city, offers more clues for resolving the striking infant-mortality differential. According to the Birmingham Health Authority, Pakistani babies are three times more likely to die in infancy than those of other races. The physician who administered the survey, Dr.

Mamoona Tahir, reported that 40 percent of these deaths resulted from genetic illnesses known to increase in frequency with first-cousin marriages.

In addition to the higher infant-mortality rates, inbred Pakistanis in Birmingham are thirty times more likely to inherit congenital diseases like phenylketonuria (a metabolic condition that can cause progressive mental retardation, brought about by unmasking a recessive allele) and the blood disorder thalassemia.[26]

In the British city of Bradford, half of all children are born to Pakistani parents, and 70 to 85 percent of Bradfordian Pakistanis marry their cousins. The proportion of consanguineous marriages in Bradford may be so high because most of the city's Pakistani population emigrated from the Kashmiri village of Mirpur. The construction of a dam inundated their birthplace in the 1960s, displacing the Mirpuri population.

Researchers have identified a range of 150 rare, congenital illnesses in Bradford. They include blindness, deafness, and skin diseases. According to the British Paediatric Surveillance Unit, 8 percent of neurodegenerative conditions among British children born since 1997 come from Bradford, even though Bradford represents only 1 percent of the population. The health system in Bradford has even found one extended family with six children who all share the same genetic condition. This disease tragically condemns them to a likely death before their twentieth birthday. In another case, a Bradfordian couple's apparently healthy five-year-old and seven-year-old were diagnosed with a devastating neurodegenerative disease in the same week.[27]

The global statistics on cousin marriage are a bit less dramatic than these examples. Professor Alan Bittles, at Edith Cowan University (in Perth, Australia), is an expert on the genetics of consanguineous marriages. According to Dr. Bittles's data from around the world, the children of first-cousin marriages only have, on average, a 3.7 percent higher mortality rate and a 3 to 4 percent higher rate of morbidity (including birth defects) than the general population, as measured up until puberty.[28] These statistics can vary according to the population. The different risks depend on the gene pool and on how much ancestral inbreeding has occurred in the past.

Why do hundreds of millions of people continue to marry their first cousins despite these higher risks? Do economic and political considerations override more evolutionarily sound behavior, as in the case of the Spanish Habsburgs? Or do some societies tolerate the risks of inbreeding depression as a tradeoff for some type of inbreeding enhancement?

Any possible "enhancement" from inbreeding or outbreeding would mean an increase in evolutionary fitness. To evaluate fitness, the gold standard is to measure how many babies are made, survive, and reproduce a third generation.

Fertility

We know that consanguineous marriages produce children with higher infant-mortality rates; but do these families compensate for the higher death rate with a higher fertility rate? How many babies do these closely related couples have, as compared with more distantly related ones?

The best insight into this question comes from a natural experiment—from a group of people called the Hutterites. The Hutterites are a population united by an Anabaptist sect of Christianity that originated in the Austrian Alps of the sixteenth century, under the leadership of Jakob Hutter. The Hutterites practice adult baptism and a communal lifestyle, inspired by the New Testament books Acts of the Apostles, and 2 Corinthians.

The Hutterites believe in absolute pacifism. This strict pacifist conviction has forbidden them from serving in armies—and even paying taxes to finance wars. On account of these principles, the young Hutterite sect soon faced persecution from its home state. So its members fled east to Transylvania. In the 1770s, the Hutterites began enjoying a century of religious tolerance in the Ukraine. But by the 1870s, the freedom ran out and the Russian draft caught up with them. To escape the imperial army, about nine hundred Hutterites migrated to what would soon become the US state of South Dakota.

Within a little over a century, the original nine hundred South Dakotan Hutterites multiplied into 135,000 descendants. They accomplished this reproductive feat through an extremely high natural fertility rate and a religious proscription against contraception. Because they had careful genealogical records tracing back their ancestry to fewer than ninety original progenitors who lived during the 1700s, the Hutterites have provided an ideal natural laboratory for studying the possible effects of inbreeding on fertility. This is why the South Dakotan Hutterites attracted the interest of Carole Ober, a reproductive geneticist from the University of Chicago, and her colleagues.

During a fifteen-year study, this research team focused on a subset of the South Dakotan Hutterites from the Schmiedeleut lineage. These families descended from only sixty-four ancestors and have had some of the Hutterites' highest lifetime fertility rates, with an *average* of over ten children during the early 1960s.

The researchers divided the married Schmiedeleut women into two groups; the first group contained the quarter of the women whose parents were most closely related to one another, and the second group included the remaining Schmiedeleut women. Those included in the first group had parents who shared around one-twelfth or more of their genes from a common ancestor. That is, their mothers and fathers were related to one another at least at a level somewhere between first cousins and first cousins once removed.

In this fast and furious reproductive race, would either group have an advantage? Any proven advantage, however small within one generation, would be greatly magnified over vast periods of time. Imagine how much a small difference in compound interest would make on a ten-thousand-year loan—if money reproduced geometrically.

The researchers compared the elapsed length of time between births for both groups. The median interbirth interval was significantly longer (by 3.2 months) for the women with the most closely related parents. The geneticists concluded that it took these women longer to get pregnant, possibly because they had lower conception or implantation rates, which ultimately could have been caused by recessive alleles.

Did the difference in the time it took for women to get pregnant affect their completed family sizes? Between 1901 and 1920, when the Hutterites had their largest family sizes ever, the quartile of women with the most closely related parents had an average family size of "only" nine children; the other women, in contrast, had a significantly higher average of eleven children. After the 1920s, however, when fertility began to decrease for both groups, the significant difference in completed family size disappeared. The researchers suggested that the first group might have tried to have more children to catch up to an "optimal" family size, even though this "reproductive compensation" wouldn't have eliminated the fertility issue caused by recessive genes.[29]

Outbreeding Enhancement in Humans

The endogamous Hutterites have very high fertility rates in general—higher than the average exogamous population (we'll learn why they have this reproductive advantage in the following chapter). But when the Hutterites reproduce *too* closely, their daughters' fertility rate drops. This drop indicates that a depression in fitness occurs toward the far end of the inbreeding spectrum.

What about the opposite scenario? Can extreme outbreeding ever enhance fitness in humans, as it has done for European house mice? When ancient humans left Africa to migrate into other continents, they came into contact with Neanderthals and Denisovans in Europe and Asia. These other hominins had already lived outside Africa for several hundred thousand years; therefore, these populations had likely evolved beneficial adaptations to their regional habitats. The thousands of generations spent by these hominins outside Africa meant that their immune systems had evolved to better contend with local pathogens. Extreme outbreeding between these hominins and humans, then, could have helped the human migrants to quickly adapt to the new environments. If ancient humans did benefit immunologically by hybridizing with other hominins, then we might find traces of this genetic history in the genes that code for parts of our immune system. Do we?

Biologist Peter Parham of the Stanford School of Medicine has studied a family of genes in the human immune system called human leucocyte antigens (HLAs—we'll learn all about them in chapter 11). HLAs essentially help the body to recognize what belongs to it, and what could be a nonself, harmful invader. When modern humans left Africa, these long migrations would have caused people to pass through successive population bottlenecks due to disease, famine, conflict, or additional migrations. These bottlenecks almost certainly led to the loss of genetic diversity, including that of HLA genes. So outbreeding with the Neanderthals and Denisovans who had already been there could have helped reverse this potentially disastrous trend.

Parham has discovered, in fact, that up to half of modern-day Europeans' variants of one class of HLA gene come from outbreeding with these hominins, compared with 80 percent of Asians' variants and 95 percent of Papuans'.[30] These percentages are much higher than the genome average of Neanderthal genes in today's Eurasian populations, which is only 1 to 4 percent. Therefore, it's very likely that these cross-species matings benefited early human migrants by helping their offspring better adapt to their new pathogen environments in Europe and Asia.[31]

In this chapter, we've learned that inbreeding and outbreeding form a continuum (much like Openness, left-right political self-placement, and RWA scores). On one side of this continuum, too much inbreeding in a human population leads to higher rates of infant mortality and genetic illnesses. Populations that reproduce with closely related individuals have higher birth rates in general, but too much inbreeding causes a drop in fitness. Together, these negative consequences explain the evolution of a force that encourages outbreeding.

This force is exerted by the first personality cluster that underlies human political orientation; specifically, xenophilia, secularism, and sexual tolerance act as a centrifugal (outward-pushing) force that ensures that some, more liberal proportion of the population will be attracted to mates who are less closely related. Perhaps like the Great Tit and many other animals, these xenophilic people will travel further from home to find more distant mates.

At a certain point, however, wandering further geographically and genetically does not increase the fitness of one's offspring indefinitely.[32] The next chapter examines why excessive outbreeding can also depress evolutionary fitness—and why some inbreeding enhances fitness. This centripetal (inward-pulling) force explains how tribalism and racism came to exist.

10.

WHEN INBREEDING IS FIT AND OUTBREEDING ISN'T

> **Inbreeding's benefits result from its *conservatism*, from a capacity to faithfully transmit parental genomes which have proved themselves.**
>
> **—Biologist William Shields[1]**

INBREEDING ENHANCEMENT AND OUTBREEDING DEPRESSION

The easiest way to immediately understand the idea of outbreeding depression is to think of the reproductive boundaries between species.[2] A male donkey (a "jack") and a female horse (a "mare") may be attracted to each other and mate. But conception, implantation, or the full gestation of a fetus may not occur. Even if they do manage to reproduce, all male offspring will be infertile. Only in extremely rare cases are some female "Molly mules" able to bear young. The problem is that horses have sixty-four chromosomes, but donkeys have only sixty-two. Mules have sixty-three, but their parents' chromosomes don't usually pair up in a way that creates fertile offspring.

Directly below, we'll look into the biological reasons why inbreeding enhancement and outbreeding depression occur. Then the chapter briefly considers a few examples of these phenomena in nonhuman animals. Finally, and most importantly, we'll discover how these forces impact human beings and shape our political reality.

Inbreeding Preserves Adaptations to an Environmental Niche

Within a species, normal individuals can breed with one another and produce viable young. This is the basic definition of a species. But not all fertile young are equally fit. Two parents, for example, may be genetically adapted to different environments. If they outbreed with one another, their offspring may lose important genes adapted to the parents' habitats.

Consider the case of two populations of sockeye salmon that inhabit the opposite sides of a lake. An inlet feeds one end of the lake, while the lake drains from an outlet on the other side. Staying in the lake area is adaptive for the young fish of both populations, since this environment nurtures their development. To remain in the deep water, the salmon fry on the outlet side instinctually swim upstream, back toward the lake; the young salmon on the inlet side, in contrast, swim downstream, to avoid going up the inlet.

When scientists crossed the two populations to create an experimental brood, some of the hybrids swam upstream, some swam downstream, and some switched back and forth. In a natural environment, an outbred sockeye-salmon population could lose more than half of its offspring from this one maladaptive trait alone.[3]

Oftentimes a trait depends on many different genes that work together. When populations of animals inbreed over many generations, coadapted gene complexes evolve that confer advantages in physiology or behavior for surviving in the local environment. These gene complexes comprise suites of genetic traits that inbred offspring inherit together as packages. Too much outbreeding with distant populations, though, could break apart these coadapted complexes, which would no longer work together to optimally adapt the creature to its environment.[4]

Inbreeding, on the other hand, holds these coadapted genes together. This preservation occurs because, in comparison with randomly chosen individuals, relatives are more likely to share many of the same alleles for related genes. So even during sexual reproduction between inbreeders, the recombination of different alleles occurs less frequently (or in some cases not at all); thus, the locally adapted gene complexes have a better chance of surviving intact.[5]

Once a population establishes inbreeding as a mating pattern, the deleterious recessive alleles are quickly exposed and weeded out, while the locally coadapted gene complexes take form. If the population were to later switch to outbreeding, it would run the double risk of accumulating more deleterious alleles in its gene pool and destroying the optimized gene complexes. Therefore, a self-reinforcing inbreeding pattern could gain momentum as a population's preferred mating system.[6]

Inbreeding Increases Altruism

There's another potential positive-feedback mechanism that promotes inbreeding in animal populations: inbreeding can increase altruism. Altruism is the force that causes an individual to reduce his or her own fitness while increasing the fitness of another. But what's the connection to mate choice?

Inbreeding raises the genetic relatedness of the members of the group. Since inbreeders share a higher percentage of their genes with one another, they act less selfishly toward one another.[7] By helping closely related individuals, inbreeders are helping to propagate copies of their own genes.[8] Chapter 19 explains in greater detail this particular phenomenon, which is called "kin selection."

If specific alleles associated with altruism exist in (or mutate into) a population, then inbreeding can increase the frequency of these "altruistic" alleles until they are "fixed" in the population; that is, they can become permanent, contingent on continued inbreeding.

In a random mating system, in contrast, altruism is likely to be lower because individuals share fewer genes in common with each other. In an outbred population, two nonrelatives may reproduce. Half of their offspring's DNA would come from the mother, and half from the unrelated father. Among inbreeders, however, parents already share a proportion of their DNA with one another. Therefore, the inbred offspring have *more* than half of each parent's genes. The longer a population's history of continuous inbreeding, the higher the genetic overlap becomes. Having a child that shares more than 50 percent of one's DNA is an easy way for a parent to increase the propagation of its genes, and therefore to gain greater fitness—without even expending any additional reproductive effort.[9]

Although sustained inbreeding in a human population can increase altruism among the in-group, it can also *decrease* altruism toward out-groups. This hostility may particularly occur when colonies of inbreeders live side by side but do not outbreed. Separation by a reproductive boundary would entail a sharp drop in genetic relatedness between the two groups. So altruism would be high within the groups but very low between them.[10]

Inbreeding Avoids the Costs of Outbreeding

Animals that outbreed face a number of potential dangers that could lower their fitness. In order to reproduce with "foreigners," animals must travel across unfamiliar territory to seek out genetically distant members of their species. These dispersals from a natal area may be hazardous. Even some human outbreeders living in perilous environments such as jungles, mountains, deserts, or war zones, may suffer higher casualty rates than inbreeders, and therefore produce fewer offspring.[11] Moreover, outbreeding animals have a higher chance of being infected by a pathogen carried by a foreign mate, or by a disease found in the mate's environment.

Sir Patrick Bateson, president of the Zoological Society of London, has suggested that outbreeding depression may also occur when acquired skills lose their utility in a foreign environment. For example, distantly related organisms could have incompatible parenting habits.[12]

The extra dangers faced by outbreeding may not even occur in the context of mate seeking. The proximate mechanism that motivates some individuals to mate with members of out-groups is often a xenophilic personality, which entails traits such as adventurousness, attraction to novelty, curiosity, and "the travel bug." These characteristics themselves may attract danger. A xenophilic man in a bellicose Amazonian tribe, for example, may be keener to participate in risky war parties that raid enemy villages and abduct foreign women.

As we review the concrete examples from nonhuman and human animals below, we must keep track of fertility. After all, if inbreeding enhancement is truly occurring, it should increase fertility over the long run; real cases of outbreeding depression should do the opposite.

NONHUMAN ANIMALS

Biologists have documented outbreeding depression across vastly different life-forms—from plants (such as white clovers and mountain delphinium), to marine invertebrates, to mammals.[13] They have also found intense, systematic inbreeding in a wide variety of creatures, often with few observable deleterious effects. For example, lizards, mice, and wolves frequently mate with their full-blooded siblings; deer and nonhuman primates regularly reproduce between fathers and daughters.[14]

But how do biologists identify outbreeding depression when they see it? We know that outbreeding depression occurs between different subspecies of deer mice because the offspring of these matings are more frequently infertile. Other times the young are perfectly viable but the parents don't get along. In fact, the scientist who carried out this research reported: "fighting does occur between some newly mated individuals, especially when members of the pair are of different races, and sometimes one or both animals may be injured or killed." Relative to the outbreeding deer mice, inbreeders would reproduce more successfully, thanks to their greater behavioral compatibility and kin-selection altruism.[15]

Other rodents have undergone extreme inbreeding with no depression in their offspring's fitness. One scientist inbred laboratory rats across dozens of generations of full-sibling matings. At the twenty-fifth generation, the rats showed no inbreeding depression in their fertility levels, growth rates, or longevity. In fact, the inbred line had a larger average litter size than its outbred control group (7.45 versus 6.70 pups), and the adult body weight of the inbred rats increased across the generations. The higher body weights and fertility rate indicate inbreeding *enhancement*.[16]

Larger mammals can suffer outbreeding depression as well. The ibex is a wild goat with enormous horns, which has subspecies that live in diverse environments around the world. Almost all of the ten or so races of ibex are separated by their distant habitats, scattered across Europe, the Middle East, and Central and East Asia. Consequently, they do not reproduce interracially in nature. In captivity, however, different ibex races *can* produce fertile offspring. When alpine ibexes outbred with the Nubian ibex from the Middle East, however, the hybrid goats calved during the coldest months of the year in their Swiss environment. This maladaptive trait would be disastrous for their kids.[17]

Primatologists have discovered similar phenomena among their own research subjects. Crosses between different races of owl monkeys frequently engender hybrids with lower fertility rates than inbred monkeys. Sometimes, these hybrids are completely infertile because chromosomal differences cause complications during meiosis.[18] There is also evidence that hybrid rhesus monkeys (crosses between Chinese and Indian subspecies) weigh less on their first birthday than do their nonhybrid counterparts. This weight difference alone, however, may not necessarily indicate outbreeding depression.[19]

HUMAN BEINGS

Like the ibex, all races of human beings can reproduce with one another and bear fertile young. Unlike the wild mountain goats, humans have extraordinarily sophisticated intelligence, communication, and technology that afford them a much higher degree of control over their environments. Do these attributes make humans immune to outbreeding depression?

Not entirely. Over most of human history, several factors have lowered the fitness of humans who have overly outbred. As we'll see below, infectious diseases have likely taken a higher toll on them, and so natural selection has favored a certain degree of inbreeding—especially in disease-prone areas. In addition, a lower compatibility between the immune systems of outbreeders has likely increased the incidence of several obstetric complications among outbreeding women, thereby decreasing their fertility rates.

In many of these cases, modern medicine has alleviated the weight of outbreeding depression to some extent. Nevertheless, science and technology have certainly not yet succeeded in transforming the underlying nature of our political psychology.

Infectious Diseases

The Political Psychology of Infectious Diseases

Before explaining exactly how infectious diseases can cause outbreeding depression, it's important to acknowledge the psychological connection between infectious diseases and extreme ethnocentrism. For example, Hitler's biographers have noted that he had unusual preoccupations with dirt, cleanliness, and impurity.[20] The leader of Nazi Germany also frequently compared the Jewish people to pathogens or the vectors of disease.

Ethnographer Raphael Ezekiel has identified similar characteristics among white hate groups in the United States. Ezekiel quotes Dave Holland, a leader of the Southern White Knights of the Ku Klux Klan:

> They bring all those Haitians and Cubans over here, and bring all these exotic diseases with them. . . . Praise God for AIDS! . . . AIDS is wiping out the undesirables. . . . And you know it's taking out blacks by the thousands; before long it'll completely depopulate Africa.[21]

Ezekiel reported that members of white hate groups have called out "*Praise God for AIDS!*" at their rallies, and printed this phrase on stickers, because these racist movements have perceived that the HIV pandemic is wiping out black people.[22]

According to the US Centers for Disease Control and Prevention, in 2009 African Americans accounted for 52 percent of all those diagnosed with HIV infections, although they represented fewer than 13 percent of the population. White Americans, who constituted 64 percent of the country, represented only 7.2 percent of HIV diagnoses. In 2010, the rate of new infections in black women was twenty times higher than in white women.[23]

A number of America's black ethnocentrists of various stripes joined their white analogues in the 1990s in politicizing and moralizing the AIDS pandemic rather than trying to understand it scientifically as a *natural* threat. Nation of

Islam leader Louis Farrakhan, for one, claimed that the US government had sent one billion units of HIV-infected blood to Africa to annihilate the continent's population. Filmmaker Spike Lee alleged in *Rolling Stone* magazine that AIDS was a state conspiracy designed to kill blacks, Hispanics, and gays: "I'm convinced AIDS is a government-engineered disease. They got one thing wrong, they never realized it couldn't just be contained to the groups it was intended to wipe out. So, now it's a national priority. Exactly like drugs became when they escaped the urban centers into white suburbia."[24]

The genocide conspiracy even infected the mainstream, good-willed, Presidential Medal of Freedom awardee, comedian, and Coca Cola spokesman Bill Cosby. Cosby was best known to Americans as the physician from *The Cosby Show*, Dr. Huxtable. In 1991 Cosby told the *New York Post* and a national television viewership that AIDS was "started by human beings to get after certain people they don't like."[25] Twenty years later, the *New York Times* reported that "a third of American blacks believe that government scientists created AIDS as a weapon of black genocide."[26]

The pandemic became an international political controversy in 2001 when South African president Thabo Mbeki sent a public letter to US president Bill Clinton. Mbeki, who questioned whether HIV causes AIDS, claimed that "a simple superimposition of Western experience on African reality would be absurd and illogical." In addition to exacerbating the racial politicization of the disease, Mbeki suggested that pharmaceutical cocktail treatments for AIDS—the best existing treatment at the time—were ineffective.[27]

Then, in 2005, Mbeki's former deputy president Jacob Zuma stood trial for allegedly raping an AIDS activist whom he knew to be HIV-positive. During the trial, Zuma admitted that he had *not* used a condom when having intercourse with this woman. Rather, to "cut the risk of contracting HIV," Zuma explained that he had taken a shower afterward. Eventually, the court cleared Zuma of the rape charges, and he was elected president of South Africa in 2009.

AIDS denialism has emerged in the Middle East as well—and again within an ethnically charged political context. The Fatah Party that governs the Palestinian National Authority in the West Bank has repeatedly accused Israel of conspiring to infect Palestinians with HIV. For instance, in February 2008 their official daily newspaper, *Al-Hayat Al-Jadida*, reported: "Israel exports AIDS and other sexually transmitted diseases . . . and therefore we have decided to take a medical test before marriage, concerning AIDS."[28]

Do these events represent anything more than eccentric demagoguery surrounding a particularly racially polarizing disease? Or is there any deeper relationship between infectious disease and the political spectrum?

Decades before the discovery of HIV, the political scientists at Berkeley who developed the original F-scale found that high-scoring authoritarians were more likely to fear germs.[29] Three generations later, in 2009, *New York Times* columnist Nicholas Kristof called the public's attention to research indicating that conservatives are more likely than liberals to worry about contamination or to feel the emotion of disgust. For example, conservatives sense significantly greater disgust at the thought of touching the faucet in a public restroom, or at the idea of accidentally sipping from an acquaintance's drink.[30]

This phenomenon that fascinated Kristof draws credibility from a growing body of studies. Evolutionary psychologists believe that the function of disgust is to repel people away from potentially dangerous pathogens. The interesting fact, though, is that self-described conservatives score significantly higher on a scale that measures disgust. The test includes statements such as, "Even if I was hungry I would not drink a bowl of my favorite soup it if had been stirred by a used but thoroughly washed fly swatter."

A group of political scientists from Rice University and the Universities of Nebraska and Illinois have subjected these findings to a more scientifically rigorous study. The research team, led by Kevin B. Smith, wasn't satisfied to simply correlate self-reported political orientation with conscious expressions of disgust. To determine the left-right placement of their subjects, Smith's team had their subjects respond to sixteen diverse, politically controversial items. More importantly, the political scientists measured their subjects' bodily responses to disgust by showing their test takers images of human excrement, a person eating a mouthful of worms, and an emaciated body.

When a person's brain perceives this type of aversive stimuli, the body releases stress hormones, like epinephrine, that activate the sympathetic nervous system. Numerous physiological changes transpire that are associated with a "fight or flight" reaction. Of particular convenience to psychophysiologists, the arousal of disgust also provokes perspiration. This secreted moisture, whose purpose is likely to cool down the body in its state of excitation, also affects the conductance of electricity across skin. More sweat on a subject's hand indicates a greater activation of the sympathetic nervous system, and presumably a greater reaction to a disgust stimulus.

Thanks to the experiment's design, Smith's team could determine whether political orientation correlates with an *involuntary* physical response to disgust as well as with the subjective sensation of this emotion. Using sixteen political issues rather than a simple self-placement also allowed the political scientists to see whether liberal or conservative responses to any particular type of controversial issue correlated most strongly with sweating.

Overall, the researchers found no relationship between the physiological reactions and attitudes toward the sixteen political issues. However, conservative responses to controversies on sex and reproduction *did* weakly correlate with the disgust reactions to the images of potential pathogen sources. Left-right opinions on tax cuts, however, had virtually no correlation whatsoever with the skin-conductance disgust response.[31] This study suggests that there's a relationship between fear of infectious disease and "cultural" conservatism (i.e., the tribalism personality cluster), but not economic conservatism (i.e., tolerance of inequality).

Does this mean that people with a greater fear of contagious illness are more xenophobic? Independent experiments conducted at UCLA by psychologist Carlos Navarette and anthropologist Daniel Fessler suggest that they are. The Los Angeles social scientists found a moderate correlation between xenophobia and people's perception of their vulnerability to infectious disease. Individuals who scored higher on ethnocentrism were also more likely to report washing their hands soon after shaking someone's hand.[32]

Is there any evolutionary meaning to this link between ethnocentrism and fear of infectious disease? One way to find out is to see whether ethnocentrism among the native populations of diverse locations varies according to their region's prevalence of pathogens. In 2008, biologists Corey Fincher and Randy Thornhill from the University of New Mexico teamed up with a couple of psychologists from the University of Vancouver to approach this question. They found ninety-seven regions around the world whose populations had previously been measured along an "individualism-collectivism" spectrum.

The Dutch anthropologist Geert Hofstede devised this individualism-collectivism scale as part of his "cultural dimensions" theory. Collectivism essentially corresponds to loyalty to the social group and ethnocentrism; individualism entails the freedom to choose one's affiliations, which permits greater xenophilia. Of course it would be nice to have RWA scores for over ninety regions, to compare apples with apples, but Hofstede's concept of collectivism has been more widely measured because of his theory's direct application to organizational psychology, in global corporations like IBM. In any case, the researchers from New Mexico and Vancouver defined collectivism as "ethnocentrism, conformity." Close enough.

Infectious diseases, however, can be difficult to measure; they flare up and die down from year to year, and they can be influenced by factors like weather and environmental change. What the researchers needed was a long-term, historical measure of pathogen prevalence. So the team collected data from between 1952 and 1961 on the incidence of major infectious diseases in each region.

These diseases included: leishmanias, trypanosomes, malaria, schistosomes, filariae, leprosy, dengue, typhus, and tuberculosis. They also created a regional index that measured the incidence of the top twenty-two local diseases for each specific area. After averaging and standardizing these pathogen-prevalence scores, the researchers could finally compare the pathogen dimension with collectivism (ethnocentrism).

Figure 20 displays the results, which are controlled for several potential confounding variables, such as population density. The scatter plot shows a strong correlation (0.70) between historical pathogen prevalence and collectivism.[a] Fincher and Thornhill concluded that xenophobic attitudes have protected people against pathogens by reducing exposure to out-groups.[33]

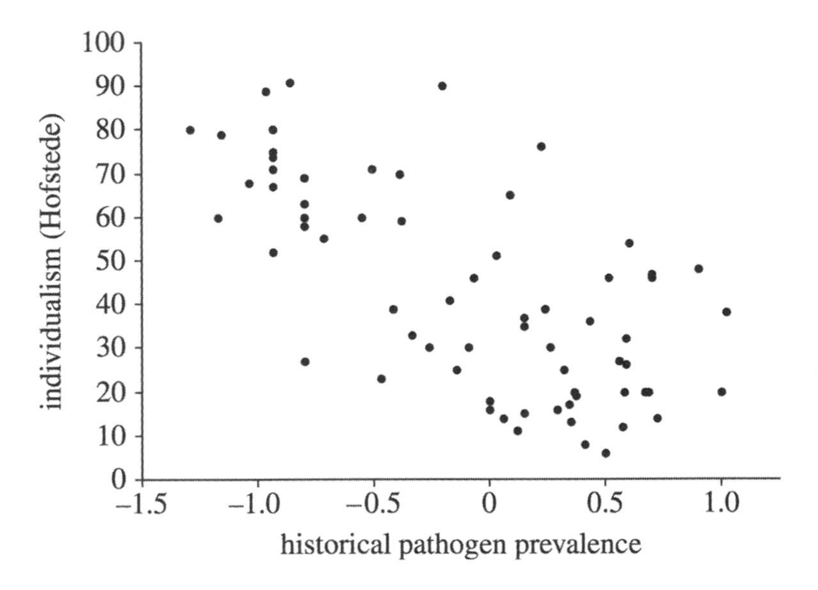

Figure 20. Correlation between Historical Pathogen Prevalence and Individualism/Collectivism (93 regions).

These findings explain why Big Five personality traits such as Extraversion correlate with distance away from the equator, as we learned in chapter 2. And they suggest why cultures native to hotter regions tend to be more conservative than cultures in cooler climes: because hot regions nearer to the equator foster a higher prevalence of infectious disease.

Mark Schaller and Damian Murray, the two psychologists from the University of British Columbia, have gone one step further. They've shown that a higher prevalence of infectious diseases not only increases ethnocentrism; it also restricts sexual behavior. In places with a higher incidence of pathogens, women show greater sexual restraint. Conversely, women in regions with fewer infectious diseases have a "greater chronic interest in new sexual partners and greater comfort with casual sexual encounters." For men, there was only an insignificant, weak correlation between pathogens and restricted sexual behavior.[34] Once again, it seems that female sexuality may be relatively more flexible than male sexuality.

Why exactly does the fear of infectious diseases, or their actual presence, make people more xenophobic, ethnocentric, and even sexually guarded? Is it simply a matter of avoiding anyone who may carry a pathogen? Why is there a need to avoid out-groupers more than fellow in-groupers?

Infectious Diseases and Evolutionary Fitness

So, why avoid out-groupers more than one's own tribespeople? No one could understand better than an Amazonian Indian in Brazil who has been called "the most isolated man on the planet."[35] He lives in the Brazilian state of Rondônia, and his nickname is "the man in the hole" because he hides from the rest of the world by running between various huts with six-foot-deep pits dug out in the center. When a government field agent specialized in contacting isolated tribes tried too insistently to approach the lone Indian, the isolated man shot an arrow into the visitor's chest.

The Brazilian government believes that "the man in the hole" is the last survivor of his tribe. The rest were killed by ranchers and loggers who have colonized their Amazonian territory—but especially by the diseases they brought with them. Although the state now has an official "no contact" policy, and has given the sole survivor more land to buffer him from the outside world, it's far too late.[36] Contact with foreigners who had extremely different immune responses to pathogens annihilated the unknown tribe. Even a common cold virus has decimated uncontacted tribes in other parts of the world.

When the Europeans first made contact with native South Americans in the early sixteenth century, the New World's indigenous population had immune systems that hadn't coevolved with smallpox. This one virus is estimated to have wiped out up to 90 percent of the Andean population by 1600—including the Incan emperor Huayna Capac.[37] At the time of his death, Huayna Capac and his full-blooded sister and queen had produced no legitimate male heirs to inherit his reign (this failure to reproduce with his sister could be an example of inbreeding avoidance or inbreeding depression). Consequently, two of the emperor's dozens of illegitimate children, whom he had begotten with less-related women, waged a civil war against one another. Thus, the Incan empire descended into chaos on the eve of the Spanish conquest.

These are extreme examples of how any contact with members of out-groups can devastate the future of a gene pool. In the next sections below, we'll see how some degree of inbreeding, under less exceptional circumstances, can enhance fitness by maintaining a population's immunity to many acquired diseases.

The Human Blood Groups and Disease

In 1919, the Polish serologist Ludwik Hirszfeld and his wife Hanka discovered that the frequencies of genes that code for the ABO blood group system vary greatly between different ethnic groups. By 1976, the British geneticist and hematologist Arthur Mourant had published a remarkably comprehensive tome describing the prevalence of dozens of different blood group systems among native populations from around the globe. Today, Mourant's encyclopedic work remains the most exhaustive record of the astonishingly complex diversity hidden within the heart and vasculature of humanity.[38]

When most people think of blood type, they think of the ABO system. If you've heard someone's blood described as "positive" or "negative," then you might also have some familiarity with the second most important blood group system: Rh. "Positive" and "negative" refer to one of the five main antigens[b] in the Rh system that are found on the surface of red blood cells (the D antigen).

Together with the ABO and Rh systems, there are thirty-one other blood group systems. Many of these systems are named after patients in whom scientists first discovered them. They are MNS, P1K1, Lutheran, Kell, Lewis, Duffy, Kidd, Diego, Yt (Cartwright), Xg, Scianna, Dombrock, Colton, Landsteiner-Wiener, Chido/Rogers, Hh/Bombay, Kx, Gerbich, Cromer, Knops, Indian, Ok, Raph, John Milton Hagen, Ii, Globoside, Gill, Rh-associated glycoprotein, Forssman, JR, and LAN. Assigned to these thirty-three blood group systems, there are about three hundred different blood group antigens.

Geneticists have found the genes that code for these antigens. And the frequencies of these genes differ between ethnic groups, since natural selection favors different antigens in different environments.[39] Mourant's book is filled with world maps that indicate the global frequencies of these antigens and their genes. Figure 21, for example, shows the geographical distribution of the *C* gene for the Rh system, among indigenous populations of the world.[40]

If you imagine hundreds of maps similar to this one, superimposed on one another, then it's easy to understand how one could "triangulate" populations that tend to reproduce endogamously. For example, the Basque population is linguistically, genetically, and consciously different from other neighboring European populations, and Basques have not historically married very freely with them. Consequently, the Basques also have quite different gene frequencies for blood group antigens.[41] After DNA, blood reveals the most about human genetic variation.

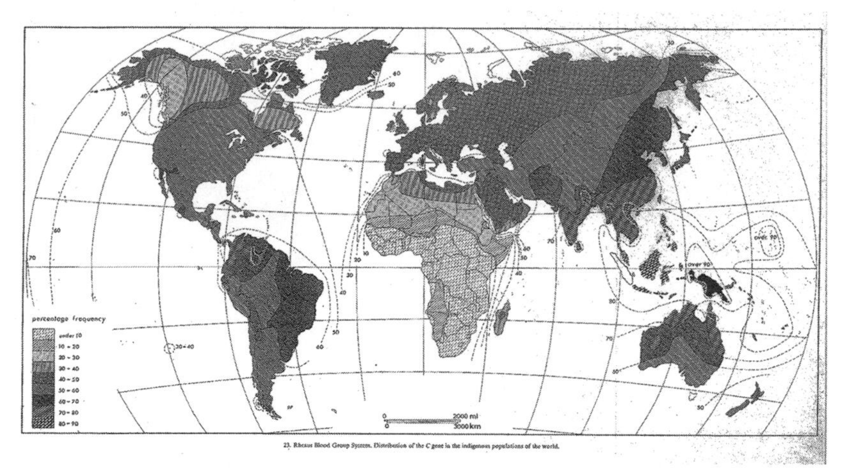

Figure 21. Rh Blood Group System. Distribution of the C Gene in the Indigenous Populations of the World.

What accounts for this great diversity running through our veins? The struggle for survival entails not only the competition for resources and mates but also the battle against pathogens. In this immunological arms race, the blood group antigens of human populations have coevolved with a microscopic world of enemies (as well as allies and neutral parties). Natural selection shapes the human immune system by favoring the configurations best adapted to a population's environment.

For their part, there are some parasites, bacteria, and viruses that have evolved antigens that are similar to those of particular human blood groups.[42] This makes it difficult for people with these blood types to produce an antibody to fight against the invading organism. Thus, the pathogen's targeted offense gives it a better chance of defeating the immune response.[43]

Disease / Microbe	**Blood Group Antigen Associated with Increased Susceptibility**
Plague	O
Cholera	O
Leprosy	A, B (lepromatous form), O (tuberculoid form)
Yaws	M
Tuberculosis	O, B
Gonorrhea	B
Smallpox	A, AB
Mumps	O
E. coli infection	B, AB
Salmonella infection	B, AB
Helicobacter pylori infection (gastric ulcer)	O
Influenza type A	A
Norwalk virus	O
Giardia lamblia	A
Leishmania	Knops

Figure 22. Associations between Pathogens (or their diseases) and Blood Group Antigens.

Figure 22 shows a list of some of the infectious diseases known to disproportionately affect people who have particular blood group antigens.[44] Pathologist George Garratty has suggested that histories of epidemics have shaped the different frequencies of these antigens around the world. These contrasting frequencies clearly appear in Mourant's maps.

For instance, the plague bacillus and the cholera bacterium target people with type O blood; those with type A_I and B blood, on the other hand, can more easily produce an antibody to defend against these pathogens. Plague and cholera epidemics, then, would kill more type-O individuals. "O" people would leave fewer descendants, reducing the proportion of this blood type in the population.[45]

The most critical diseases, from an evolutionary point of view, are those that eliminate individuals from the gene pool before they reach reproductive maturity. Victims of these pathogens cannot pass on their blood-group genes to their children. A few years before publishing his epic encyclopedia of blood-group frequencies, Mourant reasoned: "The important killing diseases of the young . . . have been the great epidemic infections. If any of these were shown on adequate statistical evidence to be more lethal to persons of one blood group than another, natural selection affecting the blood group frequencies of populations could be regarded as proved."[46] In other words, infectious diseases can discriminate against the immune systems of genetically related populations. The fitness of these groups can therefore depend on the pathogens in their environment at a given moment and across the distant past.

Consider the Norwalk virus, which also targets people with type O blood. This pathogen is responsible for 90 percent of the nonbacterial outbreaks of stomach flu around the world. As recently as 1980, before the widespread promotion of oral-rehydration therapy helped reduce the mortality rate, gastroenteritis killed approximately 4.6 million children a year; half of them were infected by the Norwalk virus.[47] This disease continues to be a major killer in developing countries.

Suppose that a small, related community lives in an environment lacking a clean water source. The dirty water could be responsible for a long history of exposure to a pathogen like the Norwalk virus, which would exert a strong force on infant and child mortality. Through natural selection, most children born in this community would *not* have type-O blood, which would make them more susceptible to this prevalent disease. If some xenophilic villagers reproduced with people from a faraway community with a clean water source and a high frequency of type-O blood, then the family might raise a greater-than-normal proportion of type-O children in the village with the dirty water. More of their children could die as a result of outbreeding depression. Since xenophilia has a strong heritable component, the genes coding for it would decrease in frequency. And the average political orientation of the village would shift to the right.

The villagers could not tell by looking at each other what blood types they have. And they would not know that this factor could impact their completed family size. Instead, they would think in terms of ethnic groups and geography, which serve as the psychological proxies for genetic relatedness. In a certain proportion of the villagers, ethnocentrism and xenophobia would motivate a conservative mate choice with members of the natal group; this endogamous reproduction would reproduce parental genomes that have historically done well in resisting the local pathogens.

Some of the villagers, however, would still be more attracted to the inhabitants of distant settlements, and perhaps even to different ethnic groups. These exogamous mate choices would create greater variation in blood types. So if a new pathogen emerged, or an existing one mutated, the resulting disease would not be able to easily wipe out a genetically homogeneous group. Therefore, it's adaptive for populations to have a certain amount of immunological diversity, but not too much. The normal distribution of the political personality helps to regulate this balance through ethnocentrism and xenophilia.

Many people today have access to antibiotics and modern medicine; these individuals may be less afraid of infectious diseases than were their ancestors or people who lack adequate healthcare. Even in today's world, though, it's easy to find diseases around which race, politics, and fear intersect.

HIV and Malaria

In the United States, as well as in other areas of the African Diaspora, HIV infects more black people than nonblacks by several fold. In 2010, Sub-Saharan Africa accounted for 68 percent of the world's HIV infections, according to UNAIDS's "Epidemic Update." Within Sub-Saharan Africa, however, not all black populations have suffered from the same infection rates. Although many factors determine the nature of epidemics, including government policy, culture, economics, population density, and numerous environmental variables, one of the most important determinants of HIV susceptibility has been almost completely unnoticed by the public: blood group antigens. Instead, much of public opinion toward the pandemic has revolved around politics and perceptions of race.

The story of HIV, race, and blood differences begins with a much older infectious disease, which has plagued human beings for fifty thousand years: malaria. In the 1920s, scientists first observed that many black Africans had evolved some intrinsic resistance to malaria. The first gene discovered to account for this resistance was the gene for sickle-cell anemia.

In parts of the continent where malaria is endemic, populations have a very high frequency of a mutated gene involved in the production of hemoglobin. Hemoglobin is an iron-containing protein that allows our red blood cells to transport the oxygen we breathe to other cells around our body. In some tribes in West Africa, up to half of the population carries one copy of this form of the gene (this "allele"), which has a single-nucleotide mutation (a change in one "letter" of DNA). These carriers of the incompletely recessive allele can produce some sickle-shaped red blood cells, but not enough to suffer from the severe symptoms of sickle-cell disease (which can include anemia, the obstruction of blood vessels, and organ damage). Being a heterozygous carrier of the sickle-cell allele confers greater resistance against the malaria parasite, and therefore a significantly reduced chance of dying from this infectious disease.

In these same West African tribes, a quarter of the population has two normal hemoglobin alleles. Since they lack the mutated allele, they do not receive any additional protection against malaria. But another quarter of the population of these tribes has two copies of the defective allele, which causes sickle-cell disease. Malaria kills many of the 25 percent who lack the mutated allele, while sickle-cell anemia shortens the life expectancy of the people who inherit two copies of the defective gene.

Geneticists call this phenomenon a "balanced polymorphism," since the different frequencies of the hemoglobin alleles depend on the level of malaria parasites historically present in the environment.[48] With life hung in the balance between malaria and sickle-cell disease, both inbreeding and outbreeding with populations that have different gene frequencies could impact the fitness of offspring.

Several of the blood group systems have coevolved in an arms race with malaria parasites. The Knops antigens, for example, exist in very different frequencies in US whites (whose ancestors come from nonmalarial zones), US blacks (whose DNA is, on average, 20 percent European), and African blacks.[49]

Approximately 90 percent of black Africans have a particular gene for the Duffy antigen system that has evolved to protect them from malaria outbreaks. Other populations usually have an allele for this gene that creates receptors on red blood cells for "chemokines." Chemokines are proteins that help the body defend against pathogens by recruiting the immune system's cells to the site of an infection. But the chemokine receptor is also the place where one of the malaria parasites (*Plasmodium vivax*) attaches to the red blood cells to infect them. In response, most black Africans have evolved red blood cells *without* this receptor, which bestows them with additional resistance to malaria.

Lacking chemokine receptors served black Africans well for millennia. But that changed in the late nineteenth or early twentieth century, when chimpanzees transferred the Simian Immunodeficiency Virus (SIV) to humans.[c] At this point, the virus became the closely related Human Immunodeficiency Virus (HIV; "H" stands for "human"). In addition to infecting the immune system's T-cells, the most common and infectious strain of HIV also attaches to the Duffy blood group's chemokine receptors on red blood cells. People who have these receptors have higher levels of chemokines in their blood, which partially protects them from HIV infection.[50] Black Africans without the receptors, however, appear to have greater difficulty in fighting off HIV in the early stages of the infection. Consequently, they are 40 percent more susceptible to contracting the blood-born immunodeficiency virus. If all black Africans had these receptors, the prevalence of HIV in Sub-Saharan Africa would drop by 11 percent.[51] And approximately 2.7 million fewer black Africans would have HIV.

Immunological Incompatibility

Infectious diseases aren't the only foreign bodies that our immune systems contend with. During reproduction, foreign DNA enters a woman's body. If she becomes pregnant, a semi-foreign body develops within her womb. Sometimes a woman's immune system may even reject her own fetus, which can result in the illness or death of the baby.

Several of these obstetric conditions occur less frequently when there is closer genetic similarity between a woman and her mate. With greater outbreeding, conversely, there is a higher chance of immunological incompatibility, which translates into lower fertility rates (i.e., a greater depression in evolutionary fitness).

Early Pregnancy Loss and Spontaneous Abortion

Pregnancy does not always occur with the deposit of semen in a woman's reproductive tract—even during the most fertile time of her cycle. To find out just how difficult it would be for them to get pregnant for the first time, over five hundred healthy, newly married women in China between the ages of twenty and thirty-four participated in a medical experiment. Fertility specialists examined the women's urine and vaginal bleeding on a daily basis for signs of conception. They determined that, over a twelve-month research period, the women conceived at a rate of 40 percent per cycle.

Of the women who *did* conceive, a quarter of them experienced "early pregnancy loss," which means a very early miscarriage within six weeks of their last menstrual period. An additional 8 percent of the newlywed women suffered spontaneous abortions after the sixth week.[52]

Among the human population in general, an average of 40 percent of all postimplantation conceptions are aborted by women's bodies. In women over thirty-five years of age, this figure likely exceeds 80 percent.[53] Any small change in this abortion rate could constitute an important gain or loss in fitness. If inbreeding were to decrease the spontaneous abortion rate, then inbreeding enhancement would occur.

Indeed, one of the major causes of early abortions is certain types of ABO blood-group incompatibilities between the mother and her fetus.[54] This fact suggests that greater genetic relatedness with their mates, up to a certain point, makes it easier for women to stay pregnant and bring a baby to term.

Hemolytic Disease of the Newborn

The best-known blood incompatibility involves the Rh system. A fetus may inherit from its father an Rh antigen that the mother lacks. In some of these pregnancies, the mother is exposed during birth to the fetus's Rh-positive blood. Since the mother is Rh-negative, she develops antibodies to her baby's paternally inherited, "foreign" Rh antigens.

During subsequent pregnancies with the same mate, the mother's antibodies may reach her younger fetuses' circulation. If these fetuses share the same problematic Rh antigen with the first baby, the mother's antibodies can attack their red blood cells. This condition, which is called Hemolytic Disease of the Newborn (HDN), can have extremely serious consequences for the baby, ranging from mild anemia to stillbirth.[55]

In addition to Rh Disease, fetuses may suffer from other types of HDNs when there is an incompatibility between the ABO blood group (as described in the previous section) and, to a lesser degree, when there is discord between the Kell, Duffy, and some other blood groups. These varieties of HDN increase a mother's risk of abortion and infertility.

Different ethnic groups have different frequencies of blood group antigens. Therefore, the more related a woman is to her mate, the less likely it is that she will face these conditions. Instead, she may have higher lifetime fertility. Thus, she would benefit from inbreeding enhancement.[56]

Until the middle of the twentieth century, Rh Disease was considered a serious public-health problem, since it caused severe disabilities and the deaths of ten thousand babies a year in the United States. Proportionally, this means that the disease impacted nearly thirty in ten thousand births in the mid-1940s, when intermarriage was much less common than it is today.

Several years after scientists began experimenting on "Rh factor," the *New York Times* mentioned the topic for the first time. On March 26, 1944, the newspaper promised: "The recently discovered Rh factor in human blood . . . need not cause infant deaths and childless marriages [anymore]." But it took another twenty-four years before a pharmaceutical company in New Jersey announced the development of a vaccine. This subsidiary of Johnson and Johnson developed the drug RhoGAM. It consists of a solution of antibodies that doctors can inject into an Rh-negative mother. The vaccine's antibodies latch onto and destroy Rh-positive blood cells that the woman may receive from her fetus. By doing so, RhoGAM prevents the maternal immune system from reacting and giving Rh disease to her future babies. RhoGAM is considered a major medical advance of the twentieth century.[57]

Preeclampsia

The most common of dangerous pregnancy complications is called preeclampsia. Preeclampsia has no known cure, and it can affect both the mother and her unborn child. The condition impacts between 5 and 14 percent of pregnancies worldwide. In some developing nations, preeclampsia effects up to 18 percent of pregnancies and is the second largest cause of stillbirths and neonatal deaths.[58]

Preeclampsia usually occurs after the twentieth week of gestation. The condition results in high blood pressure and high levels of protein in the urine. In severe cases, preeclampsia can lead to liver and kidney failure, and to eclampsia, which causes seizures, comas, and potentially even the death of the mother. The symptoms and possible causes of preeclampsia are complex and debated; however, the condition may result from the maternal immune system's reaction against foreign paternal antigens in the fetus and its placenta.

Research conducted in Turkey in the early 1970s supports this hypothesis. British population geneticists and Turkish pediatricians monitored 23,358 pregnancies at the Doğum Evi Hospital in Ankara. The researchers separated the mothers into two groups: the 17 percent of them who were married to their second cousins or closer (most were married to their first cousin), and the remainder who were not in consanguineous marriages. Women in cousin marriages had a significantly *lower* incidence of preeclampsia and eclampsia, as compared with the outbreeders. Outbreeding women were 60 percent *more* likely to experience these complications than women in cousin marriages.[59]

Physical Compatibility?

Biologist Sir Patrick Bateson believes that, in some cases, even disparities in the physical sizes of body parts can cause outbreeding depression. For example, the size and shape of teeth are strongly heritable traits. So are the size and shape of jaws. To illustrate his mandibular argument, Bateson points to the prominent Habsburg jaw, which famous painters have documented across many generations—from the Holy Roman emperor Maximilian I, to his great-great-great-grandson, Philip IV of Spain, the father of the ill-fated Charles II. Figure 23 shows these two rulers.

The problem that can arise is that teeth physiology and jaw physiology are inherited independently; they do not correlate. So if a woman with small teeth and a small jaw reproduces with a man with large teeth and a large jaw, some of their grandchildren could inherit small jaws and large teeth. Without modern dentistry, poorly fitting teeth could cause abscesses in the mouth, leading to a greater mortality rate (and outbreeding depression).[60]

Figure 23. Holy Roman Emperor Maximilian I (1459–1519), *left*, painted by Albrecht Dürer, and Philip IV of Spain (1605–1665), *right*, painted by Diego Valázquez.

Is there any evidence that greater physical similarities lead to inbreeding enhancement? One study took nineteen different measurements of couples, including their ear lengths, neck circumference, interpupillary breadth, and lip circumference. The more alike the pairs measured on seventeen of the nineteen dimensions, the more children they had.[61]

Biometrics, Personality Measurements, and Geography in Mate Choice

In light of this link between similar body measurements and fertility, it's not surprising that spouses show positive correlations in their physical characteristics in general. In addition to neck circumference and earlobe sizes, spouses tend to resemble one another, more than other potential mates selected at random, in terms of skin color, height, and weight (the same positive correlations hold true for intelligence, educational attainment, socioeconomic status, and age). But what about personality?

To answer this question, political scientists John Alford and Peter Hatemi gathered facts about tens of thousands of spouse pairs in the United States during the 1980s. In this time and place, it's safe to assume that the great majority of these unions were courtship marriages (as opposed to arranged ones) where individual preference played a key role.

The researchers discovered that spouses shared positive correlations for physical traits. But the similarities weren't very impressive. For example, height was weakly correlated (at 0.227) and weight was very weakly correlated (at

0.154). The same was true of general personality traits. The correlation between spouses for Extraversion was even a negligible 0.005.

Despite such weak correlations for general personality traits, *political* personality traits turned out to be *the highest correlations of any social or biometric variables they measured*. Support for a political party almost reached a strong correlation (at 0.596), and liberalism-conservatism scores came in at a high 0.647.[d] Specifically, spouses were most likely to share the same attitudes toward school prayer, abortion, and gay rights—three hot-button tribalism issues at the intersection of religion, sexual tolerance, and reproduction.

But how do we know that spouses didn't grow to share more similar political orientations over time, perhaps to avoid conflict? Simple: because the political attitudes of newly married spouses were very similar to those of spouses who had been married for decades. And we also know that spouses' political orientations weren't similar simply because they tended to come from the same homogeneous social or ethnic groups; *within* specific groups, marriages were occurring in nonrandom ways with regard to political personality. Among Catholics, for example, the liberals tended to marry the liberals, and the conservatives tended to marry the conservatives. The same held true for Jews and every other social group.

After eliminating these other possibilities, we can describe these coupling tendencies by their biological term: they constitute what's called *phenotypic assortative mating*, which simply means that humans and other animals mate in nonrandom ways depending on how their own traits relate to the same observable traits in others. In this case, where "birds of a feather flock together," we're witnessing *positive* assortment (negative assortment occurs when "opposites attract" more than they would at random).

But of course, not all birds are equally as likely to flock together with like-feathered birds; after all, the level of tribalism in our personalities varies substantially between individuals, and it affects mate choice. We now have a huge amount of data on human mating preferences, generated by online-dating services. In 2013 political scientists Peter Hatemi, Rose McDermott, and Casey Klofstad analyzed over three thousand online dating profiles from 313 US zip codes. They discovered that, among males, liberals are more likely than conservatives to seek partners of different body types (i.e., to assort negatively). And conservatives are more likely than liberals to prefer a partner of their own race (to assort positively).

We now have at our fingertips some very valuable information on mating and political orientation: we can calculate how people are mating nonrandomly according to political personality, and also the impact of genes on political attitudes. Hatemi, McDermott, and Klofstad took this information and built a computer simulation to see what would happen over long periods of time. They began with a normal distribution of left-right political orientations in the population (similar to the bell curve described in chapter 2). If mating were totally random (which it's not), then the political spectrum would remain the same. But when they plugged in the probabilities of assortative mating that are actually occurring today in American society, the curve began to widen substantially in the first five generations. By generations fifteen to twenty, the curve widened just a tiny bit more and then reached equilibrium, where it remained stable up to the ninetieth generation and ostensibly beyond. At this balance, the proportion of political extremists (people whose ideology is more than two standard deviations left or right of center) had increased from an original 4.5 percent up to 11.2 percent. The political moderates (who were within one standard deviation left or right of center) had dropped by 17 percent. In other words, the birds of a feather had bred a more polarized nation.

Since political orientation plays a crucial role in mating, which is so fundamental to our biology and existence, it shouldn't be surprising that concentrations of political personalities vary across geography (just as the physical traits of nonhuman animals vary across space). This probably occurs because it's easier to find a mate if one lives in a politically homogeneous territory. For a blue person to live in a blue territory would be more evolutionarily adaptive than living in a swing territory—not to mention a red one.

In America, we're in the midst of a great geographical segregation according to political personality. In 1976, just over a quarter of Americans resided in counties where presidential candidates won the election by a margin of 20 percent or more. By the year 2000, nearly *half* of Americans lived in these more politically homogeneous counties.

Geographic political disparities may be on the increase in America because it's a relatively young country with dynamic immigration and migration patterns. Still, political polarization isn't unique to America. As Alford and Hatemi point out, we know that "political divergence did not begin with the 'red state-blue state' divide, but rather is at least as old as Athens versus Sparta."

One thing that *has* changed since ancient times has occurred quickly and recently in the twentieth century: women around the world have gained higher education and political enfranchisement. Both of these developments may lead to modern women with stronger, more polarized political orientations (see appendix B) and greater freedom in mate choice. Increasingly positive assortment by education and political orientation, then, could lead to greater political polarization in the future, as compared with the dominant factors at play in earlier marriage markets (such as wealth, beauty, and the self-interest of senior family members).[62]

Fertility

Sooner or later, assessing inbreeding and outbreeding enhancement must come back to fertility. And we must answer a very basic and fundamental question: Do cousin marriages produce more children than do more distantly related mates?

The impassioned debate over marriages between close kin is not a new one. And it has always been politicized. In the second half of the nineteenth century, this contentious issue was raging in Western Europe and North America, where physicians, scientists, and politicians argued about the biological effects of consanguineous unions. Charles Darwin himself weighed in on the subject, perhaps because he had married his first cousin Emma Wedgwood.

In Darwin's 1871 book *The Descent of Man*, the grandfather of evolutionary theory expressed his irritation with the politicization of the controversy: "When the principles of breeding and of inheritance are better understood," Darwin wrote, "we shall not hear ignorant members of our legislature rejecting with scorn a plan for ascertaining by an easy method whether or not consanguineous marriages are injurious to man."[63]

First cousins Charles and Emma had eleven children together. The Darwins' fecundity was, in fact, representative of cousin marriages in their population at the time. In 1960, geneticist C. D. Darlington, at Oxford University, carried out a study of first-cousin marriages in nineteenth-century England. Darlington found that these inbreeders produced twice as many great-grandchildren as outbreeders did.[64] No wonder, then, that many modern societies still practice inbreeding between cousins, and between uncles and nieces.[65] And, predictably, people are still arguing about this extremely political issue 140 years later—including in Darwin's Britain (with respect to the marital practices of Pakistani immigrants).

Indeed, consanguineous spouses have more babies, which reflects inbreeding enhancement. But if they reproduce with individuals who share too many genes in common, as did the most closely related group of Hutterites from the last chapter, then their family's fitness drops due to inbreeding depression. Clearly, fertility is affected by many different forces, which come from different directions. These pressures vary depending on the genetic relatedness of spouses, their gene pools, and changes in their environment. When all the inbreeding and outbreeding pressures combine, their collective impact on fitness produces a curve whose peak in fertility comes to a high point at an optimal, middle ground.

In the next chapter, we map out where this optimal peak exists, and we explore how human sexuality normally guides people's behavior so that they mate accordingly. The findings are fascinating—and seldom politically correct.

11.

HOW OPTIMAL MATING HAPPENS

Evolution depends on a certain balance among its factors. . . . A certain amount of crossbreeding is favorable but not too much.

—Sewall Wright, a father of the modern synthesis between evolutionary theory and Mendelian genetics, 1933[1]

There are more than a quarter million Patels who live in Britain, which puts this surname up with "Smith" as one of the most common family names in the United Kingdom. The Patel lineage originates from the West Indian state of Gujarat. And Gujarati geography and ancestry still bear a strong influence on the mating patterns of British Patels. More than half of the Patels in the United Kingdom marry another Patel. Finding the right one is somewhat of a balancing act—an exercise in optimal outbreeding.

Patels in the marriage market attend singles events at their community centers. In addition to receiving the names and occupations of the opposite gender, guests at these gatherings can see the location of each other's ancestral villages. Patels keep track of these villages because they aren't supposed to marry someone whose forefathers came from the same one. This custom prevents excessive endogamy—and protects Patel progeny from the risks of inbreeding depression.

The ideal Patel, however, also marries within his caste and sub-caste. In addition to having slightly different cultural values and economic stations, sub-castes overlap with clusters of ancestral villages—and therefore with genetic heritage. Thus, caste restrictions also curtail outbreeding.

This balance between endogamy and exogamy, and between strictly arranged marriage and absolute love marriage, seems to be working out well for this West Indian superclan. A Patel in the United Kingdom is seven times more likely than a Smith to be a millionaire.[2]

The Patel mating pattern strives toward a middle ground. By doing so, this Gujarati kinship system avoids the inbreeding and outbreeding depression discussed in the previous two chapters.

But a number of questions remain: Is there any solid proof that some specific optimal degree of outbreeding exists for human beings? If so, what guides people's reproductive behavior so that they achieve a peak level of fitness? Finally, how do our mate-choice mechanisms compare to those of nonhuman animals?

OPTIMAL-MATING DISTANCE AND KINSHIP

Let's start with the first question: Is there any real evidence for optimal outbreeding in human beings? If some level of outbreeding does produce a peak in fitness, how could we know exactly where to find this peak? The way to answer this controversial question is to determine which couples succeed in multiplying their genes the most.

It's not enough to simply know who makes the most babies. The real question is who produces the largest number of *fit* babies. That is, who are the parents of children who survive until maturity and then reproduce the largest third generation and beyond? The fittest couples are the ones that maximize their share of the future gene pool, even long after they're dead. They are the great-great-great-great-great-grandparents of the most of us.

In 2008, a team of Icelandic geneticists decided that their North Atlantic island would be the ideal natural laboratory in which to search for an optimal-outbreeding peak. According to the team led by Agnar Helgason, Iceland would minimize the problem of confounding, nonbiological factors, since the country is one of the most culturally and socioeconomically homogeneous societies in the world. Icelanders' family size, use of contraceptives, and marital customs vary quite little.

Thanks to Iceland's genealogical record, Helgason's team also had access to such an enormous amount of information that they could research virtually an entire population rather than a sample. Their study, in fact, included all of the 160,811 known couples in the Icelandic population born between 1800 and 1965. This time span provided a genealogical depth of up to ten generations. Figure 24 shows the average reproductive success for these Icelandic couples, according to seven degrees of kinship between spouses:

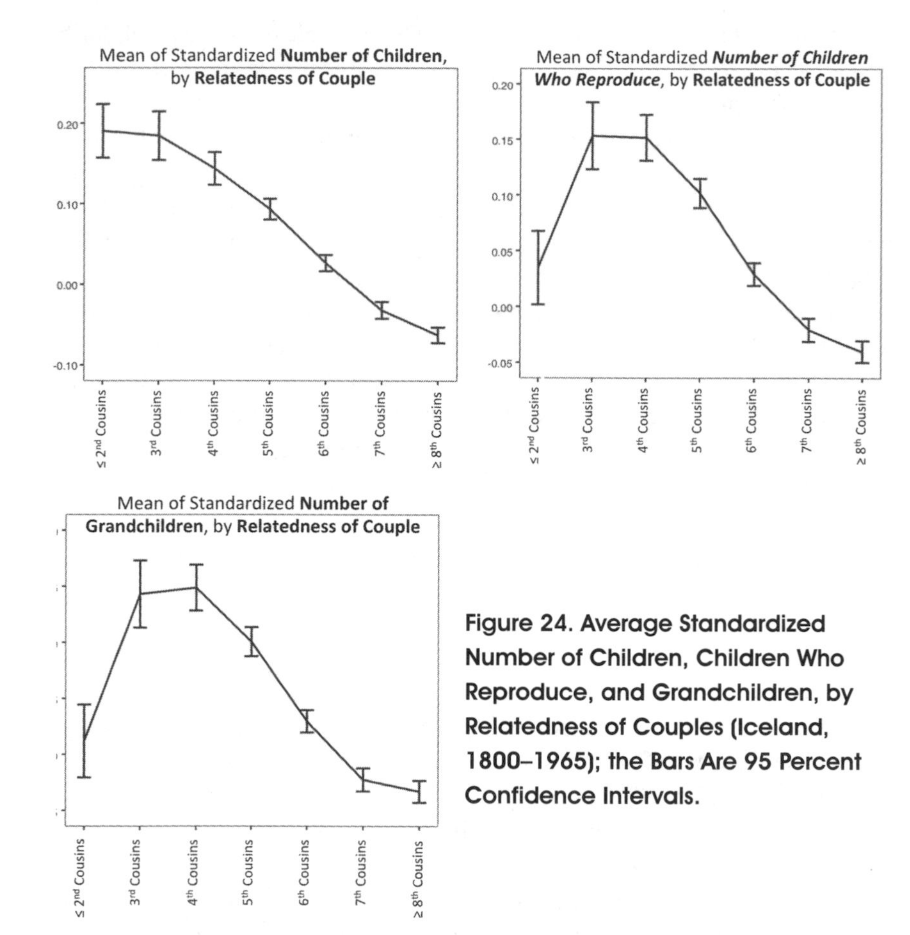

Figure 24. Average Standardized Number of Children, Children Who Reproduce, and Grandchildren, by Relatedness of Couples (Iceland, 1800–1965); the Bars Are 95 Percent Confidence Intervals.

The first graph shows that the most closely related couples had the highest number of children. The second graph indicates how many of the original couples' children reproduced. And the last graph tells how many grandchildren the original couples had.

The striking difference between the first and the last two charts concerns the spouses who were second cousins or closer: these consanguineous couples had the largest number of children, yet their children reproduced at significantly lower rates than the children of less related spouses. In other words, the original first- and second-cousin spouses had fewer grandchildren, on average, than third-, fourth-, fifth-, and even sixth-cousin couples.

During the 165-year period covered in these charts, Iceland experienced an increasing proportion of marriages between more distantly related couples, along with a substantial decline in the country's average fertility. The research team believes that the rise of modern cities, where people outbreed much further with greater ease, may have subtly contributed to this decrease in fertility. We'll look into this modern decline in birth rates in much greater depth in the following chapter. Because of this important demographic transition, Helgason's team had to carefully check the association between outbreeding and reproductive success at every twenty-five-year interval to make sure that cultural or economic variables weren't interfering; when they did, the relationships in figure 24 remained the same over the entire time span of the study.[a]

Because Helgason's team found such significant differences in fertility between small degrees of kinship, and because these same differences endured over a large span of time that witnessed substantial changes in reproductive behavior, the researchers concluded that the fitness curves they plotted had a *biological basis*. To explain the second-generation drop in fertility for first- and second-cousin spouses, the scientists presented evidence that the most inbred children suffered from higher rates of death during their first five years of life. Their increased mortality likely occurred because they had a greater chance of inheriting two copies of the same deleterious recessive genes.

Helgason's work, which was published in the journal *Science* in 2008, is truly remarkable. The Icelanders have provided, in one of the most complete population studies possible, hard data that corroborate the theories discussed in the last two chapters. Their study is likely the first to graphically illustrate the simultaneous effects of both inbreeding and outbreeding depression in humans over a substantial period.[3]

Amongst Icelanders, the peak in fitness occurs for matings somewhere between third and fourth cousins. We cannot know how much variation we may find in the optimal level of outbreeding for other human populations. Different environments and gene pools may push this peak a bit to one side or the other. But if we could plot these graphs for another population, we would likely discover a similar curve, with an optimum in the middle.

What we *can* do, in the meantime, is ask how much inbreeding occurs in different regions of the world. Figure 25 shows a world map, where the shade of each country indicates the proportion of the population that marries a second cousin or closer.[4] In the Arab world, first-cousin unions constitute between 25 and 30 percent of all marriages.[5] Further east, in Pakistan, the proportion of first-cousin marriages ascends to at least 50 percent.[6]

Yet in many other regions, such as the Americas, Europe, and North and East Asia, consanguineous marriages are under 5 percent. Why do North Africa, the Middle East, and parts of South Asia have such a high level of inbreeding? What do these countries have in common? Could the answer be Islam?

Figure 26 shows a map of the Islamic world. Of the thirteen women whom the Prophet Muhammad married, two were biological relatives.[7] One of these was his cousin Zaynab bint Jahsh, the daughter of one of Muhammad's paternal aunts. The Qur'an explicitly permits first-cousin marriages:

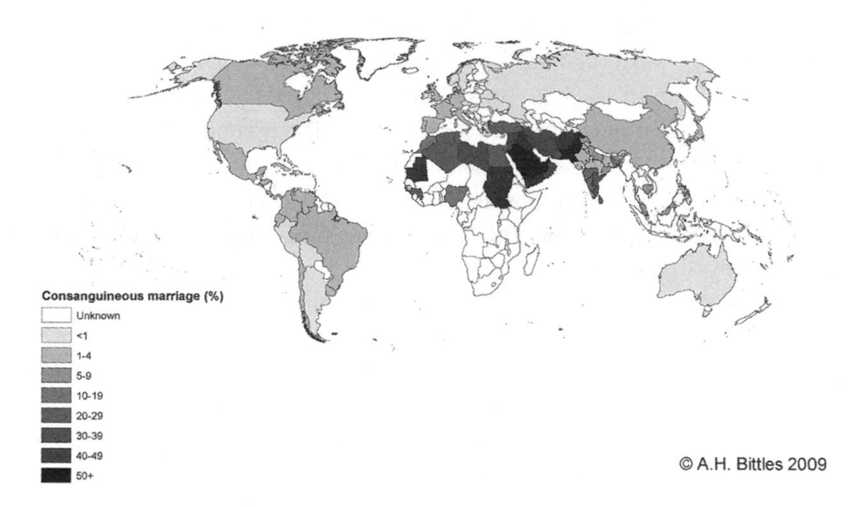

Figure 25. Probability of Marrying Second Cousin or Closer.

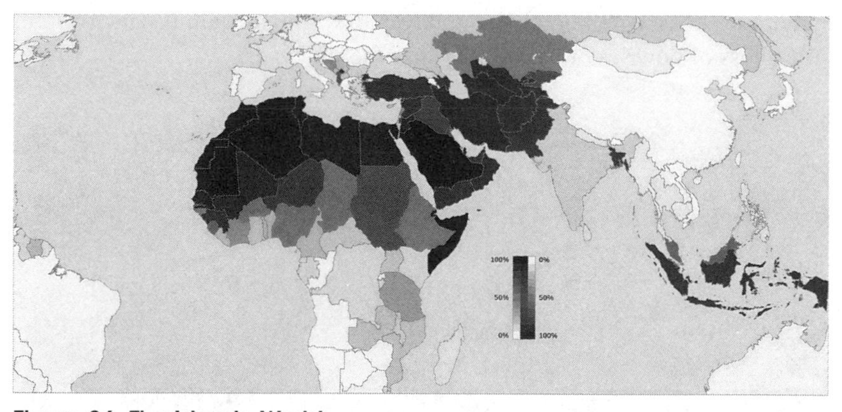

Figure 26. The Islamic World.

> We have made lawful to thee thy wives to whom thou hast paid their dowers; and those whom thy right hand possesses out of the prisoners of war whom Allah has assigned to thee; and daughters of thy paternal uncles and aunts, and daughters of thy maternal uncles and aunts, who migrated (from Makka) with thee. (Surat Al-Ahzab 33:50)

Although some emulation of the Prophet may increase cousin marriage in the Muslim world, it's unlikely that Islam is the main cause of the high level of endogamy in this region. One reason is that very few religions actually prohibit cousin marriage.

In the Old Testament, several of the Jewish patriarchs were cousins, including Isaac and Rebekah, as well as Jacob and Rachel. The Catholic Church has banned first-cousin marriages during most of its history, although kissing cousins could request dispensations. Most Protestant sects permit first-cousin marriage. Buddhism has no prohibition against the practice. Hindu marriage laws are extraordinarily diverse and complicated; nevertheless custom and law permit some types of cousin marriage—especially in South India. Among this region's Dravidian population, first-cousin and uncle-niece unions constitute approximately 30 percent of all marriages.[8]

The other problem with the notion that Islam is the primary cause of consanguineous marriages is that many non-Muslim minority religions in West Asia and the Middle East have similar marriage customs. These endogamous groups include indigenous communities of Jews, Parsees, Buddhists, Christians, and Druze.[9]

What is it, then, that causes people in this region to marry their cousins at higher rates than those in the rest of the world? A likely explanation is that much of the geography here consists of desert, rugged mountains, or other territory with quite limited resources. Inbreeding would allow these relatively poorer communities to keep more property and wealth within the family, and to circumvent the loss of dowries or bride prices to other lineages.

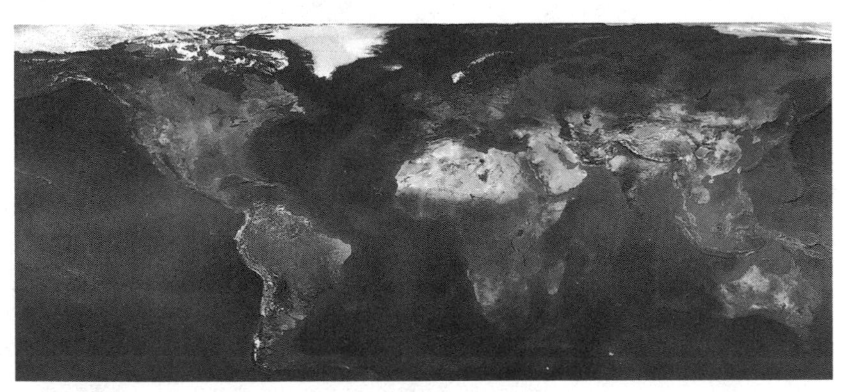

Figure 27. Barren Areas of the World, Seen from Space.

Indeed, the highest level of cousin marriages occurs among the poorest and most rural families—even within the highly consanguineous countries.[10] Figure 27 shows that these barren regions of the earth are clearly visible from outer space. Why some religions have flourished in these regions more than in others is another fascinating question.

OPTIMAL-MATING MECHANISMS IN ANIMALS AND HUMANS

In the case of human beings, the optimal genetic distance for reproduction may be between third or fourth cousins, at least from a fitness-maximizing perspective. Although some unusually harsh environments foster closer inbreeding, most human populations across most of our species's history have somehow managed to inbreed a little bit, but not too much. How have we accomplished this optimal outbreeding?

Human beings essentially use the same kinds of optimal-outbreeding mechanisms as do nonhuman animals. These mechanisms fall into two categories: (1) dispersal, and (2) kin recognition.

Dispersal

Dispersal from a natal territory allows many different kinds of creatures to avoid inbreeding with close kin. Since species that disperse do so to a *limited* extent, their behavior also prevents outbreeding depression.

In a perfect environment, animals could simply move away from their birthplace by a fixed radius. Traveling this geographic distance would put the limited outbreeder in contact with potential mates that share an optimal genetic relationship with the newcomer. In reality, however, nature isn't so simple. The environment usually has an uneven distribution of resources and threats. For example, the best nesting sites and mates may cluster around a particular area, while predators may lurk in other places—including along dispersal routes. Overcrowding and competition in the best areas can also push animals out into worse ones.[11]

In the case of many human beings living in historical times, the best economic environments have existed in cities. Cities, in a sense, exert the opposite force of dispersal; instead, many unrelated people from diverse places *converge* into a relatively small area. Coupled with modern transportation technology and globalization, today's cities allow for extreme outbreeding to occur on an unprecedented scale.

What does dispersal normally look like in nature? In most species of birds, the females fly away from the natal area. Nevertheless, there are always exceptional species. Among snow geese, the males leave. In most species of mammals, males disperse.[12]

Dispersal among a given animal species is typically more of a tendency than an absolute rule. Observant biologists have noted how even littermates have different "behavioral profiles," which influence the likelihood and extent of their dispersal. These personalities may cause one animal to disperse, and another one to stay at home, even when both face the same level of aggression from a parent.[13]

Among nonhuman primates, one sex or the other usually leaves the natal group at puberty. Occasionally, both sexes disperse.[14] Although primatologists find a few individuals of the emigrating sex that stay at home, all nonhuman primate species disperse for a very important reason: to avoid inbreeding—including with their nuclear family members.

Among chimpanzees, females are the sex that disperses at puberty. Primatologist Jane Goodall has found that, prior to their first heat, female chimps present to their brothers in reception for sexual activity. With the onset of puberty, however, this behavior suddenly reverses; sexually mature females attempt to chase their brothers away. Female chimps hate mating with their mature brothers because the females would have to pay for all of the energy needed to gestate and raise a child that could suffer from inbreeding depression. The males, on the other hand, expend minimal energy during copulation. Having a few genetically disadvantaged children would hardly impact a father's fitness. So sometimes conflicts arise.

Goodall once observed a male chimpanzee that she calls "Goblin" copulating with his sister at least twenty-six times, using force to overwhelm her resistance.[15] Harvard primatologist Richard Wrangham describes this conflict, which affects reproductively mature females that stay in their troops:

> When it comes to having sex, a female chimpanzee is not normally very picky. She finds most males attractive, or at least tolerable. One kind of relationship, however, stops her in her tracks. She doesn't like to mate with her maternal brothers. Even when those males court elaborately, with shaking branches and rude stares and proud postures, female chimpanzees refuse their brothers. Normally, the female's reluctance to mate with her brother marks the end of it. But occasionally a brother can't stand being denied. She resists and avoids him. He becomes enraged. He chases and, using his greater size and strength, beats her. She screams and then rushes away and hides. He finds her and attacks again. He pounds and hits to hold her down, and there's nothing she can do. Out in the woods, there's a rape.[16]

This strong aversion to inbreeding with their known, maternal brothers causes mature females to disperse to foreign troops. There, they can avoid the cost of inbreeding depression.

Unlike our closest living relative, the chimpanzee, *most* human beings do not attempt to copulate with their sisters. And hardly any do so with their mothers (despite the common expletive that suggests otherwise). Nevertheless, incest does occur. Arthur Wolf, an expert on kinship systems and the incest taboo at Stanford University's Department of Anthropological Sciences, estimates that 10 to 20 percent of the population has experienced "an incestuous encounter."[17] Unlike other primates, humans usually reach sexual maturity long before leaving home (especially in the case of males). Therefore, human dispersal likely serves more of an optimal-outbreeding role than an incest-avoidance function.

Like chimpanzees, the great majority of human societies traditionally disperse their females, while males stay at home. This system is called "patrilocality." A minority of human societies practice matrilocality, whereby males disperse from their natal home to live in the village of their wives. In some nomadic societies, both genders disperse and establish a "neolocal" residence away from both of their families. Neolocality has also become a predominant dispersal system in the modern economic environment of many Western countries.[b]

The distance that people disperse for marriage is not only a matter of culture, economics, and residence; dispersal distance is also a matter of biology. In fact, the geography of mate choice has very real consequences for optimal outbreeding, even in a modern Western country.

Like Iceland, Denmark keeps an almost perfect population registry. Geneticist Rodrigo Labouriau, at Denmark's University of Aarhus, set out to determine whether dispersal distance had any effect on fitness. To do so, he focused on all of the 42,165 females born in Denmark in 1954 who survived until their fifteenth birthday (i.e., until the beginning of their reproductive lives). Then Labouriau tracked these girls' fertility through the end of their reproductive careers, until age forty-five.

Once he had established the completed family size for each woman, Labouriau looked at birth records to measure the distance between the parish where she was born and the parish of her husband's birth. Figure 28 plots the relationship between the marital radius and the number of children that the couple produced. Labouriau controlled for numerous demographic variables, including education, family income, urban/rural location, mother's age at first birth, and even the distance between a mother's birthplace and the birth parishes of her own parents. Nevertheless, the relationship between women's number of children and their marital radii remained steady. The highest fertility levels occurred at a marital radius of seventy-five kilometers. But fertility dropped at closer and further distances between the birthplaces of spouses. Labouriau concluded that both inbreeding and outbreeding depression appeared to affect human fertility.[18]

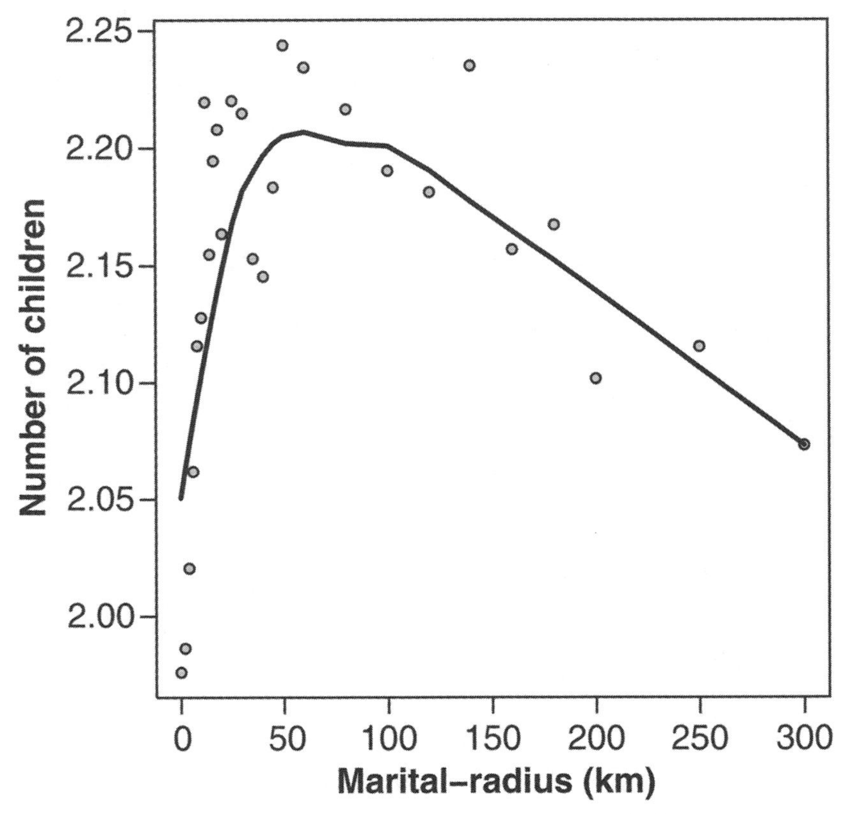

Figure 28. Relationship between Marital Radius and Lifetime Fertility for Danish Women Born in 1954.

The graph in figure 28 shows that most Danish mates dispersed between 20 and 120 km to find each other. Why did some Danes disperse much less and some disperse much greater distances than normal? Since Labouriau controlled for factors such as income and education, it's unlikely that economic factors produced this outbreeding curve. Could personality have any influence over marital radius? After all, biologists have observed that different personalities seem to determine the likelihood that some animals will disperse from their natal territory.

We already know that the human personality varies greatly in terms of ethnocentric and xenophilic attitudes. It's quite plausible that people with more ethnocentric attitudes stay closer to home, while more xenophilic people disperse more than the average distance. But can we dig deeper than attitudes? Could there be a physiological mechanism that creates an incentive for the xenophiles to travel further? Perhaps some kind of variation in the brains of the dispersing Danes causes them to travel differing distances.

Dopamine, a neurotransmitter found in the brains of a wide variety of animals, serves many different functions. Dopamine is particularly implicated in motivation and gratification. Human beings have at least five subtypes of dopamine receptors in their heads. The fourth kind of dopamine receptor (D_4) may shed some light on the dispersal mystery.

Geneticists have located the gene that codes for D_4 on the short arm of human chromosome 11. But not everyone has the same form of this gene. Individuals with a longer *D4* allele have brains that collect dopamine less efficiently (it seems like dopamine does not bind as strongly to their D_4 receptors). Consequently, these people tend to take greater risks and seek more novelties in order to get the same dopamine reward as people who live a more sedentary lifestyle.[19] The novelty seekers have a lower tolerance for monotony. So they search for new experiences to alter their mood.

James Fowler, a medical geneticist and political scientist at the University of California, San Diego, wanted to know whether variation in the brain's processing of dopamine has any relationship to political behavior. So Dr. Fowler turned to the National Longitudinal Study of Adolescent Health, which had tested a sample of 2,574 people to see which alleles for the *D4* dopamine receptor gene these individuals had.

Fowler knew that people with the long allele tended to be novelty seekers. He also knew that novelty seeking is related to "Openness" (one of the Big Five personality traits). Openness, in turn, is associated with political liberalism, as we learned in chapter 3 (see figure 6). Could Fowler show that the *D4* gene contributes to liberalism?

Sort of. Among people who had the novelty-seeking *D4* allele, the number of friends they had in adolescence correlated with a liberal political self-placement in early adulthood. However, Fowler found no *direct* relationship between the novelty-seeking allele and the number of friends people had, or between the allele and ideology. Rather, the combination of the allele and the social environment together influenced political liberalism. For people who had two copies of the novelty-seeking allele, having ten friends in adolescence pushed them about halfway from conservatism to centrism, or halfway from centrism to liberalism, on a five-point political spectrum.[20]

Perhaps novelty seekers who succeeded in finding diverse friends had a more positive appraisal of human nature than those who weren't able to win as many friendships with others. We'll look more into the relationship between perceptions of human nature and political orientation in part V. For now, though, is there any connection between liberal novelty seekers and dispersal?

Yes. A team of psychologists and an anthropologist from the University of California, Irvine, calculated the allele frequencies for the *D4* dopamine receptor gene among thirty-nine populations around the world. They found a very strong correlation (.85) between the novelty-seeking gene and populations that had undertaken long-distance migrations. Sedentary populations, on the other hand, had much lower frequencies of the *D4* allele.[21] So the gene determines brain physiology, which in turn motivates dispersal and probably outbreeding. The long-term reproductive success of these novelty-seeking groups would affect the frequency of this "liberal" gene.

Kin Recognition

Besides dispersal, the other major type of optimal-outbreeding mechanism relies on the recognition of kin. Kin recognition, in turn, comes in two types: learned and innate.

Learned Kin Recognition: Sexual Imprinting

The Canada Goose knows which individuals to mate with, and also which ones *not* to mate with, through a learned process called sexual imprinting. If a goose is raised with its parents and siblings, it will not mate with these family members—even if they're the last geese left in the world. However, when scientists separate two geese from the same brood before their eggs hatch, and then reunite these geese later in life, a biological brother and sister might mate with one another.[22]

Since most Canada Geese aren't separated at birth in nature, young siblings are exposed to one another. This exposure allows for "sexual imprinting" to occur, which prevents them from mating later in life. Thus, nuclear geese families don't inbreed—and for good reason; if they did, their lineage would die out after only a few generations of sibling matings.[23]

Sexual imprinting, as the term suggests, also leaves an impression on certain species of birds that affects which individuals they *do* choose to mate with. For instance, when Japanese Quail are reared together with their siblings, they prefer to mate with first or second cousins more than with siblings, third cousins, or unrelated birds.[24]

What about people? Do human mating preferences depend on childhood experiences? In the late nineteenth century, the Finnish sociologist Edvard Westermarck first proposed that sexual imprinting also occurs in our species. In his 1891 book *The History of Human Marriage*, Westermarck observed "an innate aversion to sexual intercourse between persons living very closely together from early youth."[25] Westermarck believed, unlike some of his contemporaries, that inbreeding produced harmful effects; therefore, he argued that natural selection had "molded the sexual instinct so as to meet the requirements of species."[26]

Several famous anthropological experiments around the globe have validated the Finnish sociologist's theory, which became known as "the Westermarck effect." One experiment took place in China, where the traditional kinship system is patrilocal and patrilineal. That is, men have generally stayed closer to their clan's home and have inherited membership in their father's lineage. Women, on the other hand, have been dispersed to other locations and have become members of other clans (provided they bear a male heir).

Having a daughter meant investing resources to raise her, only to lose her to another clan. Therefore, families who could not afford to raise a daughter sometimes practiced a now-extinct, alternative form of marriage known as *sim-pua*. This term means "little daughter-in-law" in the Hokkien dialect. According to this custom, which was once prevalent in southern China, poor parents could sell a preadolescent daughter to a richer family to become a servant and often a caretaker for a younger boy in the adoptive home. After reaching puberty, the girl would marry the rich family's son.

The Communist Party outlawed sim-pua marriage after taking power in 1949 because of the Communist ideology's strong intolerance of inequality. In Taiwan, however, this alternative form of "minor marriage" endured until the 1970s, when rapid economic growth finally antiquated the practice. Thanks to the late survival of sim-pua marriages in Taiwan, as well as Taiwan's comprehensive population registry, anthropologist Arthur Wolf was able to study the phenomenon extensively for forty years.

In his analysis of 14,200 Taiwanese minor marriages, Wolf discovered that women who married a childhood acquaintance from their adoptive family were three times as likely to have extramarital affairs, compared with women who married men whom they met as adults. Moreover, sim-pua marriages produced almost 40 percent fewer children, on average, than did conventional marriages. The more exposure sim-pua brides and their adoptive brothers had to one another during their first thirty months of life, the more the Westermarck effect inhibited sexual attraction.[27]

While poverty and culture conspired in this natural Chinese experiment to reveal sexual imprinting in humans, ideology created a parallel but very different laboratory in twentieth-century Israel. Some of the first waves of Jews to flee persecution in the Diaspora and return to their ancestral homeland established collective communities called *kibbutzim*. A strong sense of socialism pervaded these communities, which collectivized everything from agriculture to child rearing. Because children of no significant genetic relation were communally raised in age groups, however, they refused to marry one another when they grew up.

This sexual aversion to childhood acquaintances disappointed their parents, who had hoped that they would marry to perpetuate the community. In the small minority of kibbutzniks who *did* marry within their peer group, one of the spouses had usually joined the kibbutz after the age of six. By this age, the time window for sexual imprinting had closed. The children who grew up together in age groups before the age of six, on the other hand, experienced the Westermarck effect.[28]

Like the sexually imprinted birds, the Westermarck effect imprints the image of early childhood associates somewhere in the minds of human children. This imprinting de-eroticizes their relationships in adult life. The evolutionary purpose of this phenomenon is to avoid inbreeding depression. In the case of the Japanese Quail, though, the sexually imprinted image of brood mates also serves as a "template," directing the birds toward an optimal mate that somewhat resembles the prohibited prototype. Do humans have "templates," too?

According to biologist Sir Patrick Bateson, "people unconsciously choose mates who are a bit different from those individuals who are familiar from early life but not too different."[29] In other words, a learned kin-recognition mechanism, like the sexual imprinting found in quails, may motivate some level of inbreeding as well.

Bateson's explanation of optimal outbreeding sounds quite similar to Sigmund Freud's description of the Oedipus complex. According to Freud, a man chooses "his mother as the object of his love, and perhaps his sister as well, before passing on to his final choice. Because of the barrier that exists against incest, his love is deflected from the two figures on whom his affection was centered in his childhood on to an outside object that is modeled upon them."[30] Freud believed that women undergo the same process with regard to their fathers and brothers.

Is there any scientific way to test whether humans "model" their mate choice after people with whom they closely associated in childhood? A group of Polish anthropologists set out to answer this question in 2007 with a group of women and two sets of faces. One group of faces belonged to young men, whom the women ranked for attractiveness. The other faces belonged to the women's fathers. Unbeknownst to the women, the anthropologists had measured the facial dimensions of all the men, as shown in figure 29. Overall, the social scientists found no significant relationship between the facial proportions of women's fathers and the men they considered the most attractive.

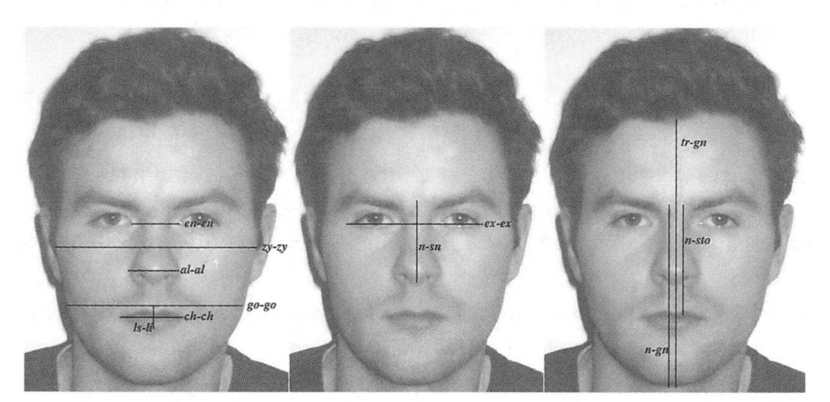

Figure 29. Measurements Taken on Faces of Female Subjects' Fathers and Stimulus Males.

But everything changed when the women evaluated the emotional investment that they received from their fathers as girls. For the half of the women who rated their relationship with their fathers *less* highly, no relationship appeared between the facial dimensions of their fathers and the males they considered handsome. For the half of the women who rated their relationship with fathers more highly, however, a moderate correlation emerged between the facial dimensions of fathers and preferred young males.

Not every set of measurements in figure 29 correlated significantly. The most important dimensions were located in the central area of the face, including the size and shape of the nose. The researchers speculated that they had found evidence of inbreeding enhancement; they cited phenomena mentioned in the previous chapter (including the benefits of maintaining coadapted gene complexes and increasing the relatedness of families).[31]

Social scientists have found similar results for male subjects, provided they had good relationships with their mothers as children. The same effect also occurs between women's *adoptive* fathers and their husbands, which shows that sexual imprinting is learned and not genetically determined.[32]

Below are some other fascinating observations of human sexual imprinting in action:

- Children of mixed marriages in Hawaii have a tendency to marry into the ethnic group of the opposite-sexed parent more frequently than they marry into that of the same-sexed parent.[33]

- Women who have older fathers have a slight tendency to choose older husbands.[34]
- The love objects of teenage girls have an eye color that matches their father's eyes more often than it does their mother's.[35]

In light of sexual imprinting, it makes sense that people in many different ethnolinguistic groups call their lovers by the same kinship terms used for "mother," "father," "sister," "brother," and "daughter"—often in a diminutive form. If mate selection were as simple as finding the fittest individual in the absolute sense, regardless of relative genetic distance, then there would be far less of a sense of destiny, romanticism, superstition, and idiosyncratic preferences surrounding the issue. There wouldn't be a Taiwanese folk belief that women and their fathers were lovers in a previous life. And no Tibetans would believe that karma attracts a soul to the opposite-sexed parent during conception. In the West nobody would speak of an ideal marriage being made in heaven between two—and only two—halves.

If sexual imprinting influences human mate choice, how exactly does it work? Are there any signs of a physiological mechanism at work in the brain? Research conducted by radiologists and psychologists at Stanford University may shed light on these questions.

People actually have a specific part of the brain that is extremely sensitive to human faces. It's called the fusiform face area (or the FFA). The FFA is usually located in the fusiform gyrus region. Imaging studies show that the FFA responds specifically to faces—more than to other human body parts, such as hands. If the fusiform gyrus suffers a lesion, the owner of the brain might develop a condition called prosopagnosia, which impairs their ability to recognize familiar faces.

The Stanford researchers wanted to find out if the fusiform face area showed any sensitivity to the race of a face. So they showed images of black faces and white faces to a group of black and white participants while imaging their brains with an fMRI machine. Afterward, the Stanford professors showed the subjects a new group of faces, half of which they had seen earlier. This "game" tested each individual's ability to recognize faces based on the genetic similarity or difference of the face.

During the brain scan, the fusiform region showed a greater response to faces of the subjects' own race in comparison with other-race faces. As for the memory test, people generally remembered same-race faces better than other-race faces. But when the psychologists divided the participants by race, they found that this was especially true for the white people. That is, black people could remember white faces better than white people could remember black faces.

Since African Americans are a minority group in a white-majority country, this discrepancy suggests that racial imprinting of the fusiform facial region likely depends on the differential exposure of one group to another, probably at an early age. Perhaps the African Americans had more white teachers, or saw more white people on television as children, as compared with their white counterparts. Unfortunately, the study is too limited to resolve this question without speculation.

It's worth noting that, even within both racial groups, not every participant had better memory for same-race faces; roughly 20 percent of both blacks and whites had better recognition of the *other*-race faces.[36] Still, the number of participants was quite small due to the high cost of fMRIs, so it's difficult to know with certainty how reliable these proportions are and what type of early experiences may have biased the recognition abilities of each individual. We can assume that environmentally acquired imprinting plays a large role in the recognition of faces of different ethnic groups. We cannot rule out, however, that the variation in racially biased face recognition *within* each ethnic group may be partially attributable to innate personality predispositions toward ethnocentrism or xenophilia.

In psychology, the well-known facial recognition bias is called the *cross-race effect*. A meta-study of 182 experiments on this phenomenon has revealed that minority groups normally recognize the majority ethnicity's faces better than vice versa. Over 80 percent of these experiments, which were conducted around the world, concluded that people can recognize emotional expressions on the faces of in-group members a little bit better than on the faces of out-group members. Averaging the experiments together, in-group members could interpret each other's facial expressions 9.3 percent more accurately than cross-cultural facial expressions.[37]

In summary, it seems that learned kin-recognition mechanisms in humans (1) center on the face, (2) guide sexual attraction toward optimal outbreeding, and (3) likely function through a highly specialized region in the brain. The shape of a face, then, can signal the genetically relative fitness of a potential mate. Other bodily differences, such as symmetry, secondary sexual characteristics, and signs of health and strength, on the other hand, likely display more universal forms of fitness.

The phenomenon of environmentally acquired sexual imprinting helps account for the large degree of relativity and change that occurs over time in the definitions of in-groups and out-groups. The *structure* of human psychology still remains the same, in that there are always in-groups and out-groups, and a variation in attitudes toward them. But the identities of these groups can and do change along with demographic trends.

For example, massive migrations in the past two centuries brought Catholic Irish and Catholic Italian immigrants to the United States—and into more frequent contact with one another. At first, the "otherness" of the groups may have made intermarriage socially "undesirable." But after long-term exposure, mixed marriages between the two groups became more socially acceptable. Likewise, Ashkenazi, Sephardic, and Mizrahi Jews in Israel didn't mix very much in the beginning of the twentieth century, when they first reunited en masse; but by the beginning of the twenty-first century, intermarriage became the rule more than the exception (particularly among the more secular majority). These examples suggest that it takes time for new generations of previously separated groups to sexually imprint and for social taboos to melt away.

Perhaps the structure of our mate-choice psychology emerged from our ancestral hunting-and-gathering environment. It could be that sexual imprinting evolved during a Pleistocene childhood to lead people to reproduce with their third or fourth cousins. Or maybe in those days the average distance traveled in dispersions, or to trade mates with other bands, brought about this same degree of reproduction.

However much our modern environments have changed, the basic structure of human psychology seems to remain much the same. Natural variation in xenophilia, novelty-seeking behavior, and dispersal still exist—even as air travel, greater economic prosperity, and mega-diverse cities create conditions where increasing numbers of people can outbreed further than ever before.

Innate Mechanisms

But not everything is relative. Some kin-recognition mechanisms may involve innate preferences determined by our genes.

To understand why this is so, let's return to the endogamous Hutterites in South Dakota—the community with the extremely high fertility rates. In addition to the relatedness of a wife's parents, the reproductive success of Hutterite spouses depends on an interaction between a part of their respective immune systems. Specifically, the couples that share too many "HLA" antigens take longer to get pregnant and experience more miscarriages.

HLA is the name of a gene complex located on chromosome 6 (see figure 30). HLA is the human version of the gene system that encodes for a tiny but crucial protein in our immune system called the Major Histocompatibility Complex (MHC).

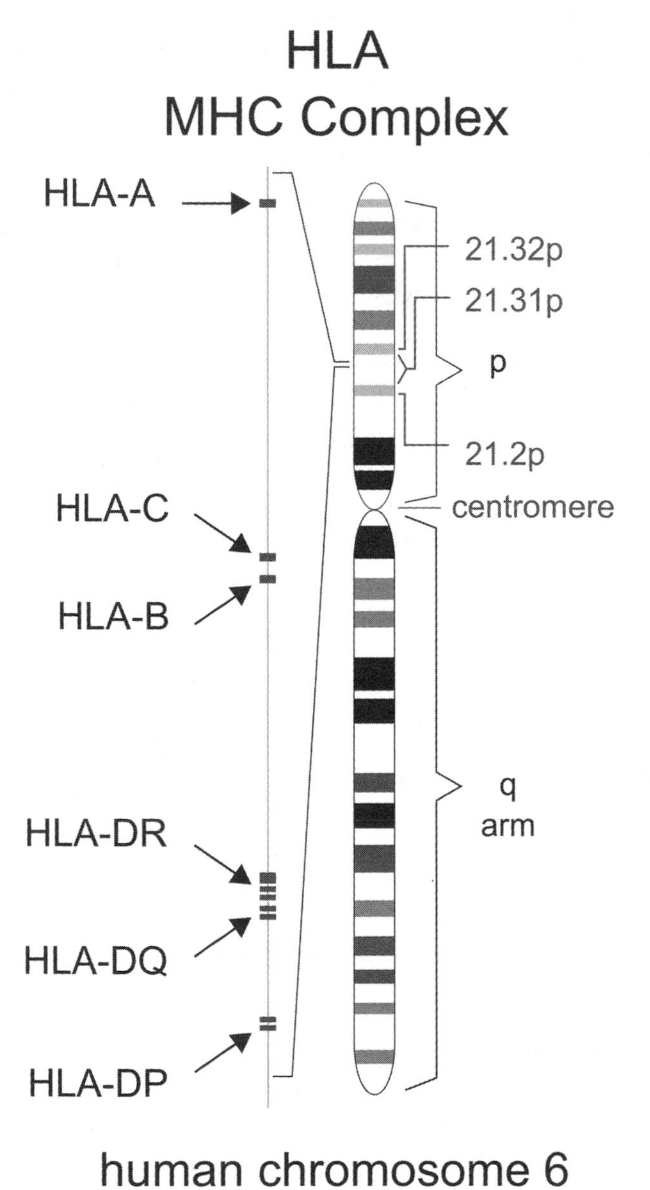

Figure 30. The Location of the Human Leukocyte Antigen Genes on Chromosome 6.

When our body finds a foreign object that may be a threatening pathogen (such as a virus), the pathogen may get broken down into smaller parts. MHCs grab digested molecules from the pathogen. Then the MHCs present these digested pieces to T-cells, which try to vanquish the outside invaders.

Essentially, the role of MHC is to help recognize what belongs to the body and what does not. Because of this difficult self-recognition task, the gene complex that encodes for MHCs is the most diverse region in our genome. In humans, this gene complex consists of about 3.5 million base pairs, and it can come in millions of unique combinations.[38]

The average South Dakotan Hutterite husband and wife are related to each other at about the same degree as first cousins once removed. And much like any other couple, the average Hutterite woman miscarries about 16 percent of pregnancies. However, some Hutterite couples share more MHC genes in common than others. For example, if a couple shares the same MHC genes at the B locus[c] of the complex shown in figure 30, then the fetal-loss rate increases to 23.4 percent. And if a couple shares all sixteen of the MHC genes sequenced by the Chicago researchers, they will lose fetuses at a rate of 33.3 percent.[39]

These findings on the Hutterites build on similar studies in different populations, where other researchers have shown that mates who share more MHC genes have more recurrent spontaneous abortions, and therefore lower fertility.[40] When these couples do give birth to live babies, their children may also have a lower birth weight.[41]

Why is it so problematic for spouses to share too many MHC genes? Because nature prefers a certain amount of diversity in immune systems. If an entire inbred community had a very similar immune system, then they'd be sitting ducks for viruses, bacteria, parasites, and other pathogens that could rapidly evolve to target them. On the other hand, populations that produce offspring with diverse immune systems create a "moving target" for pathogens.[42]

The moving-target idea is known as the "Red Queen's Hypothesis." This colorful name refers to a character in Lewis Carroll's book *Through the Looking-Glass* (1871), the sequel to the more famous *Alice's Adventures in Wonderland.* The Red Queen says: "It takes all the running you can do, to keep in the same place." Applied to biology, the Queen's words mean that an animal species needs to work hard to keep a minimum level of diversity, just to protect itself from enemy pathogens, which coevolve to catch up.

Returning to South Dakota, the Chicago team discovered another fascinating fact: Hutterites do not reproduce randomly with respect to the MHC genes that individuals carry. A significantly lower number of couples matched for this region than could be expected by chance.[43] So the million-dollar question is, "How do Hutterites know how to choose spouses with different MHC genes?" In other words, what is their optimal-outbreeding mechanism?

Perhaps the Hutterites sense immune differences between prospective mates in the same way as mice: through smell. Female mice prefer to mate with males who have different MHC genes. Not only can mice sniff out the carriers of dissimilar immune genes, they can even distinguish between individuals that are almost genetically identical, except for differing by a single gene in the MHC complex.

How could these rodents possibly smell the difference between base pairs arranged in one order or another? Scientists believe that the different MHC proteins produced by these genes allow particular strains of microbial flora to flourish inside the mouse. These floras would be associated with particular concentrations of volatile acids, which mice could smell.[44] Another possibility is that soluble MHC molecules produce different odors themselves. These molecules would have a close relationship with variation in olfactory receptors.[45] Even human noses are sensitive enough to distinguish the smells of mice with different MHC genes.[46]

MHCs also influence the mating of at least one type of nonhuman primate. Primatologists decoded the DNA of a group of mandrills, including mothers, offspring, and sires (the adult males that impregnate the mothers). The more dissimilar the MHCs between a given mother and an adult male, the more likely the male was to have sired the mother's child. Males with more diverse MHC complexes also tended to enjoy greater reproductive success—independently of the strong effects of rank and dominance.[47]

How about human beings? What allows the Hutterites to mate so nonrandomly with respect to MHCs? Again, smell might be the answer. In a premodern world without deodorant, men produce stronger odors (thanks to the bacteria that MHCs may allow to colonize men's armpits). And the more-investing, choosier sex (women) have a keener sense of smell.[48]

In 1995, a group of Swiss biologists published an early study on human odor preferences and MHC. The researchers had men sleep for two nights wearing the same T-shirt. Then they gave the sweaty shirts to women to smell. The women found the odors that came from donors with *dissimilar* MHC genes to be "more pleasant." In fact, T-shirts worn by these donors were twice as likely to remind the women of the smell of a current or former lover.[49]

Do women always prefer the odors of men with more diverse MHC genes? Thanks to the work of the Chicago geneticists, we now have an incredibly precise answer to this question.

First, these researchers discovered that women can smell the difference between a single contrasting gene in two men's MHC complexes. They also found that a woman's preference for a man's odor depends on the MHC genes that she inherits from her father—but not on maternally inherited MHC genes. Knowing these facts, the researchers decided to find out precisely how many genes differed between a woman's own MHC complex and that of her favorite male odor donor. So the geneticists harvested odors from men of diverse ancestry (most of these men were of mixed heritage; collectively, they had ancestors who were Dutch, English, German, Jewish, Polish, Scottish, Sikh, and Spanish).

After all of the smelling was over, women's odor preferences depended very much on their own MHC genes. In other words, there was no smell that the entire group of women favored. Rather, the odors that women liked best came from men who shared a particular number of MHC genes in common with them (2.3 on average, to be exact). The odors that women *disliked* the most came from T-shirts worn by men who shared *fewer* MHC alleles with them (1.5, on average). That is, they preferred more similar men.

Why did the first study find a preference for more diverse MHCs while the second one found a penchant for less immunological diversity? Because the second experiment had women smell the scents of ethnically diverse men whom they didn't know. The first one, in contrast, focused on individuals who were probably more closely related and known to the smellers. So the results are consistent: women favor an intermediate number of matching MHC genes much more than sharing zero matches or sharing all of their MHC genes with a man. Thus, these innate odor preferences, which women inherit from their fathers, guide women toward an optimal amount of outbreeding.[50]

Unfortunately, Hutterite men did not participate in this last, extremely precise study. If they had, the geneticists could have easily known the genetic distances, in kinship terms, between smellers and odor donors. Hopefully, future research can tell us how closely related a mate women prefer by following their noses. Would a third or fourth cousin, on average, smell the best? How much variation in preferences would we find? And would there be any relationship with political orientation?

In reference to this last question, the curious results of another study bear mention: a separate group of geneticists led by Peter Hatemi recently carried out a genome-wide analysis to search for genetic markers linked to left-right orientation. Among the markers most closely linked to political attitudes were numerous genes related to our sense of smell. On chromosome 9, for example, there are various genes that code for lipocalins. This group of proteins is linked to olfaction and pheromone receptors, and they're believed to play a role in odor transport and reproduction. For instance, the odorant-binding lipocalin proteins OBP2A and

OBP2B likely select and/or transport odorant molecules through the nasal mucus to the olfactory receptors. Hatemi has pointed out that lipocalins and pheromones guide mate selection in humans and other mammals. And mate choice, he has perceptively observed, is intimately related to political orientation.[51]

The last three chapters have explained the evolutionary roots of ethnocentrism and xenophilia; these two halves of the ethno-religious political spectrum motivate inbreeding and outbreeding, which in turn affect fitness. Although fitness depends on innumerable factors, the metric for evaluating fitness is fairly simple: generationally sustainable fertility. It makes sense, then, that the reproduction, survival, intermixing, and migration of different ethnic groups are inherently political topics. And the news media cover them with great interest.

Inbreeding and outbreeding, however, aren't the only evolutionary roots of tribalism. As we learned in chapter 8, attitudes toward gender inequality are also closely linked to ethno-religious political orientations (the left favors more gender equality, and the right, relatively less). In the next chapter, we'll discover how gender roles stem from biology and yet depend greatly on dynamic environmental factors. We'll also come to understand why gender inequality has varied so greatly across human history—and why it's one of today's hottest issues.

12.

WHY GENDER INEQUALITY AND FERTILITY CHANGE ACROSS HUMAN HISTORY

On October 27, 2010, a Fatah-controlled Palestinian television station aired the following public service announcement. The Arabic-language commercial takes place late at night, in the home of a couple:

Wife: (*in pain*) Ahmad, Ahmad! Come, take me to the hospital, I'm about to have the baby. Ahmad!
Husband: (*watching TV*) Yes.
Wife: Ahmad, I can't, I feel like I'm really about to give birth.
Husband: (*approaches wife*) Hold on until the morning. How can I take you now, in the middle of the night?
Wife: (*grimacing*) Take me now!
Husband: Allah will help you. Hold on a bit more. What are we hurrying for, for a fifth girl?

(*The wife collapses and falls over on the bed. An ambulance takes her to hospital.*)

In the hospital:

Husband: Doctor, what's happening?
Doctor: Allah willing, it'll be fine. Your delay in getting to the hospital affected the health of the mother and baby. Thanks to Allah, we were able to save the mother, but we weren't able to save the baby. May Allah compensate you with another child.
Husband: Bless you, doctor. It's not so bad; she has four more sisters.
Doctor: She has four sisters? But the baby was a boy!
Husband: (*with shocked expression*) A boy?

Doctor: Yes, a boy.

Husband: What are you saying, doctor? Are you really saying it was a boy? (*The husband collapses on the floor, holding his head in sorrow.*)

At the end, the following text appears:

"The Prophet said: Whoever has three daughters and he remains patient with them, provides for them and clothes them, they shall be a shield for him from the Hellfire."[1]

This public-service announcement was produced by an NGO called the Palestinian Association for Family Planning and Protection, which received funding from the United Nations Millennium Development Goals Achievement Fund. Through their partnership with this NGO, the UN surely intended to promote three of the eight Millennium Development Goals: to promote gender equality and empower women, to reduce child mortality rates, and to improve maternal health.

Ironically, the television spot founders on the first goal, and seems to make the second two goals conditional on the male gender of unborn babies. Why do so many cultures prefer baby boys over baby girls? Why does gender inequality differ so drastically in today's Western Europe, where the ad never would have appeared on television, from places like the Middle East, where it did?

Let's begin with the question of why "Ahmed" valued the life of a male baby much more than that of a female. This sex bias prevails in many conservative parts of the world for cultural reasons: males often inherit property, the family name, and the lineage itself, while females join other clans.

But there are also deeper, evolutionary reasons why many societies favor male babies. The logic is easy to understand if you imagine a family with two children—a boy and a girl. Suppose the two of them set off on a race to see which one could reproduce the largest quantity of their parents' genes. Under ideal circumstances, assuming that both children were highly fit, the male could easily win the reproductive race. Why? Because his sister could only give birth about once a year, during a more limited reproductive window. Her brother, in contrast, could impregnate more than one woman per year, over a longer period of time.

In light of this discrepancy in reproductive potential, nature has molded the behavior of men and women differently, for each sex to leverage its own interests. This chapter explores the evolutionary roots of the power balance between men and women, and why it has shifted so much across human history, altering both fertility and political reality.

THE EVOLUTIONARY GOALS OF THE SEXES

The scientific explanation of the fundamental differences between the sexes comes to us from Robert Trivers, one of the most influential evolutionary theorists of all times. Trivers first proposed his theory of "parental investment" in 1972. Since then, over eight thousand publications have cited his original article.

Put simply, parental-investment theory makes the following observations: Among sexually reproducing animals, species vary as to how much parents invest in their offspring. Moreover, one sex often invests more of its own fitness in the offspring than the other; these sacrifices include time, energy, and the ability to care for other offspring that have already been born. A hefty parental investment can also impose a large opportunity cost on a parent's ability to produce additional offspring. The highly investing sex is usually female, but not in every species.

If one sex invests much more in offspring than the other—as human females do in the lengthy processes of gestation, lactation, and childrearing—then this sex should be adapted to be choosier in selecting a mate. The fitness of the highly investing sex depends not so much on access to the less-investing sex, but rather on securing the best genes, territory, limited resources, and sometimes co-parenting, possible.

The less-investing sex (males, in the human case) is adapted to attain many matings. Because of the mating supply-and-demand incongruity between the sexes, parental-investment theory expects the less-investing sex to compete more fiercely for sexual access to the more discriminating one.[a]

There's another crucial fact that follows from the economics of mating: since the less-investing sex contends with greater competition to mate with the choosy sex, the great majority of the more-discriminating sex will have an opportunity to reproduce. On the male side, the fitter members of the more promiscuous sex will reproduce a lot, but a substantial proportion may not reproduce at all.[2]

To translate Trivers's theory into simple terms for the human context, the naked objective of men would be to maximize copulations, and the naked objective of women would be to maximize the resources that they can invest in their offspring. In reality, of course, women wouldn't be attracted to men who were pure copulation maximizers, so female interests have selected for varying degrees of loyalty in males. By the same token, men wouldn't be too interested in women who only wanted to mate once every few years to reproduce, so women are capable of enjoying nonreproductive sexuality as well.

Still, some copulation-maximizing essence remains in men. And some resource-maximizing essence remains in women. This leads to an important imbalance in the sexual economy: the male demand for sex outweighs the female demand. Consequently, *women have higher erotic capital than men.*

Men, on the other hand, have a strong incentive to acquire resources to attract larger numbers of women. Men's comparative advantage, then, would be the ability to build up "commitment capital" to invest in offspring (through women).

In fact, human males arguably do have a stronger drive to acquire resources than do females. Consider, for example, a group of individuals who represent a pinnacle of economic power: the CEOs of the Fortune 1000 companies. In the United States in 2012, 96.5 percent of these individuals were men—even after decades of gender-equality activism.[3] Of course, sexism could explain some part of this imbalance. But many women who are perfectly capable of becoming Fortune 1000 CEOs do not do so because the time, energy, and risks required come at an enormous opportunity cost to the sex that typically invests highly in offspring. To acquire the best resources for their children, women in traditional marriage markets have often sought to invest their erotic capital in wealthier men (by marrying them). However, modern economies and ideals oblige women to earn substantial financial resources, while they simultaneously invest a disproportionate amount of time and energy in children.

In any case, male wealth remains attractive. For a man who wants to maximize copulations, his best bet is to invest most of his time consolidating economic power (rather than spending all day wooing women). Evidence shows that unmarried wealthier men have more sex with potentially fertile women than do poorer men.[4]

Men also seek out other forms of power disproportionately to women, for the same reason. Consider political power, through which individuals gain greater control over public resources and influence with wealthy individuals: according to the United Nations' 2009 Human Development Report, 84 percent of the world's national legislators are men. Of the 196 member states surveyed, no country without a quota system had fewer than 58 percent male legislators (this lowest case was Finland). Indeed, patriarchy is a universal feature of human societies.[5] Some 70 percent of all societies have *only* male political leaders, while in the rest the most powerful positions are more likely to be held by men. No society has ever been found in the history of our species where women have occupied a disproportionate number of elite political positions.[6]

WOMEN AND THE COST OF TRIBALISM

To better understand our political world, it's helpful to call attention to a species-specific aspect of Trivers's parental-investment theory. In the particular case of human beings, there's an additional imbalance between the sexes in their investment in reproduction. Not only do women pay a higher price in time and energy to reproduce their genes by the same quantity as their male mates; human females also incur the risk of maternal mortality, which men do not. The more babies a woman has over her lifespan, the faster her chance of dying of reproductive complications appears to rise.

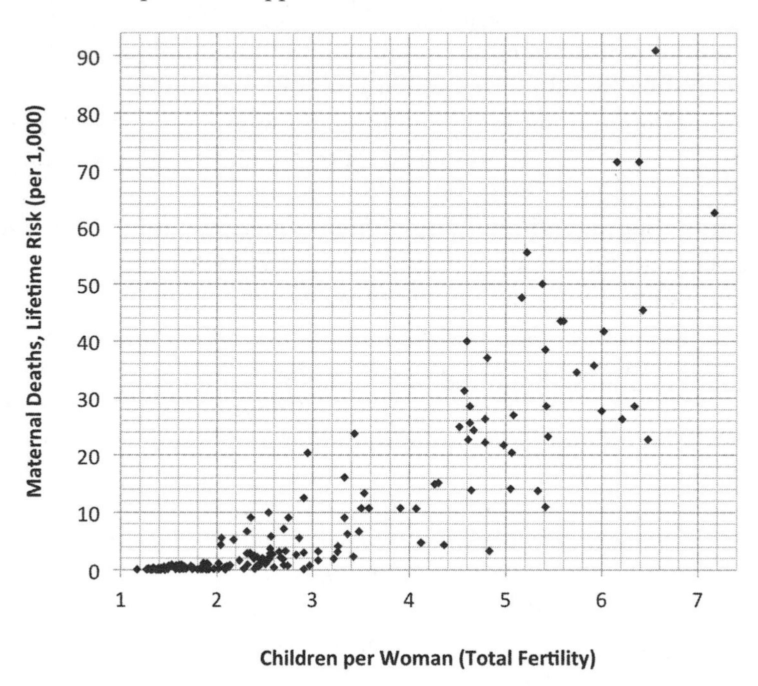

Figure 31. Lifetime Risk of Maternal Death, by Total Children per Woman.

Figure 31 shows this curve for countries around the world in 2008.[7] The highest data point represents Afghanistan. In 2008, the average Afghani woman could expect to have 6.6 children during her lifetime. She also had a 9 percent chance of dying of reproductive complications. Because of the shape of this curve, Afghani women who have an *above*-average number of children, such as eight or nine, may suffer a dramatically higher maternal mortality rate. In fact, the average children per woman in 2005 was only 0.4 higher than in 2008 (seven children), but lifetime maternal mortality was 3.5 percentage points higher.

Why is childbirth so dangerous for human women? One reason is because of heads and pelvises. The adaptation that our species has evolved to walk on two legs caused a narrowing of the pelvis. But human babies have large heads to accommodate large brains. These bulky brains allow humans to occupy an intelligence niche.

For a baby to fit through a woman's pelvic opening without causing harm, a number of processes must unfold correctly: hormones relax ligaments between the mother's pubic bones, the plates on a baby's skull must remain flexible, and the position of the baby with respect to its mother is also important. If any number of complications arises, extreme bleeding and hemorrhaging may occur, which is one of the top causes of maternal mortality.

In light of these human imperfections, it's illuminating to revisit the Judeo-Christian story of human creation, in which God banishes the primordial couple from the Garden of Eden for eating the forbidden fruit (Genesis 3:15–16). God says to Adam: "And I will put enmity between you and the woman, and between your seed and her seed; it shall bruise your head, and you shall bruise his heel." Then God turns to Eve, and says: "I will greatly multiply your sorrow and your conception; in sorrow you shall bring forth children; and your desire shall be to your husband, and he shall rule over you."

From the beginning of human history, then, according to the Bible, conflict has existed between the genders, and males have imposed their dominance over females in many areas. Childbirth has been painful and dangerous, and heads have gotten bruised.

Indeed, *women pay a much higher cost to equally reproduce the genes of both parents*.[b] And the more children a couple has, the woman's sacrifice to raise her and the father's fitness grows increasingly greater. Therefore, gender inequality goes hand in hand with high birth rates. The countries with the largest power imbalances between men and women tend to have the greatest number of children per woman. In countries with greater gender equality, on the other hand, women generally have fewer children and lower maternal mortality rates. Thus, ethnocentrists within every group favor relatively more inequality between the sexes—because male dominance will multiply their tribe more quickly. But why are gender inequality and fertility so high in some societies and so low in others?

ECONOMIC ENVIRONMENTS AND INEQUALITY

Great variation in gender inequality clearly exists between different regions and eras. If a man refused to take his wife to the hospital in many Western countries, he might be sent to jail instead of wished compensation from God in the form of another male child. As testified by its inclusion in the Millennium Development Goals, gender equality is one of today's hottest issues. Promoting it around the globe has become a central priority, a rallying cause, and an article of faith in UN agencies, multilateral development banks, and North American and European foreign-aid agencies. These institutions are spending tens of millions of dollars on a small army of gender consultants. And yet few people understand what actually causes variation in gender inequality. Most people explain it away as a cultural idiosyncrasy or a vestige of times past, to be overcome through ideological willpower. But this phenomenon also has very specific material causes, which run deep through our natural history.

The rest of this chapter searches for these ultimate causes of gender inequality. We begin by finding the determinants of social hierarchy and egalitarianism among our closest nonhuman relatives; then, we trace these determinants through the major phases of human history because they are the same factors that explain changing gender relations and birth rates.

Social Inequality among Animals

In the social animals, two factors establish how egalitarian or exploitative relationships are between individuals of different status: *mutual dependence* and *exit strategy*.

What is mutual dependence? The more that members of the group rely on one another to survive in their environment, the better they must treat each other. Even stronger or higher-ranking individuals must treat weaker ones more equally, because they need them to cooperate.

"Exit strategy" refers to how hard it is for weaker individuals to escape from exploitative relationships by leaving the group. Greater difficulty in leaving a community, or less need for cooperation, creates more hierarchical social relations.[8]

Both of these variables depend greatly on the environment. A higher risk of predation, or greater difficulty in foraging or storing food, or finding mates or nesting sites, can increase the cost of an exit strategy; these conditions force animals to tolerate more exploitation from dominant group members.[9]

Consider the Splendid Fairywren. This bright blue bird lives in dry and shrubby areas of Australia. The Splendid Fairywren's habitat is quite patchy and limited, yet the birds overproduce a large number of offspring. Since the offspring have a high rate of survival, and because the species does not undertake long-distance migration, the birds can easily saturate their habitat. Without an exit option, the social behavior of the Splendid Fairywren is quite exploitative.

In the hierarchical society of these blue wrens, many of the reproductively mature birds cannot leave the group. But neither can they attain the resources they need to start their own family. Therefore, they become "helpers at the nest." Helper birds usually work for their biological parents, but not always. Their babysitting job includes feeding, rearing, and defending the younger broods.

Other hierarchical social animals that use "helpers at the den" include the dwarf mongoose and some species of wild felids (large cats) and canids (dog relatives). Environments that require greater interdependence or that provide greater exit options, however, nurture much more egalitarian social behavior, even in closely related species. For example, one can find communal suckling in the banded mongoose as well as communal cub care and suckling among lions.[10]

Our closest living relatives are the chimpanzee and the bonobo (formerly known as the pygmy chimpanzee). DNA testing shows that our ancestors parted ways with theirs about six million years ago.[11] Chimpanzees and bonobos, however, became separate species only around 875,000 years ago.[12] Few closely related species could exhibit such striking political contrasts. Gaping inequalities characterize chimpanzee societies, while bonobos practice extreme social and sexual egalitarianism.

The most likely cause of these differences is that each species has adapted to quite distinct environments. In his outstanding book *Demonic Males* (1997), Harvard primatologist Richard Wrangham explains that chimpanzees live north of the Zaire River, while bonobos live to the south. Since the river shapes an enormous arc around the bonobo habitat, both species live within a similar range of latitudes, and in similar forests that provide comparable foods. The main difference between the two environments, according to Wrangham, is that the bonobo habitat lacks gorillas.

Although the presence of gorillas on the chimpanzee side of the river seems rather trivial, these giant apes transform the forest ecosystem by eating up the nutritious young leaves and stems on the ground. On the south side, bonobos have evolved herbivorous adaptations on their teeth for eating this plentiful food source. Bonobos can easily split up to forage for food during the day, but they return in the night to sleep among their relatively sedentary group. Compared with chimpanzees, bonobos live in opulence. Their environment would allow them an easy exit option, but they are highly social and prefer to live in large groups.

North of the Zaire, however, chimpanzees have to work much harder to feed themselves. They travel long distances in search of mostly fruits, but also seeds, resin, honey, insects, and other difficult-to-acquire items. In addition, chimpanzees form hunting parties to kill red colobus monkeys and other medium-sized game. Since chimpanzees do not acquire their high-quality foods easily, and their main staple (fruit) is vulnerable to shortages, chimps are extremely territorial. To defend and exploit their territories, chimpanzees split up into smaller groups. In fact, their population density can drop to only one chimpanzee per square kilometer. If a chimpanzee wanted to leave his group and its territory, it would be subject to aggression (especially in the case of males) from other groups defending their own scarce resources; therefore, the exit option could be extremely costly.

The more chimpanzees there are in a group, the more they pressure the precious food supply. So having extra individuals around is a burden, and mutual dependence is quite low. As a result, hierarchic exploitation pervades chimpanzee societies.

Within a group, high-ranking males often fight with low-ranking males to maintain an unequal distribution of mating privileges with females. These instances of aggression sometimes lead to killings. In addition, chimps occasionally kill unrelated infants within their communities to shorten the interbirth intervals of females, gaining a potential reproductive advantage. Female chimps also suffer from violence, and both sexes are subject to competitive dominance hierarchies. Strong chimpanzee communities also wage war against weaker ones (which we'll learn more about in the next chapter).

The social and political life of bonobos is the polar opposite of chimpanzees—even though both species are equally related to humans. In contrast with the more hierarchical chimpanzees, bonobos have evolved, according to Wrangham, a "threefold path to peace": bonobos "have greatly reduced the level of violence in relations between the sexes, in relations among males, and in relations between communities."[13]

The renowned Dutch primatologist Frans de Waal believes that bonobos have accomplished these decreases in aggression by evolving a "sex for peace" approach to conflict resolution. When territorial disputes arise at their borders, they do not kill each other like chimpanzees. Instead, female bonobos cross the line to copulate with the foreign males. Or they engage in genito-gentital rubbing with other females (also known as GG-rubbing or *hoka-hoka*). Before long, the two groups peacefully groom each other, without any blood spilt.[14] Sex serves as the glue that binds all bonobo relations together. They do it to make friends, to make up, and to alleviate stress. They even do it out of excitement for finding good food to eat.

But to simply say that bonobos "do it" hardly describes the athleticism of these sexual omnivores. Bonobos can copulate "dozens of times a day," according to Wrangham. Both sexes perform heterosexual and homosexual acts, with their genitals, hands, and mouths. They kiss mouth-to-mouth and use numerous copulatory configurations, including the face-to-face "missionary" position.[15] Male bonobos even hang from branches, facing each other with their members erect, and engage in what De Waal calls "penis fencing."[16]

In summary, chimpanzees and bonobos have evolved astoundingly different political behavior, although the two species are separated by less than a million years. Chimps live in hierarchical, xenophobic, bellicose, heterosexual, and male-dominant patriarchies. Bonobos, in contrast, are more egalitarian, xenophilic, relatively peaceful, fully bisexual, and even arguably female-dominant. No wonder humans have politicized them!

In 2007, a journalist from the *New Yorker* wrote a story that questioned whether bonobo society truly differs so radically from chimpanzee behavior. Even though the world's top bonobo experts defended primatology's research record, the *Wall Street Journal* and some notable conservatives jumped at the opportunity to "debunk" the "bonobo myth." Dinesh D'Souza, an advisor in the Reagan White House, member of conservative think tanks, and author of the book *The Roots of Obama's Rage*, expressed surprise that "the Democratic Party hasn't changed its symbol from the donkey to the bonobo." D'Souza also urged liberals to "put their bonobsession on hold" because bonobos, he argued, couldn't possibly act the way that "liberal" scientists had reported.[17] D'Souza fiercely rejected bonobo society because bonobos are our other closest relative, and perceptions of human nature were at stake (as we'll see in part V, these perceptions form the third personality cluster underlying people's political orientations).

The basic ecological principles that determine the level of egalitarianism or hierarchy in bonobos and chimpanzees also apply to humans. Economic environments that demand a high level of mutual dependence or that permit easy exit options foster social equality—including between the sexes. In these more egalitarian societies, women tend to have fewer children. Economic environments that require less cooperation or that offer less attractive exit strategies, on the other hand, are generally more hierarchical. Gender inequality and fertility are high in these cultures.

Below, we'll track the development of human societies, from hunter-gatherers, to agricultural and pastoral peoples, up to contemporary, postindustrial "developed" countries.

Gender Equality and Fertility among Hunter-Gatherers

Hunter-gatherer economies demand a high interdependence between men and women. Although they occupy a great diversity of habitats around the world, anthropologists have calculated meaningful averages for the economic participation of the sexes. Overall, men furnish 62 percent of the calories in hunter-gatherer societies' diets, while women provide 38 percent. In some groups, such as the Gwi bushmen and the !Kung, both of which live in the Kalahari Desert, the average percentage of calories contributed by men drops down into the mid-40s.

Most of the calories that men in these societies provision come from hunting or fishing. So in many hunter-gatherer groups it's not unusual for men to contribute up to 90 percent or more of the protein eaten by their families and band. These valuable nutrients and fat, however, come from vertebrate meat. And hunting requires many years of investment to master; even then, return rates are often low and quite variable. Among contemporary foraging groups, vertebrate meat can range from anywhere between 30 percent and 80 percent of the diet. So women's work is extremely important, since women extract more constant and predictable food resources, including roots, nuts, seeds, invertebrate meat, and plant fibers that require a large amount of effort to process.[18] Therefore, both sexes depend very much on one another to divide their labor between the staples needed to survive and the high-value nutrients required to thrive.

In addition to high levels of cooperation, many hunter-gatherer societies have relatively convenient exit options. Consider the Machiguenga tribe of the Peruvian Amazon. This Arawakan-speaking Amerindian group easily subsists from wild foods and small gardens provided by their lush jungle home. This environment affords the Machiguenga a high degree of independence. UCLA anthropologist and psychiatrist Allen Johnson has even described them as having an "intransigent autonomy" that "often leads single families to break away from their extended-family hamlets and live in virtual isolation for years at a time." Johnson also affirms that the Machiguenga "place the highest value on equality and noninterference."[19]

Anthropologists, in fact, have characterized hunter-gatherer societies in general as highly egalitarian—both politically and in terms of gender relations.[20] If a domineering male alienates some of the subordinate males in a small band, the latter have many options. They can tease the would-be dominant male, refuse to comply with his orders, murder him, or simply leave the group.[21]

An egalitarian ethos also pervades the culture of the !Kung people, a nomadic group of hunter-gatherers who inhabit the Kalahari Desert in Southern Africa. Even though social hierarchy does exist between the sexes and different families, the !Kung actively deemphasize any perceived differences. If a !Kung individual tries to convert a situational leadership role into one of permanent authority, the would-be subordinates simply migrate to another camp to escape the bottom rung of a hierarchical relationship.[22]

The fertility of many hunter-gatherer women is lower than women in less egalitarian societies who lack modern birth control. !Kung women, for example, have an average of 4.5 children. One way that !Kung women living in the bush keep their fertility low is by prolonged and frequent breastfeeding. Nursing an infant contracepts the birth of a subsequent child.[23]

Another reason why !Kung women have fewer children relates directly to their society's economic dependence on women. Because a !Kung woman spends much of her day walking about in the bush foraging for food, she can't easily carry two infants on her back and efficiently gather nutrients at the same time. In fact, studies show that the children of !Kung women who reproduce too frequently have a higher mortality rate. Therefore, longer birth intervals actually *increase* a !Kung woman's fitness.[24]

A similar situation faces the Aché, a population of hunter-gatherers in Paraguay. For Aché women, shorter inter-birth intervals also increase child mortality—but not quite enough to reduce the fitness of the mother. Still, Aché women *choose* to have fewer babies than the optimal, fitness-maximizing number. This choice is actually common among natural-fertility women in egalitarian foraging societies.[25] Perhaps !Kung and Aché women try to protect their own health. Or maybe they have fewer children so that they can invest more in their families and bands, which depend heavily on their labor.

Ethnographic evidence from Peru's Machiguenga tribe supports the labor theory. Allen Johnson has recorded instances—in both Machiguenga folklore and real life—of older women prohibiting younger women from having children to avoid losing the fertile woman's labor. Johnson reports two cases of infanticide, which occurred when older women (a mother and a co-wife) told young mothers: "I want you to work only for me." The Machiguenga use a natural drug called *potogo* to induce abortions.[26]

Other nomadic societies sometimes practice infanticide when a child is born too early. The ideal birth interval for foraging women is typically around four years, which allows the last child to learn to walk on its own.[27]

Gender Inequality and Fertility among Agricultural and Pastoral Societies

About ten thousand years ago, Neolithic peoples in the Middle East first developed large-scale agriculture. This invention completely transformed the economic environment of human populations around the world.

When hunter-gatherers became sedentary agriculturalists, their livelihood tied them closer to their land. Although they couldn't wander as far to hunt, they could produce and store surplus grains. With this extra food, they domesticated animals that provided easy access to dairy products, eggs, and meat. Animal manure helped fertilize crops, and cattle could plow open land that wouldn't otherwise be arable.

In his Pulitzer Prize–winning book *Guns, Germs, and Steel* (1997), UCLA professor of geography Jared Diamond explains how the domestication of plants and animals impacted human fertility. Because sedentary people don't have to hike around the bush or the jungle, they don't have to carry children with them. Women could stay at home more often and specialize in childcare, while men worked in the fields. Men could produce more of the calories of a family's diet, while women could raise all the children the family could nourish. An acre of land could feed ten to one hundred times as many farmers and herders as it could hunters and gatherers. In practice, women in many farming communities have children twice as frequently as women in hunter-gatherer societies (every two years, on average, instead of every four years).[28]

These massive changes in the economy of preindustrial sedentary peoples impacted the two factors that determine the equality of social relations. Farmers and herders became less dependent on each other for the extraction of resources from the environment. Specifically, men needed less cooperation from women to produce food on a daily basis. As for an exit strategy, people found themselves tied to their land, in an environment where the human population was booming. Leaving the group was no longer such a feasible option. Consequently, social relations became extremely unequal.

At the same time, the land's yield of food increased by one or two orders of magnitude. This new efficiency created a large food surplus that could feed many more people than those who produced it. Nonfarmers began to specialize in nonagricultural fields, becoming craftsmen, merchants, healers, priests, and so on. When the population grew too large for decision-making by elders or a tribal council to be practical, the need for full-time governors, administrators, and bureaucracies arose.[29]

These sedentary people, unlike nomadic foragers, also had the ability to store food and animals, to trade surplus food for durable goods, and to accumulate a greater degree of wealth.[30] Accumulated wealth, in turn, created the temptation for stealing and the need to defend valuable resources. Warrior castes arose to serve this function, while governments raised taxes to finance defense and other services. In addition to defending wealth, armies could extort tribute from other populations, or conquer them completely. Hierarchy soared. So did the inter-generational transmission of inequality.[31]

Castes, guilds, and taboo systems emerged. These social structures often created reproductive barriers—even within the same ethnic group. Perhaps people instinctively needed to break down the mushrooming population into smaller units, more akin to the size of the hunter-gatherer environment in which human psychology and mating preferences had evolved.

Among agro-pastoral peoples, gender inequality reached the highest point in human history. This culmination of gender inequality is glaringly obvious at the highest level of governing elites. Rulers of the Babylonian, Egyptian, Indian, Chinese, Roman, Aztec, and Incan empires had harems of up to several hundred women. The members of the harems were young, fertile women prohibited from other men. In the case of the T'ang Imperial Court (China, 618–907 CE), conjugal visits with the ruler coincided with the women's peak level of fertility.[32]

Gender inequality increased among commoners as well. Because survival depended on inheriting privately owned land, animals, and wealth, parents gained unprecedented control over when children could marry and with whom they could reproduce, particularly in the case of daughters.[33]

If children didn't receive this investment from their parents, they risked facing a grim future in a world without social safety nets.[34] Thus, young adults had enormous incentives to obey their parents, to conform to an extremely hierarchical relationship with them, and to accept arranged marriages.

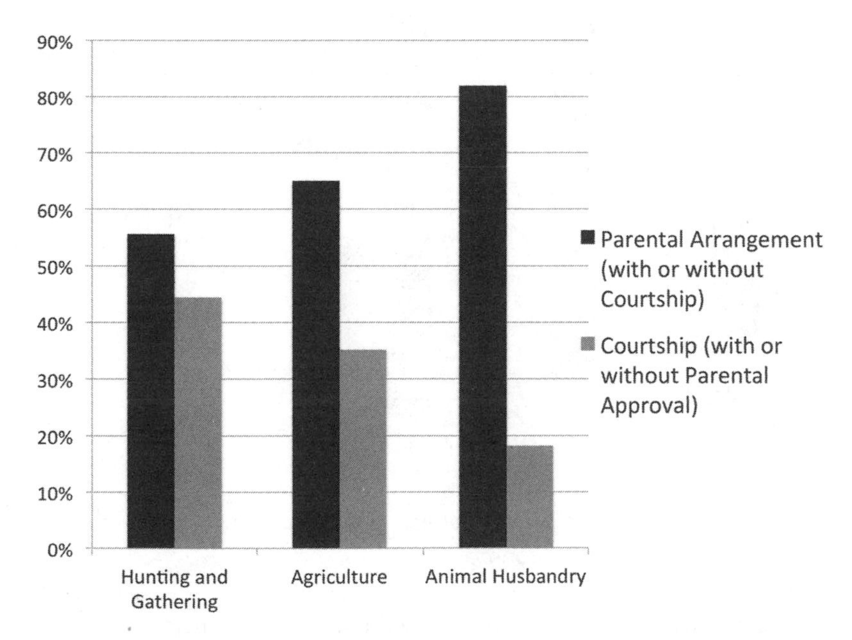

Figure 32. Marriage Types for Females, across Various Economic Environments (in 186 preindustrial societies).

It's easy to visually appreciate how much gender inequality increases from hunter-gatherer societies to agro-pastoral ones. Across this transition, the percentage of societies that practice arranged marriage for their daughters rises, while the practice of courtship marriage drops (see figure 32).[35] Gender inequality is also much higher among herders than among farmers. This increase likely occurs because pastoralists tend to live in more barren regions. These resource-poor environments make children even more reliant on inheritance or dowries from their parents.

The next chart shows the subset of cases where parents arrange marriages for their children without giving them courtship freedoms. In these most hierarchical of kinship systems, the economic environment impacts gender inequality with respect to the *parents'* decision making. In most of the hunter-gatherer societies that arrange marriages for their children, senior women's voices are heard as or more loudly than men's in making these decisions (see figure 33). But older males have a much greater say in their children's mate choice in farming societies—and more control still in pastoral cultures.

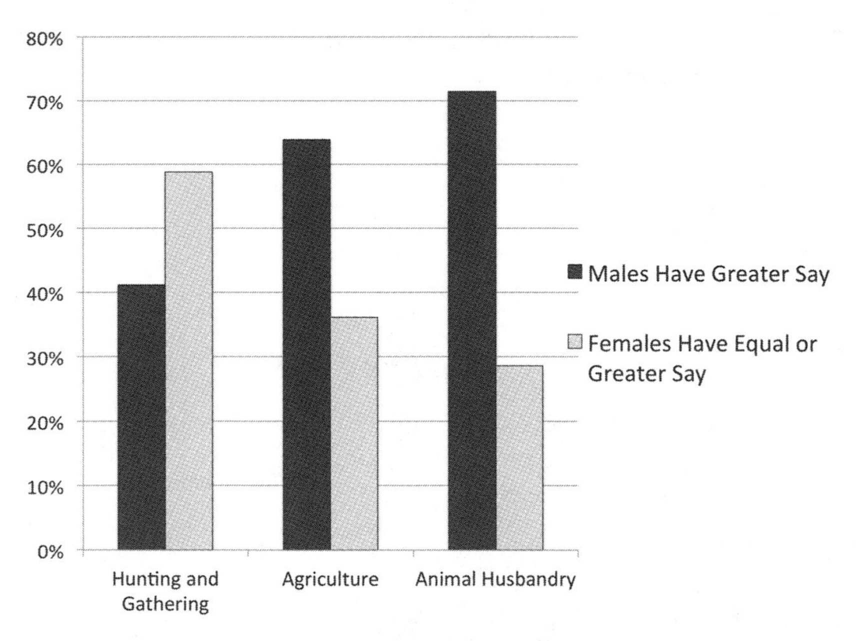

Figure 33. Gender of Decision Makers in Arranged Marriages without Courtship, by Economic Environment.

The evolutionary psychologist who analyzed these data, Menelaos Apostolou, concluded that males extract more wealth from their environment in agro-pastoral societies, which gives them a greater say over their children's mate choice.[36] Since women specialize in tending to the home, their husbands depend less on them economically than hunter-gatherer husbands depend on their wives to provide staple foods. Therefore, gender inequality is greatest among preindustrial sedentary peoples.

With high gender inequality, the birth rate increases because males don't need female labor as much in the workplace. The agro-pastoral woman essentially becomes a baby-making machine. The more hierarchical the relationship between the sexes, the higher the fertility rate climbs, and the higher price in health a woman pays to increase her genes and her husband's genes by the same amount.

Extremely high fertility is not simply a matter of birth-control technology not existing, but rather a consequence of gender inequality. After all, abortifacient plants, abstinence, lactational contraception, and infanticide had existed from the days of more egalitarian hunter-gatherer societies, which had lower birth rates. Gender inequality, then, is why Afghanistan—where one cannot easily see the faces of women in public—has one of the highest fertility rates in the world.

Since greater gender inequality reproduces an ethnic group more quickly, ethnocentrists favor relatively more of it. Within groups, conservatives and liberals argue over this topic. Between nations, the liberal developed countries seek to promote gender equality in more conservative agro-pastoral countries with higher birth rates.

Modern Societies and the Demographic Transition

While traditional farming and herding communities still exist around the world, many "developed" countries have undergone "demographic transitions." This paradoxical phenomenon occurs when the overall availability of resources in an industrialized society increases, yet fertility rates drop substantially. In many wealthy countries, in fact, women aren't even having enough babies to keep the population at a constant size. Figure 34 shows the demographic transition in action during the year 2009. The dotted line indicates the fertility rate that the average woman would have to meet to replace the population. This replacement rate is a bit *above* two children, mostly because of child mortality. Why are rich countries shrinking? Are they sabotaging their own evolutionary fitness? Let's rewind time to see what happened.

In the year 1800 in the United States, 94 percent of the population lived in rural areas, and about 85 percent of Americans worked in the agricultural sector. In this agro-pastoral society, the average white woman gave birth to seven children over the course of her lifetime. Having that many children was useful because a farmer in the early 1800s had to spend around three hundred hours to produce one hundred bushels of wheat.

Over the next 150 years, however, a series of spectacular technological innovations completely transformed the country's economic environment. In 1890, a horse-drawn machine could produce one hundred bushels in only forty-five hours. And by 1930, a simple tractor could accomplish the same feat in about eighteen hours. Thanks to the increased efficiency of machines, America needed far fewer farmers to feed itself. By 1940, only 43 percent of the population lived in rural areas, and the average white woman had two children—a fertility level *below* the replacement rate.[37]

As machines replaced more and more people, the demand for farm workers dropped. Towns and cities gained a greater labor force, and child labor was banned in factories. Instead of working, children had to attend school to prepare for the demands of the more competitive open markets of an increasingly skill-based economy.[38]

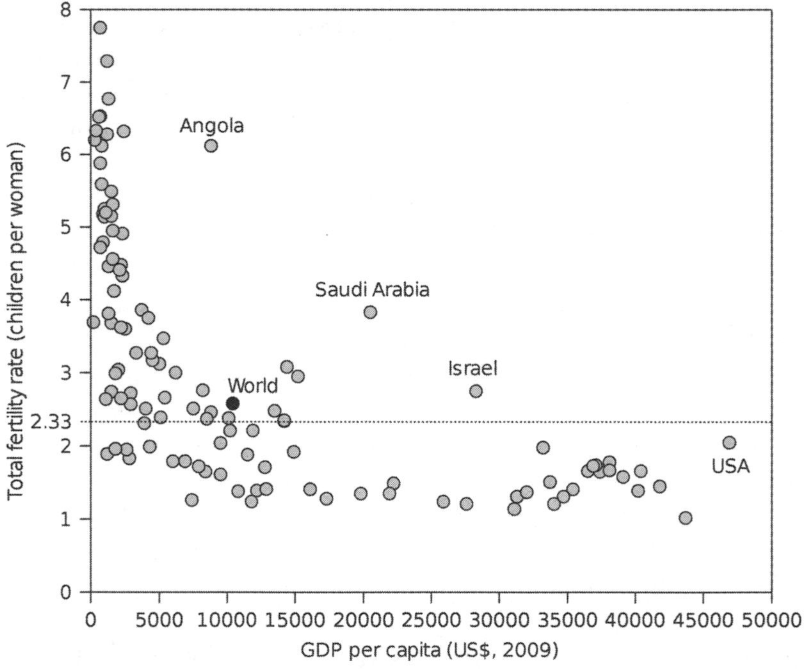

Figure 34. Total Fertility Rates, by GDP per Capita, in 2009. Source: *CIA World Factbook*.

Educating children instead of sending them to work required a longer-term, more expensive investment from parents. To afford raising their nonproductive kids, women eventually began to enter the workforce—especially during the two world wars. With more women in the workplace, mutual economic dependence increased between the sexes. Families relied more on women's income to pay for the increasingly expensive investment in children.

Livelihoods in industrial societies also depended much less on particular plots of land. Therefore, the exit option became relatively easier. This freedom to leave raised equality between the sexes even further.

Hillard Kaplan, an anthropologist at the University of New Mexico, emphasizes that fertility levels had already dropped tremendously in places like the United States *before* the introduction of the birth-control pill in 1960. He argues that scientists developed modern contraception *in response* to the demand for lower fertility brought about by a postagricultural economy.[39]

In any case, the availability of the oral contraceptive in the second half of the twentieth century gave women even greater control over their bodies. Deciding to postpone reproduction became easier than ever, so more and more women delayed marriage and childbirth to extend their education. Higher levels of education allowed women to extract greater amounts of resources from the modern economic environment, which further increased their independence from men. Divorce (an exit strategy) increased, as did gender equality.

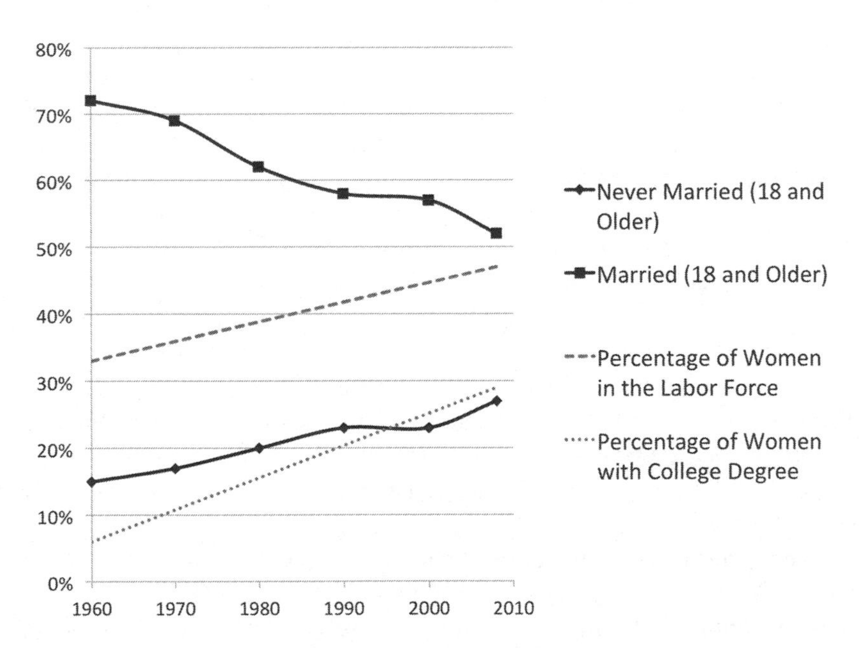

Figure 35. Marriage Trends Compared with Women's Education and Labor-Force Participation (1960–2008, United States).

The bottom grey line in figure 35 shows how the percentage of US women with a college degree multiplied by fivefold between 1960 and 2008. Since the 1990s, the majority of new college graduates have been women. The top grey line shows a similar rise in women's participation in the labor market during this period. At the same time, the proportion of married people in US society shrank substantially, since more women were delaying marriage and depending less on men for financial support.[40]

Unlike in agro-pastoral societies, women in modern economies also depend less on their parents to inherit land, animals, or dowry; instead, children of both genders can postpone marriage until after their education, when they achieve financial independence. Free mate choice is therefore much more common in postindustrial societies.[41] Here, individualism and liberalism increase.

It's difficult to overemphasize the magnitude of the shift in gender equality that has taken place in many developed countries. In the United States, for example, one of the largest changes ever recorded in public opinion concerns approval "*of a married woman earning money in business or industry if she has a husband capable of supporting her*." In 1937, only 18 percent of Americans accepted this proposition; by 1990, 82 percent approved.[42] The greater the proportion of a society's women who work, the fewer the babies born. And the more hours women work, the lower their fertility rate.

By the time a few generations pass, the demographic transition can even change cultural expectations. Between 1960 and the end of the twentieth century in Flanders, Belgium, the proportion of women who attend college rose from 20 percent to nearly 100 percent. The proportion of women in the labor force doubled. And women *wanted* fewer babies. The average number of children they preferred plummeted over this time period from 2.9 to 1.2 babies—enough to quickly shrink the population.[43]

Not all regions of today's world have reached the final stages of the demographic transition. Yet even in the countries with the highest fertility rates, these rates began to plunge by the 1980s. Countries whose fertility rates have remained relatively higher are concentrated in Africa, the Middle East, and South Asia. These are also the same countries that have had the highest levels of gender inequality.

The particular way in which the demographic transition will unfold in very conservative countries remains uncertain. In India, for example, wealthier, more educated families are having fewer children, as expected. However, these are the same families that also have greater access to ultrasound equipment, which can determine the sex of an unborn child. Although the rich, educated segment of the population desires only a couple of children, so that they can invest heavily in them to give them an edge in the modern economy and marriage market, cultural gender discrimination still exists. Among these families, sex-selective abortions of girls have increased the most dramatically—but only when the first-born child is female. In other words, the rich are going to extreme lengths to ensure that they have only two children, and that at least one of them is male.[44]

Whether cultural change will catch up to economic and demographic change in every part of the world, and how long it might take, is an open question. Gender inequality and fertility, however, are inherently intertwined with tribalism and will surely remain politically sensitive issues into the indefinite future.

In the next chapter, we'll explore the roots of the most extreme facets of tribalism: war and genocide.

13. THE BIOLOGY OF WAR AND GENOCIDE

The Prussian military theoretician Karl von Clausewitz (1780–1831) famously wrote that war is the continuation of politics by other means.[1] If war furthers the pursuit of political motives, and if politics reflects deeper evolutionary conflicts, then is it possible to contemplate war from a biological perspective?

The scientific body of the United Nations has declared that there is no need to consider this possibility; war, it argues, is a purely human invention, with no biological roots whatsoever. In 1986, the Spanish Commission for UNESCO convened scholars from around the world for a Colloquium on Aggression. Three years later, UNESCO finally adopted the "Seville Statement on Violence" at a general conference in Paris. According to the Seville Statement, which is backed by twenty academic signatories, war is completely avoidable because it has no natural history. The statement includes the following points:

- "*It is scientifically incorrect to say that we have inherited a tendency to make war from our animal ancestors. . . . Warfare is a peculiarly human phenomenon and does not occur in other animals. The fact that warfare has changed so radically over time indicates that it is a product of culture.*"
- "*It is scientifically incorrect to say that humans have a 'violent brain.'*"
- "*It is scientifically incorrect to say that in the course of human evolution there has been a selection for aggressive behaviour more than for other kinds of behaviour.*"[2]

It's certainly *politically* incorrect to say these things. But would it really be scientifically wrong to explore these possibilities—especially if rephrased in less absolute terms?

Before undertaking this challenge, it's crucial to keep two philosophical fallacies in mind. The authors of the Seville Statement on Violence committed a textbook case of the "moralistic fallacy": they declared that the way they believed the world *should* be is the way the world actually is. The rest of this chapter explains why the world actually *isn't* the way the Seville Statement claimed it is/should be.

We must also be careful not to commit the inverse error, which is called the "naturalistic fallacy." According to this equally problematic misconception, the way the world actually is, is always equivalent to what is morally good. To avoid this trap, it's imperative to recognize that an explanation of *why* something occurs is not necessarily the same as morally justifying it. In the words of political philosopher Fred Alford, "one must draw a sharp distinction between rationality and rationalization."[3]

Indeed, the great majority of crime is perfectly rational from an evolutionary perspective, assuming the perpetrator does not get caught and punished. Crime occurs when people cheat the established rules of their society in their fitness-enhancing effort to acquire resources or to multiply their genes. Additionally, people commit crimes of aggression against those who limit their resource acquisition or reproductive interests.

War, from this perspective, occurs when groups commit group crimes (such as those described above, as well as others predicated on political ideology), or when they defend themselves from others seeking to do so. An evolutionary account of war does not morally justify it—but it may explain *why* war occurs.

Finally, an evolutionary explanation does *not* imply genetic determinism, or any other kind of inevitability. War depends on geographic, environmental, economic, historical, cultural, political, technological, and myriad other factors. Moreover, individuals may very well have free choice over their decisions. Still, if we look at large groups of people over large periods of time, war is a question of "how" rather than "if." And evolutionary theory best explains the ultimate "why."

An evolutionary analysis of war comes from a cold, somewhat narrow place—one that is distantly removed from the indescribable emotions of the millions of individuals who have experienced war firsthand. This chapter examines war from this dispassionate point of view. It seeks to understand the underlying logic of this human behavior—without moralization, naturalization, or overdeterminism.

If war did have deep evolutionary roots, in spite of UNESCO's insisting otherwise, how would we know? The following four propositions would be true:

- War would be a universal phenomenon among human societies;
- War would certainly occur among human groups living in the most primitive economic environments;

- Nonhuman animals would engage in warlike behavior; and
- War would alter the fitness of groups; that is, it would impact the reproduction and destruction of gene pools, as well as the natural resources available to them, in nonrandom ways.

These criteria are quite similar to those that we've used to examine political orientation. And the last point, if it turns out to be true, would link war to the biology of tribalism.

Let's begin with the first point: Is war universal? In the year 1993 alone, at least sixty-three armed conflicts raged around the world—in Europe, the Middle East, both North and Sub-Saharan Africa, Central and East Asia, the Pacific islands, and the Americas.[4] Over the course of the twentieth century, wars killed over 100 million civilians. And more than 210 million people perished from the genocidal actions of states.[5]

Could these horrific statistics amount to some unnatural side effect of modernity? Perhaps people living a lifestyle closer to that of our early human ancestors have been less violent.

To the contrary, 90 percent of hunter-gatherer societies have been documented to engage in warfare; 64 percent of them fight wars at least every two years.[6] In fact, hunter-gatherers die from collective violence at a rate *twenty times higher* than that of Europeans and Americans during the twentieth century—even including all of the century's major wars.[7]

How about the third bullet point listed above? Do nonhuman animals practice anything akin to warfare?

THE PRIMATOLOGICAL ORIGINS OF WAR AND GENOCIDE

Our closest living relatives are chimpanzees. We shared a common ancestor with them some six million years ago, and we still share all but 6 percent of our DNA.[8] Despite this close kinship, many human beings are amazed to learn that chimps are more similar to us than we once assumed. One particular discovery made over three decades ago still shocks people years later.

In the early 1970s, primatologist superstar Jane Goodall was conducting her famous research on chimpanzees at Gombe Stream National Park in Tanzania. Long before Kofi Annan named Goodall a United Nations Messenger of Peace in 2002, she was studying the aggressive nature of East African chimp communities. Richard Wrangham, who worked under Goodall as a graduate student, tells the enthralling story of what happened at Gombe in his book *Demonic Males*.

According to Wrangham, the trouble began in 1970, when a personal conflict between two dominant individuals flared up. Two subgroups within the chimpanzee community formed, led by these two alpha males. A north-south geographical rift opened up between the feuding subgroups. By 1972, they had become two isolated communities.

Wrangham recalls how groups consisting mostly of males would set off from both communities—in what appeared to be hunting parties—to patrol the border areas of their respective territories. The patrols would shout out to one another, trying to gauge from these vocal calls how many individuals were in the enemy party. If they perceived a substantial discrepancy in the number of their adversaries, a patrol would attempt to attack an isolated individual from the other group.

In 1974, the occasional incidents of inter-group aggression had escalated into something different. A raiding party of six individuals from the northern community silently slipped into the southern group's territory. There, they located a lone straggler named Godi. The northern raiding group then hollered and shrieked as they bit and pummeled Godi to death. They even used a stone as a weapon.

Less than two months later, another unfortunate southern chimp suffered the same fate after trying to escape up a tree. Before leaving him to die from his wounds, the aggressors from the north chased away two southern male bystanders and abducted a young southern female.

By 1977, the northern community had systematically annihilated all of the male rivals from the southern group. Although a few southern females were killed, the northern group absorbed the rest into their community. As soon as the aggressors had annexed the southern territory, however, a third community from even further south began to launch lethal raids against the northern victors.

Wrangham estimates that approximately 30 percent of adult male chimpanzees at Gombe died from the aggressive behavior of members of their own species. Experienced, independent primatologists at other locations in Tanzania, and in Senegal and Ivory Coast, have observed similar raiding behavior. Primatology now considers lethal inter-group raids to be a normal, species-wide pattern in chimpanzee populations across Africa.[9]

A fascinating detail that Goodall observed in this behavior might be relevant to human beings. Although she discovered an eagerness among young male chimpanzees to form lethal raiding parties, Goodall emphasized that this eagerness varied greatly between individuals.[10] In other words, she believed that chimps have distinct political personalities that affect their aggression toward enemy communities.

Some of her colleagues claimed that Goodall overly anthropomorphized her primate subjects. For example, she gave them names instead of numbers. If Goodall was right about the variation in chimpanzee personalities, though, it's interesting to note that human beings also show pronounced individual differences in analogous attitudes. For instance, in the months leading up to the 2004 presidential election in the United States, Zogby and Gallup polls found that not all the voting human primates felt the same way about America's lethal raiding of an enemy territory. Specifically, 77 percent of Republicans thought the Iraq surge was improving the situation, compared with only 28 percent of Democrats.[11]

How common, across the animal kingdom, is the lethal-raiding behavior that primatologists have found in chimpanzees? Do other creatures engage in warlike behavior? An exceedingly few number of animals, in fact, deliberately seek to kill adults of their own species. According to Wrangham, spotted hyenas, lions, and wolves are among the only other animals that sometimes do so.

What is the common denominator that each of these species share with one another? They are all social predators that hunt together but sometimes get caught alone, off guard in enemy territory.

A resource-poor environment increases territoriality. So chimpanzees have a "motive" to eliminate the competition from other groups when they can easily use superior numbers to gang up on a hapless out-grouper.[12] Indeed, the great evolutionary theorist William Hamilton speculated that the origin of the "war party," in both chimpanzees and human beings, was "an all-male group, brothers and kin, practiced as a team in successful hunting and at last redirecting its skill towards usurping the females or territory of another group."[13]

Bonobos, on the other hand, live in a comparatively resources-*rich* environment. So they don't bother to hunt and eat the difficult-to-attain chimpanzee delicacy of colobus monkeys. Nor have primatologists ever documented bonobos engaging in war raids.

Still, the bellicosity of chimpanzees and human beings is far greater and more lethal than that of the handful of other animals that occasionally intend to kill adults of their own species.[14] Why?

Chimpanzees, you may remember, are patrilocal. This means that males stay on the territory where they were born, while females disperse to other groups to avoid inbreeding. Male chimps within a territory, then, are close relatives of one another. Unlike the females, who can leave and join an unrelated group, the males must hold onto their territory to survive. Another group of chimps would have no use for foreign males, who would only consume the scarce food supply and compete for matings with females. So foreign males would kill males without territory and absorb their females.[15]

For the victors, having extra females around increases the fitness of the male chimpanzees; additional copulations cost males little—and they invest practically nothing, compared with many human fathers, in parenting. So it's well worth their while to incorporate the females of vanquished groups.

This incongruity between the sexes in parental investment also explains why inter-group aggression among humans and chimps is by far the most lethal, compared with the handful of other animals where this behavior occurs. Spotted hyenas, rhesus macaques, and savanna baboons are matrilocal; males disperse to neighboring groups, while genetically related females inherit and defend their territory. Among these species, Wrangham explains, females are burdened with the duty of fighting, while their males have much less at stake.

A species like the hyena, then, has *matriotism* (as opposed to the patriotism of humans and chimps). Hyena wars, however, are far less genocidal than the wars fought among the two patriotic species. Since hyenas live in *female*-dominant groups, the matriarchs' interest as the more-investing sex is focused on maximizing resources rather than copulations. Having additional male hyenas around would *not* increase a female hyena's fitness, because it's not difficult for her to get pregnant. The effort that she invests is in gestation, lactation, and childcare. Female hyena warriors seeking resources only have to kill enough of their enemies to drive them off their land; they don't have to exterminate the females and abduct foreign males.

Even though the world has 8.7 million species,[16] Wrangham believes that humans and chimpanzees are the only two patrilocal, male-dominant, intensely territorial, lethal-raiding social predators. Warfare and genocide, then, occur only in human beings and chimpanzees. They do *not* occur, however, in bonobos, our other closest living relative.[17]

DEATH AND REPRODUCTION IN HUMAN WAR

Does human war follow the same gender logic as chimpanzee lethal raiding? Like chimps, humans generally practice a patrilocal residence pattern—but not overwhelmingly so. Of the world's *hunter-gatherer* societies, 63 percent are patrilocal, 16 percent are matrilocal, and 16 percent are bilocal (meaning they practice both types of residence patterns[a]).[18]

Because humans are generally patrilocal, and because we likely evolved from territorial social predators, people have patriotism and male warfare—as opposed to matriotism and female-dominated war. Accordingly, human war is a mostly-male behavior. Only on rare occasion do ideologies or extreme conditions permit the participation of women on the battlefield. During World War II, for example, some 8 percent of Soviet combatants were women. This female involvement was due partly to the Marxist ideal of sexual equality, and partly to the dire necessity that arose with the German invasion.[19] Another exception in today's world can be found in Israel, where military service is compulsory for women (who can serve as combat soldiers). But the Israeli Defense Forces is the only army in the world that continuously requires female service. This unusual need arises from the country's tiny territory, which is surrounded by much larger states that have historically sought to destroy their smaller neighbor.

But the important principle, as political psychologist Jim Sidanius points out, is that "there's not a single recorded event in human history of women organizing and constituting armies for the purposes of conquest or intergroup predation."[20] In nearly every known human society around the globe, the fighting of war is an exclusively male activity.[21]

The two sexes even *feel* differently about war, even though their opinions about most other controversial issues do not differ dramatically. According to political scientists Benjamin Page and Robert Shapiro:

> In practically all realms of foreign and domestic policy, [American] women are less belligerent than men. They are more supportive of arms control in peaceful foreign relations; they are more likely to oppose weapons buildups for the use of force. . . . On these issues, over a twenty-year period, there has been an average opinion difference of about 9 percentage points.[22]

This gender gap in attitudes toward war is extra-ordinary. When measured as large groups, men and women typically diverge in political opinions by a much smaller spread—even with regard to issues such as feminism.

How Violent *Are* Human Males?

Human beings aren't actually descended from chimpanzees. They're just our "first cousin" species. But so are the peaceful bonobos. Both chimps and bonobos are patrilocal, yet the former engage in phenomenally more violent behavior than the latter. Where does the human male fit into the picture?

Our ancestors likely evolved into modern humans while living in a woodland savanna. Wrangham believes that this environment would not have supplied these prehuman apes with the vegetation necessary for foraging in large, stable groups like bonobos. Instead, the savanna woodlands would have provided sparse, high-value foods (such as fruit and hunted meat). These items would have required a chimpanzee-like fissioning of animals into small, highly territorial groups.[23] From this social and ecological configuration, one would also expect a chimp-like pattern of inter-group violence. Does the ethnographic record confirm this prediction?

Wrangham argues that the warfare of human groups living in primitive economic environments compares remarkably well to chimpanzee lethal raiding. To make his case, he revisits anthropologist Napoleon Chagnon's famous studies of the Yanomamö. This indigenous group of some twenty thousand people is native to the Venezuelan and Brazilian Amazon. Like most peoples, the Yanomamö are patrilocal; men remain in their territory while women marry into other villages.

According to Chagnon, a Yanomamö village can only reach a population of about three hundred before a conflict arises. Whether conflicts originate with allegations of witchcraft, or a fight over women, or over resources (which the Yanomamö actually deny), they can quickly spin out of control.

The Yanomamö don't have a developed rule of law, but they do have customs for adjudicating disputes. The process essentially consists of a series of increasingly dangerous rituals to resolve a clash between individuals. But if the escalating duels do not extinguish the enmity, then the parties involved may recruit allies for war.

One of the main forms of Yanomamö war craft is the raid (*wayu huu*), in which ten to twenty men stealthily approach an enemy territory to kill a particular individual target. In practicality, these missions may end "successfully" with the death of any man from the other village. If the warriors are able to ambush an isolated individual at the edge of the enemy village, they fire lethal curare-tipped arrows at him, run away, and prepare for retaliation against their village.

Before returning home, the raiding party may try to abduct a woman from the enemy group. If they succeed, according to Chagnon, all the participating warriors rape her. Then the rest of the men in their home village rape her before they give her to one man as a wife.[24]

The Yanomamö live in a chronic state of inter-village conflict and war raids. Although a raid may only result in one or two casualties, the violence is so pervasive that some 30 percent of all men die violently at the hands of their own ethno-linguistic group. Wrangham emphasizes that the same percentage of the Tanzanian chimpanzees at Gombe National Park are killed by members of their own species.[25]

How representative of primitive war are the Yanomamö? To draw conclusions about such a controversial facet of human nature, we should demand hard data from comparable societies around the world, collected by independent researchers. Figure 36 shows the percentage of deaths caused by warfare among indigenous groups living in independent, pre-state societies. These tribes are native to South America, Papua New Guinea, and Australia (the data come from Lawrence Keeley's book *War before Civilization*[26]). With the exception of the Yanomamö, a different researcher collected the casualty rates for each group. The chart also includes, for comparison's sake, Wrangham's Gombe chimpanzees as well as the war casualties of the United States and Europe during the twentieth-century.

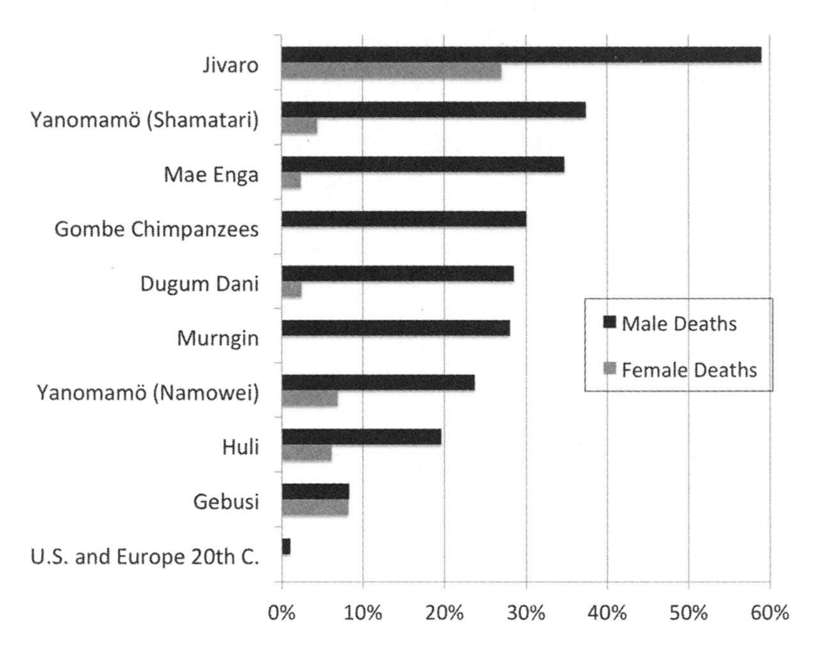

Figure 36. Percentage of Deaths Caused by War.

The Jivaroan group whose male war-casualty rate approaches 60 percent is the Waorani of the Ecuadorian Amazon. Even though highland Quichua Indians who colonized regions of the jungle had guns and the Waorani only had spears, the former were so afraid of this tiny but hyperbellicose group that five hundred Waorani once had twenty thousand square kilometers of homeland.[27] The Quichua word for the Waorani is *auka*, which means "savage."

Before outsiders finally succeeded in pacifying them, the Waoranis' exceptionally high level of violence acquired the tribe an enormously disproportionate amount of natural resources per capita. But the price of the land was steep: a nearly 60 percent war-casualty rate.

The Waorani are actually an outlier; in fact, they are the people with the highest war-casualty rate ever discovered. Still, even at half the Waorani rate, why do such a high proportion of males die from warfare in these independent, pre-state societies?

War as a Male Reproductive Strategy

Imagine that you are a perfectly peaceful Yanomamö boy minding your own business. You may not hate the village across the river, but one day a group of men from that village kill your favorite uncle. And then three years later your village's enemy kills your cousin and kidnaps your sister. By the time you're old enough to fight, you don't like the neighbors. You also realize that you have a one-in-three chance that they will shoot a poison arrow at *you* one day. What do you do?

You could decide not to fight, but your people wouldn't respect you. It might be hard to find a wife, because the best warriors take more than one. To make matters worse, even if you're lucky enough to find a woman not taken by a warrior, Yanomamö culture permits other men to seduce your wife if they think you don't contribute your fair share to war efforts. In a small community, everyone can easily keep track. Your future as a Yanomamö pacifist isn't a very bright one.

If you become a skilled warrior, on the other hand, you may take part in the killing of a target from the enemy village. With this feat, you would join the 40 percent of adult Yanomamö men honored as *unokai*s. On average, unokais have more than two and a half times as many wives as non-unokai men; and unokais produce over three times as many children![28]

Under these circumstances, engaging in war would be the most rational decision. Evolutionary pressures would naturally select for more bellicose Yanomamö very quickly.[29] In light of such facts, primitive warfare could be considered "a male-coalitional reproductive strategy."[30]

If the reproductive patterns of most of human history looked more like the Yanomamö pattern than the one we find in our contemporary, state-level societies, then many men would not ever reproduce. But most women would. It's easy to test this theory. All we have to do is examine the human genome. This genetic record can show how many female ancestors humanity has had, compared with how many male ancestors engendered our contemporaries.

Geneticists at the University of Arizona did the counting. To do so, they took genetic samples from populations around the world. Then, they analyzed Y-chromosome DNA, which sons inherit from their fathers. They also looked at mitochondrial DNA, which mothers pass down to their children. When the geneticists contrasted the diversity found within these two types of DNA, they discovered that the maternally inherited DNA was twice as diverse as the paternally inherited DNA.

The popular science press jumped on the finding to claim that, throughout human history, twice as many women have reproduced as men. However, some other issues could partially explain the reduced diversity of the male DNA. For example, older men often reproduce with younger women, which would make male generations longer than female ones. Still, the geneticists who conducted the study believe that only a substantial differential in reproductive success rates between the sexes could explain their findings. Simply put, most women in our species' history have reproduced; but for men, many have not reproduced at all, while others have done so disproportionately.[31] So the genetic record suggests that males have had an evolutionary incentive to engage in warfare to increase their reproductive success.

Why, then, is the war-casualty rate so much lower for twentieth-century Americans and Europeans than for the Yanomamö? To begin with, the American and European societies are not polygynous. Even in modern states, countries with legalized polygyny have more civil wars.[32] Secondly, the populations of these modern states were enormous compared with a Yanomamö village. The death of one Yanomamö man in a village of one hundred would be the equivalent of 1.4 million American casualties in 1945. So the genetic stakes are far higher for the Yanomamö political unit. In addition, the twentieth-century states had sufficiently large populations for a small subset of men to specialize as full-time soldiers; among "primitive" peoples, in contrast, all men may be generalists, and expected to fight.

But most importantly, the modern states had . . . *states*. They existed precisely because they had the political complexity and technology necessary to exert a monopoly in coercive force over enormous populations of people. Moreover, they had a much more developed rule of law. This combination of effective force and rule of law helped them to reduce conflicts within states—and occasionally even between them. Political scientist Francis Fukuyama's book *The Origins of Political Order* (2011) tells the story of how these and other institutions have developed over the course of human history, leading to the emergence of modern democracies.[33]

Although today's modern states have reduced the chronic internal warfare typical of societies like the Yanomamö, the male predisposition for violence remains—even during peacetime. In the United States, American men are approximately nine times more likely to commit murder than are women.[34]

We can also find traces of biological adaptation to war in women. Consider the wartime dilemma faced by countless females over the ages, as exemplified by the biblical story told in the fourth book of the Judeo-Christian Bible. In Numbers 31, the Lord commands Moses to take revenge on the Midianites. Midianite gods and women had attracted the Israelites and their men, which had led to a plague:

> [The Israelites] mounted an attack against Midian, as the Lord had commanded Moses, and they killed every male. . . . The children of Israel took the Midianite women and their small children captive, and they plundered all their beasts, livestock, and all their possessions. . . . Moses said to them, ". . . So now kill every male child, and every woman who can lie intimately with a man you shall kill. And all the young girls who have no experience of intimate relations with a man, you may keep alive for yourselves." (Numbers 31: 7–18)

At the end of the war, the Israelites took thirty-two thousand female Midianite virgins. They killed the nonvirgin women. Chimpanzees, in contrast, typically kill the children of females who give birth shortly after abduction from another troop but leave most females alive. Like chimpanzees, though, the Israelites killed all the enemy males.

Now let's see how this war would have impacted the fitness of males and females very differently. After the battle, the female Israelites would reproduce at a normal rate, but the males would have multiplied their rate of reproduction because they had divvied up the conquered Midianite females among themselves. The fitness of the Israelite males turned out to be higher than that of the smitten Midianite males.

For the Midianites, on the other hand, the gender advantage was reversed. Midianite females had greater fitness than their male counterparts because females were the only ones to survive the war. If a Midianite woman were conceiving a baby on the eve of her people's defeat by the Israelites, then it would clearly be better for her to have a female child, which might survive to pass on its genes.

The extreme impact that war can have on the fitness of the genders is still relevant to this day. On September 11, 2001, Americans woke up to a foreign terrorist attack. The nation braced for the possibility of war on its own soil. People felt vulnerable and nervous. Across the fifty US states, pregnant mothers involuntarily miscarried 12 percent more male fetuses after the twentieth week of gestation than during the average September. Mothers continued to give birth to fewer boys three to four months after the terrorist attack.[35]

Scientists aren't sure what exactly reduces the male birth rate, but they suspect that male fetuses are more sensitive to corticosteroids released by mothers experiencing intense stress.[36] If a pregnant Midianite woman aborted a male fetus during the war, perhaps she would have had a better chance of surviving and continuing her reproductive life with a foreign husband. Or, in the case of a long siege, she could get pregnant again and give birth to a daughter, who could survive a defeat.

War and Peace

Pre-state peoples like the Yanomamö don't always engage in "external" wars, where the objective is to annihilate the male population of an enemy group and to take over its resources and women. Often their wars are more "internal," or regulatory.[37] The objective of a regulatory war might be to recalibrate a political alliance between villages, to change a boundary or the use of natural resources, or to adjust reproductive patterns between populations.[38]

Exogamous populations that marry each other's women typically fight regulatory wars. After all, it wouldn't make sense to wage a genocidal campaign against a village where your nephews and nieces live.

If pre-state peoples fight wars in a pattern similar to chimpanzee raids, they often resolve regulatory conflicts like bonobos—by offering women to the "enemy." If war entails reducing the enemy's fitness by taking away his life and women, then peacemaking entails the opposite: increasing his fitness by giving him the women of one's own group, even in a symbolic gesture. Enemies destroy genes, while allies exchange genes.[39]

Among the Kiwai people of Papua New Guinea, a former adversary lends its own women to the host of a peacemaking feast "to put out the fire."[40] The Maring tribe (also from Melanesia) exchanges women during peace talks. The number traded by each side is typically commensurate with the number of men killed in the other group.[41] Some Australian peoples send out women to their would-be attackers to give the aggressors the option of sexual intercourse in lieu of fighting.[42]

Western Europeans also used sexual relations to form alliances and make peace. Like the Spanish Habsburgs, many royal families traded women with one another between the eleventh and the nineteenth centuries.

Exchanging genes for peace, however, can't always prevent war. When war does break out, a violent clash of forces and wills alters which economic, political, and military resources are available to which populations. In a distinct subset of wars, mass killings of civilians systematically and asymmetrically destroy the *genetic* resources of particular groups. These events irreversibly impact evolutionary fitness on a large scale and transform political reality.

GENOCIDE: THE ULTIMATE POLITICAL CRIME

In March 2011, 350,000 people had become fans of a Facebook page calling for a third Palestinian *intifada*, or rebellion. The organizers of the movement had hoped to use social media to harness the rage coursing through the Middle East during the Arab Spring, and to turn this current against Israel.

Nevertheless, Facebook made the highly unusual decision to remove the "third intifada" page because of the following sentence that appeared on it: "*Judgment Day will be brought upon us only once the Muslims have killed all of the Jews*."[43] Did this genocidal provocation originate with an individual political radical who suddenly found 350,000 followers? The controversial incitement to commit genocide ultimately comes from *Sahih Muslim*, which is considered the second most authentic of the six major collections of Hadith of Sunni Islam. In book 41 of *Sahih Muslim*, verse number 6985 reads:

> Abu Huraira reported Allah's Messenger (may peace be upon him) as saying: The last hour would not come unless the Muslims will fight against the Jews and the Muslims would kill them until the Jews would hide themselves behind a stone or a tree and a stone or a tree would say: Muslim, or the servant of Allah, there is a Jew behind me; come and kill him; but the tree Gharqad would not say, for it is the tree of the Jews.

This verse appears fairly often in the context of the Israeli-Arab conflict. Palestinian politicians use it to motivate a holy war to destroy the world's only Jewish state. Article seven of the Hamas Covenant, which has been the charter of this political organization since 1988, includes the same verse.

Even the more secular Fatah Party, which occupies the center of the Palestinian political spectrum, quotes the lines above. On January 29, 2010, a Fatah-controlled television station broadcast a sermon that paraphrased this same provocative line. The imam explicitly politicized the verse; he left little doubt as to how he wanted his followers to apply the meaning to the contemporary conflict with Israel:

> The Jews, the enemies of Allah and of His Messenger! Enemies of humanity in general, and of Palestinians in particular. . . . Our enmity with the Jews is based on faith; our enmity with the Jews is a matter of faith, more than an enmity arising from occupation and the land. . . . The Prophet [Muhammad] says: "You shall fight the Jews and kill them, until the tree and the stone will speak. . . ." Thus, this land will be liberated only by means of Jihad.[44]

The imam's words beg the question of whether he incited genocide as a religious duty, or to gain land, or for both causes. How significant a role do material motives play in genocides, compared with extreme ideological hatred? What is the relationship between genocidal incitement and genocide itself?

There ought to be a systematic way to study genocides, because they have occurred throughout time and across the globe. Genocidal behavior has been recorded from Julius Caesar's destruction of the Germanic tribe of Eburon in 53 BCE, to Genghis Khan's nearly complete obliteration of the population of Herat (in present-day Afghanistan, 1220), to the Prince of Yen's politicidal slaughter of tens of thousands of Ming-dynasty Confucian officials and their families in 1402. The list goes on and on in any region, into modern times.[45]

Although humans have lived through long periods of peace, the chronicles of history document the chilling universality of genocide. While it doesn't happen in every war, genocide does occur in a distinct minority of armed conflicts.

What is the logic of this horrific phenomenon? War, as we learned above, is the continuation of politics by other means. If the logical extreme of war is genocide, then genocide is the ultimate political crime.

The Subtypes of Genocide

Few people have better demonstrated the political nature of genocides than political scientist Barbara Harff. To properly categorize these phenomena, Harff had to coin the word "politicide." Politicide refers to a mass murder where the perpetrators target their victims based on perceptions of their political orientation.[46] Harff's contribution of this term is especially important because the international community officially lacks the concept of politicide.

In the wake of World War II, the United Nations emerged as a renewed effort to prevent another world war (which its predecessor, the League of Nations, had failed to do). One of the United Nations's early accords was the Convention on the Prevention and Punishment of the Crime of Genocide, adopted in 1948. The final document defined genocide as "the intent to destroy, in whole or in part, a national, ethnical, racial, or religious group" through killing, maiming, sterilizing, or transferring away the children of the group in question.

The first draft of the convention also protected populations defined in terms of political orientation. But objections from Russia and a few of its allies removed this political criterion from the final historical document.[47] Presumably, Stalin wanted to avoid international scrutiny of his purges, or political mass murders. He used these politicides to persecute "reactionary" groups believed to resist assimilation into the dominant society, as well as to eliminate the old aristrocracy and other "class enemies." For instance, in the 1930s Stalin liquidated and deported eight million *kulaks* (independent, wealthy peasants of an "antisocialistic" class). Throughout history, regimes on *both* extremes of the spectrum have committed politicide to eliminate ideological adversaries and to expropriate their resources.

Having rightly introduced a term for politicides, Harff examined the death tolls of the dozens of genocides and politicides that had occurred between the end of World War II and the turn of the twenty-first century. Between twelve and twenty-two million civilians perished in these mass killings. This figure exceeds all the military victims of the wars fought during the same period. In most cases, the state or its agents enabled genocides and politicides to occur.[48] The extreme concentration of power in the modern state has made these crimes against humanity feasible on previously unthinkable scales.

Next, Harff wanted to know what percentage of mass killings were genocides, and what proportion were politicides. To answer this question, she took as her sample the thirty-seven genocides and politicides documented between 1955 and 2001, which occurred on every continent. Of these events, Harff considered 32 percent to be pure politicides. These include the forced disappearances of leftists and the stealing of their babies in Argentina (1976–80) and similar patterns of persecution in Chile (mid-1970s to mid-1980s), and El Salvador (1980–1990).

The largest category of mass killings was "politicide with communal victims," which made up 35 percent of the sample. Saddam Hussein committed this atrocity when he massacred Iraqi Kurds who supported antigovernment parties; but Hussein didn't systematically murder the Kurds working in his regime whom he considered loyal.

Harff categorized 19 percent of the mass killings as "combined genocide and politicide," as exemplified by the Guatemalan Civil War. In Guatemala, the thirty-six-year conflict involved a left-right dimension between Communist and anti-Communist belligerents, as well as a racial conflict between Guatemala's large indigenous population and white-mestizos.

Only 14 percent of the events were classified as pure genocides, such as the wars between Hutus and Tutsis in Rwanda and Burundi.[49] In other words, fully 86 percent of the events had a *substantial ideological dimension beyond ethnic conflict*. One could easily argue that all the events—including the "pure" genocides—had political elements as well. In the case of the genocide in Rwanda, for instance, the Hutus didn't target only Tutsis; they also killed Hutus whom they thought stood "too far" to the left in the Hutu political spectrum. In fact, the first killings targeted liberal members of the government, including the president of Rwanda's Constitutional

Court. The Hutu militias also targeted liberal civil rights activists, priests, and journalists. The lack of resistance from the more xenophilic, assassinated Hutu leaders surely made it easier for the perpetrators of the genocide to hack to death with machetes eight hundred thousand people in only one hundred days.[50]

In summary, it's difficult to tease apart the genocidal aspects of mass killings from the politicidal ones. Still, Harff's categorizations make useful distinctions. Her work emphasizes that only a minority of mass killings are "pure" genocides without a substantial amount of overt, political discrimination.

Genocides have asymmetric impacts on the gene pools of different ethnic groups. They also change the amount of territorial and economic resources available to them. Therefore, genocides clearly alter the evolutionary fitness of both populations and individuals, in relative and absolute ways, with long-term repercussions.

The end result of *politicides*, on the other hand, is mainly the transfer of economic resources from one class to another, as well as the elimination of political adversaries. The motives of governments to commit politicides, then, would apparently be the relatively more short-term goals of enrichment, redistribution, and/or reducing the opposition by physical elimination and fear.

One cannot help but wonder, however, if politicides may have long-term affects as well. If a politicide were to accurately target people based on their ideological leanings, and if a substantial component of political orientation is heritable (as chapter 2 suggests), then it's conceivable that a politicide could alter—however subtly—a population's average political disposition.

The Causes of Mass Killings

Barbara Harff has also made an enormous contribution to the scientific study of the causes of genocides and politicides. To determine in a rigorous way what triggers these events, Harff worked with the US government's State Failure Task Force to compile a list of the 126 major cases of internal war and regime collapse that started between the years 1955 and 1997. Of these 126 instances, genocides/politicides ensued 28 percent of the time. What precisely sparked them to occur when they did—and only then?

This research identified six factors that were present more often than not in the cases of internal wars and regime change that *did* lead to mass killings. On average, this six-factor model could predict a genocide/politicide with an accuracy rate of 74 percent. If all six of the factors occurred, then the probability rose to 90 percent.[51] Harff's six factors that predict mass killings are:

(1) An elite mostly constituted by an ethnic minority;
(2) The presence of an autocratic rather than a democratic regime in power;
(3) The occurrence of past genocides;
(4) Extreme political upheaval;
(5) A country's level of trade openness with other states; and
(6) Extremist, exclusionary ideologies.[52]

Even though numerous of the sixty-six candidate variables that Harff's team tested related to resource scarcity (including the ratio of population to arable land, damage due to drought, famine, and so forth), only one of the final six key predictive factors even pertains to material factors (number five). The five other factors relate to distinctly political conditions. Factor six even concerns political orientations themselves.

Since this book is a natural history of political orientations, let's have a closer look at the sixth factor. Extremist, exclusionary ideologies are especially likely to increase the chance of mass killings when espoused by elites who come to power in the absence of democratic safeguards. Among Harff's set of 126 countries that had undergone state failures, the countries where the ruling class adhered to exclusionary ideologies had two-and-a-half times the chance of committing genocides or politicides.[53]

Indeed, exclusionary ideologies have often preceded historical genocides. The Turks had discriminated against the Armenians for centuries before World War I. The political upheaval during and following this war served as the catalyst to put this ideology into action: the Ottoman government starved, massacred, and marched 1.5 million Armenians to death.

Similarly, Germany had a tradition of anti-Semitism that stretched back hundreds of years. This ideology, which pervaded many realms of the country's culture, stemmed from theological sources. One of them was the German priest and Reformation leader Martin Luther (1483–1546).

Luther wrote several major works demonizing the Jews in the 1540s, including *Warning against the Jews* and *On the Jews and Their Lies*. In the latter treatise, Luther argued for the burning of Jewish schools and synagogues, for rabbis to be executed for practicing their livelihood, and for Jews to be enslaved as agricultural labor.

In his book *Hitler's Willing Executioners* (1996), Harvard political scientist Daniel Goldhagen contended that "eliminationist anti-Semitism" was *the* greatest cause of the Holocaust. In other words, ordinary Germans became Nazis not mainly because of the economic depression, or by falling prey to Hitler's charismatic leadership; rather, the cultural identity of a substantial segment of

German public opinion had long been infused by a medieval, genocidal ideology. This culture, Goldhagen argues, made ordinary Germans *want* to implement Hitler's attempt to annihilate the Jews.

In addition to their theological roots, exclusionary ideologies in Germany and elsewhere can arise from the need for a scapegoat to rationalize historic blows to an ethnic group's self-esteem. For instance, a sense of lost glory or endangered superiority preceded mass killings in Turkey, Cambodia, and Argentina. The perpetrators in these cases blamed the decline of a golden age on minority ethnic and political groups.[54]

Exclusionary Ideologies Today

Some people compartmentalize mass killings away in a box of evil, separating them from the here and now of ordinary life. Yet when internal war or regime collapse unexpectedly triggers these horrific events, they're typically fueled by exclusionary ideologies that previously lay in ferment beneath the surface of public opinion for many years. Although these ideologies alone don't always lead to mass killings, they're often the first and most foreseeable step toward them. And exclusionary ideologies aren't hard to find. Below, we'll look at a few examples of these negations of identity, reported from around the world in 2010.

At this point in Myanmar (Burma), an autocratic military regime had been ruling the country for nearly fifty years. The majority of Burma's population is Buddhist, but the country's 1.5 million Chin people are mostly Christians. Burma's policy toward the Christian Chins has reflected an exclusionary ideology—to the point of denying their religious existence. The government has forced them to destroy all the crosses that once stood on the mountains of Chinland; in their place, Buddhist pagodas have been erected.

Because of the ethnic persecution and economic discrimination that have accompanied Burma's ideology toward the Chin, some two hundred thousand have fled the country.[55] As is true for so many other groups, the fractured populations of diasporas settle in other people's countries. There, they run the risk of assimilation or renewed persecution. So ethnic cleansing (the use of terror to transfer populations) can result in partial or complete ethnocide.

To the west of Burma, communal tensions between Hindus and Muslims in Northern India have periodically flared up and cooled down over the centuries. One point of contention involves a battle over religious shrines. Hindus believe that their deity Ram was born in the city of Ayodhya around 5000 BCE. An ancient temple once marked Ram's birthplace. It was one of the holiest sites of Hinduism. But in 1528, the Mughal emperor Babur ordered the temple destroyed and replaced with a mosque. The Islamic place of worship was built, and it became known as the Babri Mosque (named after Babur).

In 1949, shortly after the political upheaval of Indian independence, Hindus placed idols inside the mosque to honor Ram's birthplace. This act angered Muslims, whose religion prohibits the graphic depiction of the human form. Questions also arose about whether the mosque was built in accordance with Islamic law; a Hindu judge maintained that the Mughals had built the mosque out of the materials of the ancient Hindu temple. The temple's reused pillars depicted Hindu gods and goddesses. Muslims, however, generally denied that the Hindu Temple had ever existed.

The site remained an object of tension entangled in lawsuits for forty years. During this time, the redemption of Ram's birthplace became a banner issue for right-wing Hindu politicians. Then, in 1992, Hindu extremists destroyed the mosque. In the violent riots that followed, over two thousand people were killed. In neighboring Pakistan and Bangladesh, Muslim mobs directed their rage at the Hindu minority populations. Hundreds of Hindu homes and temples were destroyed, and many women were raped.

By 2010, eighteen Indian judges had heard the Babri Mosque case. In the latest ruling, the court ordered that the holy site be split between Hindus and Muslims. The compromise displeased many people. Two months later, an explosion in Varanasi, Hinduism's holiest city, wounded thirty-four people and killed a one-year-old girl sitting on her mother's lap. The terrorist organization Indian Mujahideen claimed responsibility via e-mail. In their explanation, the Islamist group criticized the verdict and claimed to avenge the mosque's destruction eighteen years earlier.[56]

The same exclusionary gesture of building a religious structure above a conquered people's temple has occurred repeatedly throughout history. When the Spaniards conquered the Andean region, they found the most revered temple of the Incan Empire in Cusco. The Coricancha (or "Golden Courtyard") was dedicated to the Sun God. And the sun was believed to have begotten the progenitor of the Inca dynasty. The Spaniards promptly demolished the Incan temple, and they built the church of Santo Domingo on top. But the Spaniards made sure to leave the distinctive foundation of Incan stonework visible—*beneath* the church.

The most famous example of this exclusionary phenomenon lies in Israel. The Temple built by King Solomon in 957 BCE has been considered the holiest site in Judaism for nearly three millennia. Believers consider this location to mark the resting place of the divine presence. It's also the place where God gathered the dust to create the first man. It once contained the "holy of holies"; here, the Ark of the Covenant contained the Ten Commandments, which Moses is believed to have received from God on Mount Sinai. Observant Jews have turned toward the Temple's location during prayer for thousands of years.

When the Muslim armies conquered Jerusalem in 637 CE, the Umayyad Caliphs commissioned al-Aqsa mosque and the Dome of the Rock to be built on top of the Temple Mount. Today, right-wing Jews and Muslims still harbor mutually exclusionary ideologies revolving around these religious structures. In July 2010, religious Jews marked the annual fast day of Tisha B'Av, which marks the two destructions of the Temple in Jerusalem. A poll conducted ahead of this day of mourning found that 49 percent of Israelis would like to see the Temple rebuilt.[57] Reconstruction would presumably endanger the mosques standing on the Temple Mount.

On the other side of the conflict, Palestinian political and media leaders have denied Jewish history in ancient Jerusalem for many years. The Palestinian Authority's official daily newspaper, *Al-Hayat Al-Jadida*, has repeatedly dismissed the existence of the Jewish temple. This paper has referred to the holiest site in Judaism as "the legend and fable of the alleged Temple." The Palestinian Authority's rejection of the ancient Jewish temple in Jerusalem contradicts the archeological record and even the Qur'an itself (Sura 17:2–7).[58] But the force of exclusionary ideology is sometimes stronger than historical fact and religion.

When the Umayyads first built Islamic shrines on top of the Jewish Temple in the seventh century, however, the Temple had already been destroyed for over five hundred years. After this destruction of the second Jewish Temple in 70 CE, the crushing defeat against the Romans scattered most of Judaea's Jews into exile. When the Muslims arrived in the seventh century, the region had been under the rule of the Christian Byzantine Empire. The Byzantine emperor Justinian had constructed the Church of Our Lady on the Temple Mount, claiming to have outdone King Solomon. The main religion for Islam to compete with in Jerusalem, at the time, was Christianity.

In this context, the writing inside the Dome of the Rock mosque negates the validity of Christianity. The following lines from the Qur'an are inscribed in this structure:

> That is Jesus, the son of Mary—the word of truth about which they are in dispute. It is not [befitting] for Allah to take a son; exalted is He! When He decrees an affair, He only says to it, "Be," and it is. [Jesus said] "And indeed, Allah is my Lord and your Lord, so worship Him. That is a straight path." Then the factions differed [concerning Jesus] from among them, so woe to those who disbelieved—from the scene of a tremendous Day. (Maryam 19: 34–37)

The phrase "God has no companion" also appears five times in the Dome of the Rock.[59]

In the adjacent country of Egypt in September 2010, only months before the Revolution, Christians and Muslims were clashing over these very same Qur'anic verses. Bishop Bishoy, the second highest clergymen in the Coptic Church, denied these words' divine origin. Instead of being revealed to the Prophet Muhammad through the archangel Gabriel, as Muslims believe, the bishop suggested that one of Muhammad's successors inserted the words into the Qur'an after the Prophet's death. In response to outraged Muslims, the Coptic pope Shenouda III appeared on state television and called Bishoy's comments "inappropriate."[60] That is, the pope denied his bishop's denial of the divinity of the Qur'anic verses that deny the divine origin of Jesus.

As this political cross-section of 2010 shows, exclusionary ideologies frequently negate religious identity or claims of divinity. Since religiosity is a facet of, and proxy for, tribalism, these ideologies subtly deny the existence of a tribe as well. These exclusionary worldviews can escalate into pregenocidal ideologies when they begin to explicitly dehumanize the members of an out-group.

The Predictable Link between Dehumanization and Disease

The dehumanization process almost always compares out-group members to infectious disease, or to the insect or animal carriers (the "vectors") that transmit pathogens. This fact is fascinating in light of findings from chapter 10 that ethnocentric people fear infectious disease more than xenophiles. We also discovered that pathogens likely bear some responsibility for the existence of xenophobia: by avoiding outbreeding, groups have limited their contact with novel pathogens; inbreeders have also avoided inheriting genes less adapted to their in-group's pathogen environment.

By linking out-groups to pathogens and their carriers, xenophobic people not only strip away the humanity of the object of hatred; they also compare the other to something that is much easier and "desirable" to kill than humans. The comparison simultaneously justifies, encourages, and enables the would-be perpetrators of genocide.

History offers numerous examples of pregenocidal, dehumanizing ideologies. In the 1320s in France and some of the northern Iberian kingdoms, Jews were persecuted together with lepers.[61] After the Black Death struck Germany in the 1340s, Germans blamed the Jews for the plague (which in reality, of course, was caused by the *Yersinia pestis* bacterium) and slaughtered many Jewish communities.[62] In 1542, Reformation leader Martin Luther implored the emperor and lower state authorities to expel the Jews, whom he described as "plague" and "pestilence."[63]

Four-hundred years later, German Nazis often referred to the Jews as a "rat infestation."[64] Hitler himself made this dehumanization ideology a cornerstone of his policy; he even updated the Jews-as-disease metaphor in response to the nineteenth-century founding of bacteriology. At a dinner with Heinrich Himmler (head of the SS) in February 1942, Hitler said:

> The discovery of the Jewish virus is one of the greatest revolutions that have taken place in the world. The battle in which we are engaged today is of the same sort as the battle waged during the last century by Pasteur and Koch [who identified the bacteria that cause anthrax, cholera, and tuberculosis]. How many diseases have their origin in the Jewish virus![65]

The disease metaphor has resurfaced again in recent times in central Africa. Before and during the 1994 Rwandan genocide, Hutus commonly called Tutsis "insects" and "cockroaches."[66] Cockroaches spread bacteria, including those harmful to humans, by transporting them on the surfaces of their bodies. The right-wing Hutu publication *Kangura* wrote:

> Hutus, you will be injected with syringes full of AIDS viruses because the peace accord gave the Ministry of Health to the cockroaches [Tutsis] . . . sleeping Hutus, be prepared to be killed in your beds by cockroaches.[67]

At the time, Rwanda had one of the worst AIDS epidemics in the world; almost a third of the capital's residents between the ages of eighteen and forty-five carried HIV.[68] The disease was quite real, but Hutu extremists transferred fear from the pathogen to the Tutsi people.

In North Africa in 2011, the Libyan Civil War broke out. On February 23, "only" three hundred people had died in uprisings against the regime of Colonel Muammar Gaddafi. Then, Gaddafi ordered his supporters to attack the "cockroaches" and "rats" (referring to the protestors), and to "cleanse Libya house by house."

The Libyan envoy to the United Nations, Ibrahim Dabbashi, told reporters that Gaddafi's statement was "code for his collaborators to start the genocide against the Libyan people. It just started a few hours ago. I hope the information I get is not accurate but if it is, it will be a real genocide."[69] In the next four months alone, ten to fifteen thousand people were killed. The UN Human Rights Council found evidence that Gaddafi's forces had committed war crimes.[70]

Gaddafi's civilian enemies were human beings, of course, and not really cockroaches. But real cockroaches collected from hospitals and nearby households in Libya really do carry dangerous bacteria. Of a sample of cockroaches collected by zoologists in Tripoli, some 97 percent were carrying twenty-seven types of potential pathogens. Among them were bacteria resistant to at least six different antibiotics.[71]

Explicitly Genocidal Ideology

When pathogen metaphors escalate, they turn into explicitly genocidal ideologies. The Islamist Hamas Party, which was elected to power in the 2006 Palestinian elections, has a long track record of disseminating such incitement. In December 2002, for example, a poster appeared on Hamas's website that promised the rewards of Heaven to people who kill Jews. The text of the poster read: "I will knock on Heaven's doors with the skulls of Jews." An axe cut through the word "Jews," which was placed on a stump. Crania rolled on the ground.[72]

In December 2010, Hamas's official television channel, Al-Aqsa TV, broadcast a music video with lyrics that implored Allah to vanquish the enemies of Islam and "strike the Jews and their sympathizers" as well as Christians and Communists, saying, "Allah, count them and kill them to the last one, and don't leave even one."[73] By including Communists, the video promoted politicide as well as genocide.

Both the Hamas and Fatah political elites have encouraged mass political killings. But what do members of the Palestinian public think about political violence? Some polls have shown that a substantial majority of Palestinians agree with the killing of Jews. For instance, a poll published in the *New York Times* revealed that 84 percent of Palestinians supported the 2008 attack that killed eight religious high school students in Jerusalem.[74]

The Selective Separation and Destruction of Gene Pools

Occasionally, Harff's six factors align in some combination that actually spawns mass killings. According to her statistics, genocides and politicides occurred in 28 percent of all state failures in the latter half of the twentieth century.

In some mass killings, one ethnic group seeks the absolute extinction of another. These cases, in which women and children are killed at similar rates to men, are usually genocides. In other instances, the perpetrators disproportionately target enemy males, while leaving many females alive for reproductive ends. This second scenario often occurs in politicides, but it can happen in genocides as well.

In any scenario, the evolutionary "bottom line" is clear: *both genocides and politicides dramatically, rapidly, and disproportionately change the rates of destruction and reproduction of different gene pools.*

This reproductive logic gruesomely appeared in the French Wars of Religion. In the late sixteenth century, the aristocratic houses of France battled one another over the reformation of the Catholic Church and the rights of French Calvinists (the Huguenots). In 1570, a peace accord had ended the Third War of Religion. Nevertheless, the agreement was fragile because Catholic hardliners refused to accept the terms. At the same time, the Huguenots had a strong defensive position. In an attempt to cement the peace, the Queen Mother Catherine de'Medici planned to marry her Catholic daughter to the Protestant prince Henry of Navarre.

Six days after the wedding, one of Europe's worst mass killings of the century broke out. In the St. Bartholomew's Day massacre of 1572, between ten thousand and thirty thousand people died in the outburst of violence. Catholic mobs mutilated the genitals of male Protestants and cut open the pregnant bodies of female Protestants.[75] The actions taken by the Catholic extremists show that they intended not only to kill Protestants; they also destroyed the reproductive potential of many survivors.

In the Holocaust, Nazi Germany systematically murdered six million Jews, two to three million Soviet prisoners of war, two million Poles, one million Romani, and thousands of other religious, ethnic, and sexual minorities. The massive genocide began, however, with a state policy to sterilize and sometimes euthanize Germans whom the Nazis considered "inferiors."

The policy took effect in 1933. It led to the sterilization of three hundred thousand to four hundred thousand people suffering from conditions such as epilepsy, schizophrenia, mental retardation, and physical deformations. The Nazis sought to destroy the reproductive potential of these first victims because they believed that the congenitally ill carried deleterious genes. In the context of the 1930s and the height of the eugenics movement, Hitler's scientists hoped to artificially improve the German gene pool.[76] Perhaps Nazi geneticists feared the inbreeding depression that might follow after Germans could no longer reproduce with out-groups.

Two years later, the 1935 Nuremberg Laws forbade marriage between Germanic peoples and Jews, as well as extramarital sexual relations between them. Section 3 of the laws even barred Jews from employing female domestic workers of Germanic ethnicity under the age of forty-five. The intention was clearly to prevent contact between fertile German women and Jewish men.

In 1937, the Nazi regime launched a campaign to sterilize the so-called Rhineland Bastards. These children of German women were fathered after World War I by African troops who served in the French army's occupation of the Rhineland. Hitler called their mothers "whores and prostitutes."[77] These policies show that the Nazi regime intended to safeguard inbreeding, to prevent outbreeding, and to decrease the relative fitness of non-Germanic populations.

The Rwandan and Burundian genocides also show how different ethnic groups have sought to gain relative fitness advantages over each other. The fact that Hutus and Tutsis have different origins, however, has been obscured by a politically correct discourse, especially in the West. This narrative states that, once upon a time, there was no difference between the two groups, who formed part of a single tribe. Then German and Belgian colonialists arrived and introduced artificial distinctions and discrimination, which eventually led to genocide.

The truth, however, is even less politically correct. Intermarriage between Hutus and Tutsis for four centuries has reduced differences to a certain extent.[78] Still, the Tutsis have a different ancestral history. They likely migrated to central Africa from Ethiopia, or elsewhere in the Horn of Africa, in the sixteenth century. The Tutsis integrated fairly peacefully with the Hutus, but a society emerged that has been described as "Rwandan feudalism." This caste-like system, in which the Tutsis were at the top, dominated the socio-economic structure of precolonial Rwanda.[79]

Hutus and Tutsis have distinguished between one another based on physiological traits such as height, weight, hair, and facial features. According to social psychologist Neil Kressel, an expert on genocides, the average Tutsi is four inches taller and five pounds lighter than the average Hutu.[80] These differences, however, are difficult to scientifically attribute to differing genetic histories; intermarriage and semipermeable socio-economic classes could both confound perceptions.

Genetic studies, however, do corroborate the oral histories that the Hutus and Tutsis originated from different regions. Rwandan Hutus have a proportion of sickle-cell gene carriers similar to neighboring central African peoples. Tutsis, though, virtually lack this mutation. This absence of the sickle-cell mutation suggests that the Tutsis spent many centuries in a region with less malaria (or that they have one of the other mutations that resist the disease). In addition, the majority of Tutsis have a gene that helps them digest lactose; most Hutus lack this gene. The ability to digest milk products is typical of desert-dwelling pastoralists, which further substantiates the theory that Tutsis originated in East Africa.[81]

Psychologically, Hutus and Tutsis are conscious of their different origins and the socio-economic tension between the two populations. These perceptions come across in the words of a Burundian Hutu refugee, who described the nature of his hatred of Tutsis:

> In the past our proper name was Bantu. We [Hutus] are Bantus. "Hutu" is no tribe, no nothing! . . . *Muhutu* is a Kihamite [Tutsi] word which means "servant." . . . The name means "slave." We are not Hutu; we are *abantu*—human beings. It is a name that the Tutsi gave us.[82]

In this context, Hutus and Tutsis have committed genocide after retributive genocide against one another, going back to the independence of Rwanda and Burundi.[b]

How do politicides, such as Cambodia's mass killings between 1975 and 1980, selectively alter the destruction and reproduction of genes? The Khmer Rouge's main objective was to eliminate those they believed would not conform to their political ideology—that of the Communist Party of Kampuchea. In the process, however, Pol Pot's regime disproportionately destroyed Vietnamese, Cham, Thai, and Chinese genes.

Unlike the situation in Germany, where Nazi policy prohibited reproduction with the victims of their genocide, the Khmer Rouge *could* reproduce with the majority of their politicide's victims. Consequently, the gender ratio of the Cambodian victims is highly skewed. Males were targeted relative to females at a ratio of nearly two to one. The population structure of Cambodia remained shifted for many years. This gender imbalance would give a reproductive advantage to the surviving males—including the perpetrators of the politicide.[83]

Gender-biased killings have also occurred during more "conventional" genocides. During the Bosnian War, a leaked Central Intelligence Agency report stated that the Serbs had carried out 90 percent of the war crimes, and that they were the only belligerent party that attempted to "eliminate all traces of other ethnic groups from the territory."[84] Nevertheless, the Serbian perpetrators of the genocide sometimes targeted the males of other ethnic groups more than their females.

In 1991, Serbian militias committed the Vukovar Hospital Massacre in Ovčara. The perpetrators of this crime removed hundreds of Croats from a hospital and executed 263 of their men in a woodland ravine. Only one woman was killed. The following year in the Omarska Concentration Camp, Serbians castrated Bosnian Muslim and Croat men before killing them and throwing them into the river.[85]

The most infamous instance of gender-biased killing during the conflict occurred in July 1995. During the Srebrenica Massacre, Serb forces rounded up and executed eight thousand Bosnian Muslim men and boys.[86] What becomes of the "extra" women who are left alive?

Rape: A Weapon of Mass Reproduction

In mass political killings, as well as in wars, mass rape is a well-documented phenomenon. It's true that some victims of these rapes are males, and some are infertile females. Based on this fact, some people have argued that rape in the context of mass violence—or even in general—is about power, and not at all about sex.

The aim of this subsection here is not to argue whether or not this view is valid; rather, the purpose is to look at cold statistics to evaluate any evolutionary implications that may arise from this behavior. For instance, what percentage of the rape victims in wars and genocides are fertile females? How many become pregnant, thereby multiplying the genetic material of the rapist?

Rape is a weapon of mass reproduction. UNICEF has calculated that 67 percent of the females raped during the 1994 Rwandan genocide were between fourteen and twenty-five years old. Thirty-five percent of them were impregnated.[87] In several months of mass killings, more than 250,000 women were raped. Rwandans referred to the product of these rapes as "devil's children."[88]

During the Bosnian Genocide, records show that Serbian soldiers *intended* to impregnate their enemy's women. According to the US State Department's *Seventh Report on War Crimes in the Former Yugoslavia*, two Serbian soldiers raping a Muslim woman told her: "You should have already left this town; we'll make you have Serbian babies who will be Christians." Then they forced her to eat pork and drink alcohol, which are forbidden to observant Muslims.

Independent human-rights organizations have reported the case of a woman named Sofija who was raped every night by half a dozen Serbian soldiers for several months. Another woman had suffered twenty-nine rapes in one night before she lost consciousness. In the course of the conflict, Serb combatants committed between twenty thousand and fifty thousand rapes.[89] This behavior surely multiplied the DNA of the rapists in a substantial portion of cases.

Armed conflict has long plagued the Democratic Republic of Congo (DRC). The east of the country, which borders Rwanda and Burundi, has suffered in recent decades from spillover from the Hutu-Tutsi wars. Rape in this conflict-torn region of the Congo has reached record-breaking levels. The United Nations special representative on sexual violence in the conflict has called the DRC "the rape capital of the world."[90]

She's not exaggerating. In 2009, the UN reported that fifteen thousand women had been raped in the country.[91] Still, the UN's estimate may be extremely low compared to the actual number of rapes committed. In 2011, a team of American scientists published an article in the *American Journal of*

Public Health. Their study had found that the real number of rapes recorded between 2006 and 2007 was four hundred thousand—a far, far higher number than the official estimate of sixteen thousand. The four hundred thousand rapes, which occurred at an average of forty-eight per hour, included only those committed against females between the ages of fifteen and forty-nine. This is of course the demographic most likely to become pregnant.[92]

In February 2011, after Colonel Gaddafi's use of genocidal language, the Libyan Civil War erupted. Gaddafi's forces killed thousands of civilians. And they also raped them, as a matter of state policy. Credible reports emerged from high-level military defectors that Gaddafi's regime had distributed crates of Viagra to his mercenaries. The International Criminal Court's chief prosecutor, Luis Moreno-Ocampo, said that Gaddafi ordered the use of the erection-inducing drug so that his fighters could rape hundreds of women in rebel-controlled territories.[93]

According to Arafat Jamal of the UN's refugee agency (UNHCR), Gaddafi instituted the rape policy because of the particularly destructive effects that it has in his country's culture. When a rape occurs in Libya, according to Mr. Jamal, "a whole village or town . . . is seen to be dishonored." Gaddafi's forces have often raped girls and women in front of their male family members. The act is so psychologically devastating that rape victims often commit suicide. Others face "honor killings" at the hands of their fathers or brothers. Reports have also surfaced of rapists who have infected their victims with HIV.[94]

Whether rapists are fundamentally driven by power, sex, or the desire to destroy an enemy by impregnating its women, rape ubiquitously accompanies war and mass political killings. The great majority of rape victims in these circumstances are fertile women. Many of them bring pregnancies to term.

Disregarding the evil of the crime, nature rewards the rapists by multiplying their genes more than those of the nonrapists. Therefore, it's plausible that evolution has selected for a predisposition in some men to engage in this behavior under certain circumstances.

The same logic pertains to war and mass political violence: in each case, the perpetrators steal economic and reproductive resources from other groups, selectively altering the fitness of different populations. Because of the dramatic evolutionary advantages of these behaviors, war and genocide are natural phenomena. Their basic logic runs like a thread from chimpanzees, to humans living at a primitive level of political complexity, to the participants of historical and contemporary conflicts in every region. But to condone war, genocide, and rape, or even to consider them genetically determined or exempt from free will, would be to commit the naturalistic fallacy.

PART III.
DO WE LIVE IN A JUST WORLD?

Conflicts between ethnic groups cannot explain the entire left-right political spectrum. After all, full-blooded siblings sometimes have dramatically different political attitudes. If they share half of their genes, why do siblings often disagree so much?

In chapter 5 we broke down the content of the RWA test into three clusters of personality traits (see figure 10). The first cluster, colored in grey, pertained to tribalism. The second cluster, colored in black, is the topic matter that parts III and IV now address. This black cluster concerns varying levels of tolerance toward inequality and authority. These attitudes span across a spectrum, with egalitarianism on one end and hierarchy on the other.

The chapters of part III show how these egalitarian and hierarchical attitudes toward inequality correspond to left-right political orientation. Part IV then traces the evolutionary origins of these attitudes back to conflicts between members of the nuclear family (see figure 12).

The first chapter of part III (chapter 14) describes how differing levels of tolerance toward societal inequalities influence people's placement on the political spectrum. The second chapter (15) shows that tolerance toward inequality in the nuclear family also predicts left-right orientation.

14.

ATTITUDES TOWARD INEQUALITY AND AUTHORITY IN SOCIETY

Arthur Lucas was born in 1908 in the US state of Georgia, but history remembers him mostly for where and when he died.

In 1962, Mr. Lucas sat in a jail cell, waiting. The court had convicted him of murdering an undercover narcotics agent from Detroit who was scheduled to testify in a drugs trial. Circumstantial evidence convinced the jury that Lucas had traveled from Detroit to Toronto to commit the killing. The Canadian judge who presided over his case had sentenced Lucas with the words: "You shall be hanged by the neck until you are dead."

It was Lucas's fate to be executed with a small-time Canadian criminal named Ronald Turpin (figure 37 shows their mug shots). Ten months earlier, Turpin had held up the Red Rooster restaurant in Toronto for $632. The thief made a clean getaway—until a policeman pulled him over for a broken tail light. Turpin murdered the constable but was immediately apprehended.

On their last night, both men ate a dinner of steak, potatoes, vegetables, and pie, served on paper plates. Less than six hundred feet from their cells, Lucas and Turpin could hear the shouting of activists who had come to protest their imminent hanging. The anti-death-penalty protesters stood outside in –12°C weather, holding signs that read "Public Murder." Most were teenagers and people in their early to mid-twenties, including many college students.

The young demonstrators echoed a greater clamor in the contemporary public opinion of many democratic countries for the abolition of capital punishment. Before their hangings, the two convicts were informed that they would likely be the last two people ever executed in Canada. Turpin responded, "Some consolation." But Lucas reportedly told his chaplain: "We is lucky . . . if we were on the street, I could be killed by a car and I wouldn't be ready to meet my Maker."

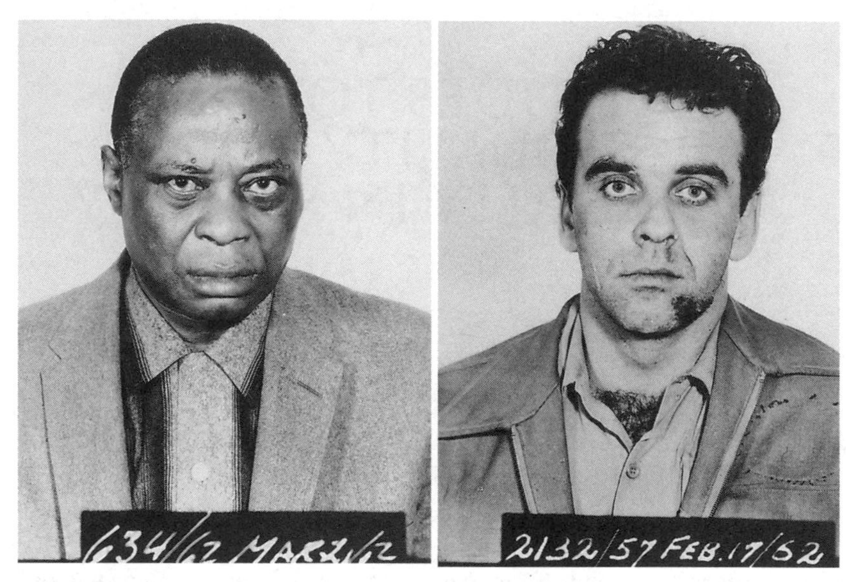

Figure 37. Arthur Lucas (*left*) and Ronald Turpin (*right*).

Two minutes past midnight on December 11, 1962, the hangman carried out the sentences. According to the chaplain, Lucas's head was "nearly torn right off," since the executioner had reportedly miscalculated his weight. After Lucas and Turpin's hangings, Canada commuted all death sentences to prison terms.[1]

This chapter's purpose is to explain why the anti-death-penalty protesters were likely liberals, and also what Lucas's and Turpin's contrasting comments suggest about each man's political leanings.

A half-century after Canada's last two executions, the majority of US states still use the death penalty to punish serious crimes. This practice has fascinated Stewart McCann, a Canadian psychologist at Cape Breton University. Professor McCann wondered why there was so much variation in the number of death sentences and executions carried out each year in the states with active capital-punishment statutes. Among these states, was there any relationship between the degree of societal threat and the number of condemned criminals?

To answer his question, McCann devised a measure he called "state threat." The degree of "threat" depended on a state's homicide rate, violent-crime rate, and the nonwhite percentage of the population. When McCann calculated the average level of "threat" for each state between 1977 and 2004, however, it had no power to predict the number of death sentences and executions.

McCann then entertained another hypothesis: that the number of inmates on death row depends on the average political orientation of a state. To determine the ideological leanings of each relevant state, McCann gathered the state-aggregated results of 122 national telephone polls of almost 142,000 individuals, carried out by the *New York Times* and *CBS News* during this same twenty-eight-year period. McCann discovered that in conservative states, higher levels of societal threat increased the number of death sentences and executions. In liberal states, however, the exact opposite phenomenon occurred: higher threat *decreased* death sentences and executions.[2]

McCann's discovery suggests that liberals and conservatives have fundamentally different views on crime and punishment. Indeed, when conflicts arise between societal authority and rule-breakers or dissidents, *conservatives tend to support the status-quo power, while the left generally sides with the non-conforming, weaker party.* The underlying principle here concerns the second universal personality cluster underlying political orientation: tolerance toward inequality and authority. The right has greater tolerance, while the left has less.

CRIME, PUNISHMENT, AND THE RWA SCALE

As shown in figure 10, one of the six content categories underlying the RWA scale elicits attitudes toward societal authorities. Altemeyer and his Berkeley-school predecessors have found that conservatives have especially negative attitudes toward weaker parties who challenge these powers or transgress their rules. Those on the right of the political spectrum are more likely to agree with the following RWA statements:

- "*Laws have to be enforced without mercy, especially when dealing with agitators and revolutionaries who are stirring things up.*"
- "*In the final analysis the established authorities, like . . . our national leaders, generally turn out to be right about things, and all the protesters don't know what they're talking about.*"

Leftists, on the other hand, advocate greater tolerance toward rule-breakers, protesters, and dissidents. They characteristically endorse RWA items such as these:

- "*It is important to protect fully the rights of radicals and deviants.*"*
- "*The self-righteous 'forces of law and order' threaten freedom in our country a lot more than most of the groups that they claim are 'radical' and 'godless.'*"*[a]

- *"It is best to treat dissenters with leniency and an open mind, since new ideas are the lifeblood of progressive change."**[b]

In addition to decades of administering the RWA scale, Altemeyer has also conducted supplementary work with the highest- and lowest-scoring quartiles of his test-takers. One such experiment involves giving both groups the chance to impose a prison term on people convicted of a given crime. Conservatives always choose longer sentences. Follow-up questions reveal that, compared with liberals, conservatives consider the same crimes to be more serious, and the perpetrators to be more "repulsive and disgusting." Conservatives think that punishment will be more effective in preventing future transgressions, and they derive greater satisfaction from punishing wrongdoers. Finally, conservatives believe that judicial leniency encourages more crime. Thus, they also tend to favor capital punishment.[3] This finding explains why conservative states sentence more people to death as violent crime rates increase.

The further one goes to the right of the political spectrum, the more pronounced attitudes toward crime and punishment become. Consider the words of a member of a radical right-wing hate group, documented by ethnographer Raphael Ezekiel. This young man from Detroit had the following opinion about murderers:

> It's sick, this is not right, you shouldn't have done it. You pay with your life. Shoot him or pull the rope that drops the trap door, and hang him, something perfectly legal. I would not be against that because I feel this man has done wrong where there's not excuses.[4]

The extremely conservative person who uttered these words would presumably have supported the executions of Lucas and Turpin. Statistically speaking, it's a safe bet that there were far more liberals than conservatives among those who did protest their hangings in 1962.

THE RIGHTS OF SUSPECTED DISSIDENTS

In real-life conflicts between governments and dissidents, it's not always immediately apparent who is innocent or guilty of criminality—or who may be a nonviolent political enemy of a regime. Sometimes a government suspects rebellion or wrongdoing but needs information to prosecute an individual. Governments have a number of high-handed methods for extracting this information. Conservative RWA scores moderately correlate with approval of these government techniques, including extrajudicial wiretaps, letter openings, and warrantless searches.[5]

A greater proportion of the political right also tolerates the extracting of information from individuals under physical duress, which the left opposes and considers torture. One such technique, known as waterboarding, simulates drowning. In his memoir *Decision Points* (2010), former US president George W. Bush defended this practice. Bush, a conservative, argued that waterboarding terrorist suspects effectively prevented attacks against Heathrow Airport and Canary Wharf in London, various targets within the United States, and American diplomatic facilities abroad. This practice also features prominently in the movie *Zero Dark Thirty* (2013), which relates how the waterboarding of terrorist suspects produced information that helped lead to the killing of Osama bin Laden. The left-leaning human-rights organization Amnesty International, on the other hand, rejects the utility of information obtained through torture, which it claims is "notoriously unreliable and inadmissible."[6]

When Barack Obama ran for president on the liberal, Democratic ticket, he pledged to shut down the Guantanamo Bay detention camp, which was associated with coercive techniques. Although Obama signed an order to close the prison on the second day of his presidency, more conservative elements of the legislature prevented the transfer or release of the jail's inmates.

TOLERANCE OF INEQUALITY, AND BELIEF IN A JUST WORLD

Of all the relationships between two human beings, few could entail as large a disparity in power as that between a torturer and a captured suspect. The torturer may work for a government that represents any segment of the political spectrum, although there's a higher chance that his or her employers occupy one of the extremes. Likewise, the suspected dissident might be a left-wing extremist, a right-wing extremist, or a centrist democracy activist. The only given is that the torturer holds an enormous amount of power, and the suspect, extremely little.

The same principle applies to the RWA items listed above. No context whatsoever surrounds the nature of the "*laws [that] have to be enforced without mercy*." We know not the political orientation of "*our national leaders*." By the same token, the statements offer no criteria for judging the grievances of the "*protestors*," "*agitators*," or "*deviants*" mentioned in the RWA statements. Nonetheless, a regular proportion of the test takers from Canada, the United States, Israel, the Palestinian Territories, South Africa, and numerous other countries sides with the societal authorities; another proportion sides with the weaker nonconformists. Regardless of the circumstances, the right and the left perceive power hierarchies through different moral lenses.

Princeton bioethicist Peter Singer has defined what he believes to be the essence of political ideology in similar, context-free terms. Singer once asked a Trotskyist, unionist, pro-Cuba, dyed-in-the-wool leftist why he had spent over fifty years working for left-wing causes. The friend responded that he was "on the side of the weak, not the powerful; of the oppressed, not the oppressor; of the ridden, not the rider." And "that," Singer thinks, "is what the left is all about."[7]

This leftist morality follows from the following premise: even though the world is filled with many unlucky souls, all people share a fundamental equality of moral worth and potential. Because leftists are more likely to believe in the innate, inner equality of all people, they attribute the world's inequalities to outer, structural injustices. In particular, the left sees many power hierarchies as unmerited and exploitative. Leftist morality is rooted in the imperative to equalize, to various extents, discrepancies in power (especially through education). Compared with conservatives, leftists have a lower tolerance for inequality.

In this leftist worldview, evil comes primarily from undeserved inequalities in strength or power: from capitalists who exploit workers, unscrupulous corporations that deceive consumers, colonialists who leach off third-world countries, soldiers and police who abuse civilians, men who mistreat women, humans who disrespect the animals and plants in their environment, and so on. From a leftist perspective, these power discrepancies do *not* necessarily reflect the inner moral worth of the parties on either end of the hierarchy.

When leftists show greater leniency toward petty criminals and dissidents, they view them as the weaker victims of more powerful forces. They would point out that Arthur Lucas was borderline retarded, with an IQ of only 63. Or that all the evidence against him was circumstantial, so antiblack prejudice could have biased the jury. And they would recall how both economically underprivileged men had to make do with an alcoholic lawyer who came to court with a hangover.[8] In short, leftists tend to focus more on outer, structural inequalities as the cause of crime and dissidence.

Finally, the left is more prone to the "moralistic fallacy" described in the previous chapter. That is, leftists sometimes assume that their moral beliefs about the way the world should be have a greater impact on reality than they actually do. For example, a secular leftist might oppose religion, and believe that the world would be a better place without it. This individual could commit the moralistic fallacy by underestimating the actual religious motivation of religious extremists in other parts of the world and overattributing the cause of their discontent to external inequalities (such as poverty or colonial history). Another example of the moralistic fallacy is when people who consider torture unethical claim that this practice never produces any actionable intelligence.

Conservatives, on the other hand, are much more likely to believe that hierarchies reflect inner, individual capabilities—be they moral, intellectual, or physical. Inequalities, according to this worldview, reveal the worth of the powerful and the weak, of "winners" and "losers." Thus, the right has a high tolerance for inequalities.

From this perspective, the poor have fewer resources because they lack the intelligence or moral strength to acquire them. People commit crimes out of moral weakness, or because they believe they can escape punishment. By this logic, harsher penalties and better law enforcement reduce crime. People like Lucas and Turpin, who expire at the gallows, sink to the bottom due to their own inner flaws. They thought that they could cheat the system at the expense of society, and they deserve their fate. The real victims, conservatives would argue, are the widow and four daughters left behind by the policeman that Turpin murdered and the FBI agent killed by Lucas.

From the rightist perspective, evil also comes from those who seek to artificially manipulate or subvert legitimate power hierarchies. University of California, Berkeley, metaphor theorist George Lakoff believes conservatism follows the logic that "The Moral Order is the Natural Order." This dictum describes an old idea formerly known as the "Great Chain of Being," whereby the most powerful and noble beings deservedly occupy the top of a moral ladder; in this descending order of greatness, they are God, men, women, children, animals, and plants. Each rung of the hierarchy meritoriously has authority over subordinates.

In this conservative moral framework, goodness comes from obedience to a higher power. Evil follows from disobedience. According to Lakoff, the right demonizes people who invert the order. For instance, atheists try to put people above God. Feminists and lesbians represent women trying to play a role equal to or greater than that of men. Homosexual men take on the role of women. Environmentalists prioritize the interests of plants and animals over those of humans and their corporations.[9]

Therefore, the political right is more susceptible to the "naturalistic fallacy" described in the previous chapter. That is, conservatives tend to believe more than most people that the way the world is, is equivalent to what is morally good.

Political psychologists call the conservative attitude toward power hierarchies a "Belief in a Just World" (BJW). The BJW asserts that "the world is a fair place wherein people get what they deserve and, often, deserve what they get." Social psychologist Melvin Lerner first postulated the just-world hypothesis in the 1960s. Lerner was originally interested in the psychological tendency of some people to blame the victims of crimes or accidents for their own misfortune. He originally theorized that this thought pattern served as a defense mechanism against the anxiety-provoking idea of living in an unjust world. If you've ever read about a horrible tragedy in the news, and then a little voice in the back of your head suggested that the victims might have done something to deserve their fate, then that would be a sign of BJW.

Individual differences in "Belief in a Just World" significantly correlate with political conservatism.[10] Someone like Arthur Lucas apparently lies on the conservative end of the BJW spectrum, as judging by his comment that he was lucky to be the last person to be killed by execution in Canada because he'd have a chance to prepare himself to "meet [his] Maker." We would have to assume that Lucas was more conservative than Turpin, who implied that being the last person to hang was like a cruel joke in an unjust world.

As we'll see below, people who score highly on a BJW scale are much more likely to be political conservatives, and to deem natural, social, political, and economic inequalities as morally just.

Acts of God

Sometimes disparities in power result from natural disasters. Other times mass political violence at the hands of humans increases hierarchies. In both cases, people with a high BJW and high religiosity justify the ensuing inequalities by attributing their cause to the will of God, or to karma.

History abounds with examples where religion sustains the belief in a just world, even in the face of overwhelming disasters. In the year 1096, European crusaders traveled to the Middle East seeking to liberate the Holy Land from Muslims. Along the way, they slaughtered Jews living in German towns. Both the Christian perpetrators and the Jewish survivors used religious thinking to maintain their belief in a just world. The crusaders blamed the victims for their own deaths by accusing the Jews of killing Jesus. A Jewish chronicle written shortly after the massacre, on the other hand, ascribed the pogrom to sins that "stretched back to the days of Moses in the wilderness."[11]

Leaders of more recent times have also attributed significant geopolitical events to the will of God. One of them is Pat Robertson, a Southern Baptist televangelist who unsuccessfully sought the Republican nomination for the 1988 presidential election. When a political extremist assassinated Israeli prime minister Yitzhak Rabin in 1995, Robertson claimed that Rabin died because he thwarted God's plans for continued war in the region. A decade later, Israeli prime minister Ariel Sharon suffered a devastating stroke that left him in a permanent vegetative state. Robertson insinuated that God had punished Sharon for his withdrawal from the Gaza Strip.

Incidentally, why did an American Christian leader care so much about divine influence over political events in the modern state of Israel? Robertson belongs to the thirty million American Christians called Armageddonites. This group vehemently opposes the division of "God's land" between Jews and Muslims. Rather, they pray for the Jews remaining in the Diaspora to return to Israel. Complete Jewish integrity and political control of the Promised Land, they believe, will hasten the arrival of Armageddon. Many Armageddonites yearn for the end of the world, when they believe born-again Christians will be "raptured" up to Heaven.[12]

Meanwhile, while the world kept turning and globally warming, Israel experienced the deadliest fire in its history in December 2010. The Mount Carmel Forest Fire caused the evacuation of over seventeen thousand people, and took the lives of at least forty-four people. Before an international effort succeeded in extinguishing the blaze, a controversial Israeli leader attributed the disaster to divine punishment. The Iraqi-born rabbi Ovadia Yosef serves as the spiritual leader of the Shas political party, which represents Orthodox Jews from diasporas in the Middle East and North Africa. During a weekly sermon, Rabbi Yosef read an excerpt from the Babylonian Talmud stating that "fire only exists in a place where Shabbat is desecrated." With these words, the spiritual leader of a right-wing religious party implied that the disaster occurred because the residents of the Carmel region had failed to keep the Sabbath.[13] Although one area of the country suffered from the fire while others did not, the religiously conservative Yosef preserved the possibility of a just world by blaming the victims for incurring God's punishment.

Wealth and Private Property

One of the greatest inequalities between human beings is the vastly different amount of wealth and private property that individuals accumulate. Leftists and conservatives have conflicting explanations for and moral judgments of these disparities, but the existence of staggering economic inequalities is an indisputable fact.

In the year 2000, the richest 2 percent of adults in the world owned over half of global household wealth; and the richest 10 percent owned 85 percent. By contrast, the poorest half of the world's adults owned only 1 percent of global wealth.[14]

Enormous economic disparities also exist *within* countries—even in rich developed states. In America, these inequalities have increased dramatically during the end of the twentieth century and the beginning of the twenty-first. In 1978, the top 1 percent of US families netted 9 percent of the country's income. By 2007, their share rose to 23.5 percent. Stanford University political scientist Francis Fukuyama credits this growing economic gap to "a period of conservative hegemony in American politics. Conservative ideas," he asserts, "clearly had to do with the rise in inequality."[15]

Indeed, conservatives accept inequality as legitimate much more readily than do leftists. Research shows that, the higher an individual's belief in a just world, the more they attribute poverty and wealth to character and internal causes, and the less they consider external factors such as structural injustices.[16]

New Yorker cartoonist Robert Mankoff published a classic cartoon on economic BJW in 1981. In his sketch, a large fish opens its mouth to eat a medium-sized fish, which in turn is about to eat a little fish. The little one is thinking, "There is no justice in the world"; the medium fish thinks, "There is some justice in the world"; and the big fish thinks, "The world is just."

Do "big fish" truly have the highest BJW? In reality, BJW correlates with the left-right spectrum, but it does *not* correspond well to the size and power of "fish" (i.e., the resources that people have at their disposal). If there is any relationship with income, it may be the reverse of the Mankoff cartoon. A 1993 study of Southern Californians found that the people of the lowest socio-economic status (the smallest fish) had the highest belief that "the world is just."[17] Since poverty increases religiosity (see chapter 7), and religiosity reinforces BJW, perhaps poverty increases BJW through religiosity.

If, hypothetically, the poor were to have the greatest belief in an *unjust* world, then the fifty smallest fish in the tank would easily destroy the two largest fish that eat half of the resources.

Because the belief in a just world corresponds to political orientation, governments on different ends of the left-right spectrum have opposite attitudes toward the ownership of private property. Regimes on the far left view economic inequalities as highly unmerited and exploitative. Therefore, socialist and Communist states often expropriate property from wealthy individuals and their private corporations. The stated goal of such expropriations is to equalize the ownership of wealth by redistributing it directly, or by nationalizing a company and later redistributing its profits. For example, the socialist government of Bolivian president Evo Morales has nationalized foreign-owned oil, gas, and electric companies to redistribute the wealth that they generate to the country's impoverished majority. Under Morales's Movement for Socialism party, the Bolivian state has seized the assets of primarily British, French, Spanish, and Brazilian companies.[18]

In Venezuela, Hugo Chavez's Bolivarian Revolution for twenty-first-century socialism has entailed what Chavez called an "economic war" against the "bourgeoisie." In 2010 alone, Chavez's far-left government expropriated over 200 businesses spanning across the financial, ranching, housing, food, and industrial sectors.[19]

Governments on the far right of the political spectrum, on the other hand, have had a much more positive attitude toward private property—along with the inequalities that it entails. Between 1970 and 1973 Chile had a Marxist government led by Salvador Allende. During Allende's far-left government, the country nationalized many industrial and agricultural properties in the name of equality. However, a coup removed Allende from power in 1973 and installed the far-right military dictatorship of Augusto Pinochet (1973–1990). One of Pinochet's key economic policies was the privatization of numerous state-owned industries, many of which Allende's government had previously nationalized. Economic inequality increased dramatically under Pinochet's authoritarian regime (as did the torture and politicidal disappearances of dissidents). But Pinochet's policies of privatization and economic liberalization also attracted an inflow of capital from multilateral and private financiers. Whether these measures contributed to the long-term economic growth known as the "miracle of Chile" has, predictably, been a controversial topic subject to much debate.

Government Spending on the Poor

Another means that governments have at their disposal to control the degree of economic inequality in their countries is the authority to impose heavier taxes on the rich (which is called "progressive taxation") and then to redistribute a greater proportion of this revenue to the poor. The amount of the tax burden on the wealthy, and the social spending on the poor, constitutes a tremendously contentious political issue.

The left favors a fiscal policy aimed to increase economic equality. That is, heavier taxes on the rich, and more social spending on the poor. The right typically supports measures that maintain or increase economic inequality (lighter taxes on the rich, and less social spending on the poor). At the 1992 Republican National Convention, the topic of progressive taxation arose. In reference to this issue, George H. W. Bush's running mate, the future vice president Dan Quayle, asked: "Why should the best people be punished?"[20] Quayle's argument against progressive taxation reflects a high BJW: if the largest incomes accrue to the most worthy individuals, those individuals ought to keep their well-earned reward.

Conservatives also argue against greater taxes on the rich because they consider wealth to be the incentive that motivates people to work. By the same logic, government assistance of the poor lessens the poor's incentive to productively contribute to the economy. Research shows that conservatives increase their opposition to government assistance to the disadvantaged even further if the beneficiaries do not show an effort to help themselves.

Liberals, in contrast, believe more in the goodwill of all social segments. In the liberal mind, unemployment results less from a lack of incentive and more from the presence of social injustice, such as discrimination. Thus, liberals in the United States favor social spending on the poor in general, and to an even greater extent when the recipients belong to underprivileged minority groups.[21]

In some cases, conservatives have worried that government social spending on the poor could increase poverty by giving the lower class an undue reproductive advantage over the middle class. Howard Flight, a right-wing member of the United Kingdom's House of Lords, once warned a newspaper reporter: "We're going to have a system where the middle classes are discouraged from breeding because it's jolly expensive, but for those on [welfare] benefit there is every incentive. Well that's not very sensible."[22]

Whether social spending causes the poor to have more children is a thorny, complex question. In any case, greater access to resources increases fitness in one way or another—whether in the quantity of children a family can raise, or in the quality of the nutrition and education that parents can provide. The important premise underlying Flight's statement, however, is that the poor are less worthy to reproduce than the middle class by virtue of their lower economic standing. This just-world assumption offended liberals in the United Kingdom, who did not believe that the poor are less worthy human beings, or even that they reproduce according to government incentives.

When Howard Flight made these controversial remarks about welfare in 2010, the liberal opposition compared them to an even more explicit denigration of the poor made by a member of the Conservative Party in 1974. Sir Keith Joseph, credited as the kingmaker of Thatcherism, uttered the following words in a speech that year:

> The balance of our population, our human stock is threatened. . . . A high and rising proportion of children are being born to mothers least fitted to bring children into the world and bring them up. They are born to mothers who were first pregnant in adolescence in social classes 4 and 5. Many of these girls are unmarried, many are deserted or divorced or soon will be. Some are of low intelligence, most of low educational attainment. . . . They are producing problem children, the future unmarried mothers, delinquents, denizens of our borstals, sub-normal educational establishments, prisons, hostels for drifters.[23]

Sir Keith Joseph left no doubt how he felt about the poor: that they are intellectually and morally inferior, and that they should not reproduce.

Some far-right governments in other parts of the world have felt the same way, and taken it into their own hands to forcibly sterilize the poor. Between 1996 and 1998 in Peru, the government of Alberto Fujimori sterilized some three hundred thousand women from the lowest socio-economic classes as an "antipoverty measure."[24]

The Justification of Rank, Social Status, and Political Authority

If believers in a just world attribute poverty to the personal flaws of the poor, they also attribute the status of high-ranking people to their individual virtues. According to this logic, the sources of success are superior intelligence, strength, or moral qualities.

In his autobiographical work *Mein Kampf*, Hitler took his far-right-wing political ideology to a logical extreme. The following excerpt reflects his radicalization of the just-world philosophy, applied to political power: "Nature does not know political frontiers. She first puts the living beings on this globe and watches the free game of energies. He who is strongest in courage and industry receives, as her favorite child, the right to be master of existence."[25] In short, Hitler argued that might equals right.

Even much more centrist conservatives are more likely than the left to admire physical, sexual, or military strength in their political leaders. In 2010, supporters of the right-wing United Russia Party produced an erotic calendar in praise of their leader Vladimir Putin on his fifty-eighth birthday. In the pin-up calendar, young women from Moscow State University posed in lingerie. One of the captions from a scantily clad coed flirted with Putin: "You put out forest fires, but I'm still burning." Miss February asks: "How about a third time?" Her question ambiguously referred to a third term in office. Images of Putin's physical strength appeal to his supporters. The Russian prime minister has appeared on television riding shirtless in the mountains and swimming in a Siberian river.[26]

A photo of Putin sitting in the cockpit of a helicopter evokes a similar image to that of his contemporary US counterpart, the conservative president George W. Bush. In 2003, Bush gave his 2003 "Mission Accomplished" speech on board the aircraft carrier USS *Abraham Lincoln*. The speech announced (years prematurely) the end of major combat operations in Iraq. For his televised arrival on the ship, President Bush posed for photographers while wearing a flight suit—a symbol of masculinity and military strength.

Believers in a just world, then, tend to admire—and show submissiveness to—strong, powerful, and high-ranking individuals. At the same time, the BJW mindset belittles poor, weak, and low-ranking people. One of the authors of the *Authoritarian Personality* was the German-born sociologist Theodore Adorno. Adorno pointed out that a German folk expression captured these characteristics of BJW individuals: the word *Radfahrernaturen* means "bicyclist's personality." The term refers to the fact that above the waist, bicyclists bow, while below, they kick.[27]

Body gestures can be quite revealing. On the far right of the political spectrum, an extended-hand salute shows deference to a high-ranking, powerful leader. This salute was adopted by Nazi Germany in the 1930s to show obedience to Adolf Hitler. Schoolchildren's socialization in those years involved learning to give it to their teacher. The same salute is now used in the Shia Islamist Hezbollah Party, which numerous Western countries and Israel have classified as a terrorist organization.

The far left, on the other hand, does not believe in a just world. Rather, its egalitarian ethic opposes hierarchy. Therefore, Marxists, socialists, and trade unionists have adopted as their gesture a raised clenched fist, which they symbolically punch up against power. One can find this raised fist all over the world—from Naxalite Communists in India who give the "red salute" (*lal salaam*), to the clenched fist in the logo of the Socialist Party of Malaysia.

CARROTS AND STICKS

At the beginning of this chapter, we learned that conservatives have a more negative attitude than the left toward "criminals," "agitators," "deviants," and "protesters." Conservative believers in a just world tend to favor the rules and leaders of the powerful state instead of the weaker challengers to government authority. The left, in contrast, shows greater tolerance toward rule-breakers and policy-challengers. Liberals take a softer stance because they more readily attribute the motives of dissidents and rule-breakers to outer, structural inequalities; conservatives take a hard stance on dissent because they see it as the product of inner moral weakness.

Consequently, those on the left and those on the right disagree about the effectiveness of punishment in resolving conflicts. Conservatives often consider punishment preferable—especially when a disparity in power exists between two parties. Leftists are less likely to find a weaker power at fault in a conflict, and they therefore advocate the settling of disputes through the balancing of perceived injustices or through positive incentives.

Former US secretary of state James Baker mentioned these two approaches once in an interview, when asked how diplomats manage conflicts between countries. His interviewer was a faux-streetwise, wannabe Anglo-Jamaican poseur named "Ali G," who had asked Secretary Baker to contribute to an educational program for young people. Unbeknownst to Baker, "Ali G" was actually a fictional satirical television character performed by the comedian Sacha Baron Cohen. The following is an excerpt from their dialogue:

> Ali G: Ain't they the same thing though?
> James Baker: No, they're two different countries, Iraq and Iran.
> Ali G: Do you think it would be a good idea if one of them changed they name to make it very different sounding from the other one?
> James Baker: No, because "Iran" doesn't sound like "Iraq."
> Ali G: Ain't there a real danger that someone like gives a message over the radio saying "Bomb Ira . . ." and the geezer don't hear it properly and bomb Iran rather than Iraq?
> James Baker: No danger.
> Ali G: How do you make countries do stuff you want?
> James Baker: *You deal with carrots and sticks.*
> Ali G: But what country is gonna want carrots even if there is like a million carrots you be givin' over to them? (emphasis added)[28]

In reality, Baker's dominant country has experienced military and diplomatic conflicts with the less powerful states of both Iraq and Iran. In the course of these clashes, conservatives tend to support the threat of physical punishment (i.e., sticks); liberals, as a group, favor more positive incentives (i.e., carrots) to "make countries do stuff they want."

Iran, for instance, has become entangled in a conflict with numerous countries over its illegal nuclear-weapons program and its support for Islamist militant organizations, such as Hezbollah. The United States, Israel, and many European and Arab countries want Iran to stop these hostile military endeavors. The right of the political spectrum in these countries believes that Iran will only respond to threats or physical force. The left hopes that carrots might convince the Iranian regime to desist from weaponizing enriched uranium.

A diplomatic cable released by WikiLeaks in 2010 quoted the conservative European Union official Robert Cooper as saying: "Iran needs to fear the stick and feel a light 'tap' now."[29] Center-left US president Barack Obama, however, began his engagement with Iran through a different approach. In the first sit-down interview of his first term (in January 2009), President Obama spoke to the pan-Arab satellite network Al Arabiya about American-Iranian relations: "If countries like Iran are willing to unclench their fist, they will find an extended hand from us," he promised. The hypothetical "carrots" in Obama's unclenched fist would ostensibly include economic aid and the lifting of sanctions.[30]

When faced with a heightened diplomatic conflict with Iran, Obama's first instinct was to reach out his hand. As a liberal politician, he made this egalitarian gesture instead of making a hierarchical threat. Indeed, liberals generally believe that force and violence exacerbate conflicts. This belief might explain why liberal US states *reduce* death sentences and executions as crime rates escalate.

When conflicts arise between politicians and their ideological adversaries at home, the right is more likely than the left to use "sticks," or to threaten the use of violent force—albeit metaphorically. Carl Paladino, the 2010 Republican nominee for the New York gubernatorial election, once bragged that he would "clean out Albany [the state capital] with a baseball bat." Paladino also threatened to "take out" a reporter from the *New York Post* who had inquired about the politician's mistress and their daughter.[31]

In 2010, Ben Quayle (the son of former vice president Dan Quayle) successfully campaigned for election to the House of Representatives in Arizona. The conservative candidate declared Obama "the worst president in history" and promised to "knock the hell out of" Washington.[32]

On the far right, Tea Party–supported political commentator Glenn Beck has discussed poisoning Democratic speaker of the house Nancy Pelosi, choking leftist filmmaker Michael Moore, and beating to death liberal Congressman Charles Rangel with a shovel.[33]

In early 2010, the Democrats passed a controversial healthcare-reform bill in the United States. Conservatives opposed the law, which they perceived would increase government spending on insuring the poor. Sarah Palin, the VP candidate in the ill-fated 2008 Republican presidential campaign, retaliated by literally targeting pro-reform Democrats standing for reelection in vulnerable districts. Palin tweeted: "Commonsense Conservatives & lovers of America: 'Don't Retreat, Instead—RELOAD!' Pls see my Facebook page." On her Facebook account, Palin had posted a map of the United States with gun-sight crosshairs marking these liberal districts.

Democratic representative Gabrielle Giffords of Arizona protested: "We're in the crosshairs of a gun sight over our district. When people do that, they've got to realize that there are consequences to that action."[34] In spite of Palin's campaign to remove Giffords, the latter won reelection in the 2010 mid-term elections. Two months later, however, a mentally disturbed gunman shot a bullet through Giffords's skull and injured her brain.

On the extreme far right of the spectrum, politicians' threats against their left-wing colleagues grow even greater. Tom Metzger, the founder of White Aryan Resistance (WAR) and a Grand Dragon of the Ku Klux Klan, ran for a seat in Congress in 2010 to represent Indiana's 3rd district. During his campaign, Metzger told a local news outlet: "I'd go to Washington and get into Congress and have a fistfight every day."[35]

Why does the right favor physical punishment more than the left as a conflict-resolution method? In the next chapter, we'll explore the use of authoritarian sticks and egalitarian carrots within conservative and liberal families—both in the United States and around the world.

15.

ATTITUDES TOWARD INEQUALITY AND AUTHORITY WITHIN THE FAMILY

In August 2005, the SurveyUSA polling firm conducted a nationwide study to find out how American adults felt about the corporal punishment of children. Figure 38 shows the proportion of adults across all fifty states that agreed that "*it is OK for a school teacher to spank a student*."

The bars colored in black show the states that voted for the conservative candidate in the 2004 presidential election; the grey bars correspond to the states that voted for the liberal candidate. The most striking feature of the graph is that every one of the twenty-five states above the median level of spanking approval voted conservatively.

These findings reverberated within academia and the media. Political scientists Marc Hetherington of Vanderbilt and Jonathan Weiler of the University of North Carolina, Chapel Hill, began a book called *Authoritarianism and Polarization in American Politics* (2009) by pointing out this very strong correlation between spanking approval and state partisanship.[1] The following year, journalist Nicholas Kristof publicized the same finding in his *New York Times* column.[2]

The solid statistical link between spanking approval and political orientation may have been newsworthy to public opinion in 2010. But the Berkeley authors of *The Authoritarian Personality* had famously made this discovery sixty years earlier. Altemeyer's continuation of their work, leading to his development of the RWA scale in the 1980s, replicated the finding that attitudes toward family disciplinary strategies have great power to predict political ideology. Right-wing test-takers are predisposed to agree with the following RWA statements:

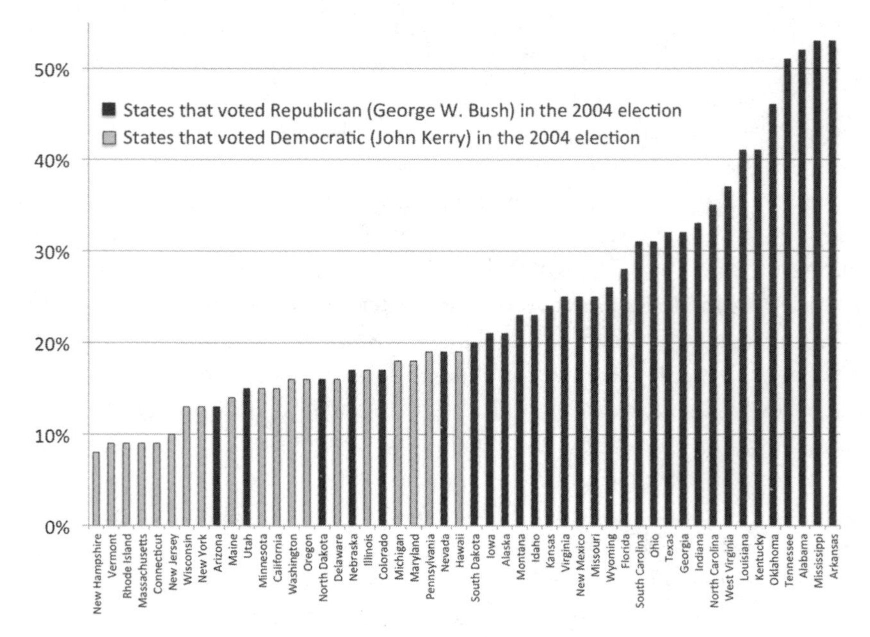

Figure 38. Percentage of Adults (18 and Older) Who Agree It's "Okay for Teacher to Spank Student" (2005), Compared with the 2004 US Presidential Election Results.

- "*One reason we have so many troublemakers in our society nowadays is that parents and other authorities have forgotten that old-fashioned physical punishment is still one of the best ways to make people behave properly.*"
- "*Obedience and respect for authority are the most important virtues children should learn.*"
- "*If a child starts becoming unconventional and disrespectful of authority, it is his parents' duty to get him back to the normal way.*"

According to Altemeyer, liberals are more likely to have parents who "modeled equality within the family" and who "nurtured a 'question and decide' approach to life rather than [a] 'memorize and obey'" one.[3] These left-of-center test-takers tend to endorse RWA items such as the following:

- "*The sooner we get rid of the traditional family structure, where the father is the head of the family and the children are taught to obey authority automatically, the better. The old-fashioned way has a lot wrong with it.*"*

- "*Students in high school and university must be encouraged to challenge their parents' ways* . . . ."*

These RWA items about inequality and authority in the family parallel the test's statements about inequality and authority within society (discussed in the previous chapter). In both cases, these questions polarize the political spectrum in the same way because they test for belief in a just world (BJW). The right, which has a higher BJW, believes that inequalities and power are generally justified; therefore, conservatives tend to advocate strict punishment for criminals and dissidents who challenge societal rules, and they also tend to support physical discipline for children who disobey family authority. The left, in contrast, believes in a world filled with unfair inequalities; consequently, leftists have a less favorable view of authority—both in society and in the family.

A comical example illustrates these contrasting attitudes toward family authority. Political scientists asked over sixteen hundred adults from the United States, United Kingdom, Argentina, and Canada the following question: "Would you be willing to slap your father in the face, with his permission, as part of a comedy skit?" They found that the thought of physically punishing a father—even under these hypothetical, approved circumstances—distressed conservatives significantly more than liberals.[4]

Let's take a closer look at where family inequalities come from and why they polarize liberals and conservatives.

THE CARROTS AND STICKS OF FAMILY DISCIPLINE

Within a nonimmigrant, two-generation nuclear family, the parents are indisputably more powerful than the children. The older generation has an advantage in size, strength, age, knowledge, intelligence, social network, and access to resources. Thus, extreme inequalities prevail in families, just as in society.

For reasons that will be explained in part IV, conflicts sometimes arise between parents and their children. Parents have different means at their disposal to deal with insubordinate children. These means come in two basic disciplinary flavors: hierarchical and egalitarian moralities. Attitudes toward these moralities, such as those expressed by the RWA items above, correlate very well with left-right political orientation.

Within the family that conservatives approve of, the parents use a hierarchical morality to coerce their children to conform to the parents' desires. Parents particularly leverage their external, physical advantages in power to resolve conflicts of interest. The key incentive for children to behave is to avoid parental anger, threats, and punishment. When parents in these families do discipline their children, punishments are more likely to have an aggressive nature. That is, they prefer the proverbial "stick" of corporal punishment and spanking. Thus, fear serves as a key motivator. In families governed by authoritarian parents, children usually feel relaxed and enjoy playing more when one or both strict parents are away from home.

Conservative families also emphasize rewards for good behavior. These positive incentives are sometimes external. They may comprise objects, money, or privileges, in addition to positive emotional feedback.

Liberals, on the other hand, prefer egalitarian families that mediate conflict quite differently, without as much regard for exterior inequalities. Instead, egalitarianism emphasizes an intrinsically equal worth of individuals based on interior qualities. According to George Lakoff, liberal adults assume that children learn and obey through emotional attachment to their parents more than through external, physical reward and punishment. In other words, leftists believe that "bonds of affection and earned mutual respect are stronger than bonds of dominance."[5]

To motivate desired behaviors, liberal parents use guilt, which they can evoke by withdrawing affection. If you have ever felt as though you were penetrated by a "judgment ray" (making you feel guilty but not fearful), that would be a sign of egalitarian coercion. In egalitarian families, children sometimes feel sad when parents are away from home because the physical absence of parents can simulate the withdrawal of affection.

The last RWA item above implies that liberals approve of children *challenging* their parents. Why? According to Lakoff, open two-way communication is essential in an egalitarian home: "If parents' authority is to be legitimate, they must tell children why their decisions serve the cause of protection and nurturance." This logic makes children's questioning a desired trait.[6]

Hierarchical disciplinary morality, in contrast, discourages excessive questioning by children. Instead, children are expected to accept their parents' views about what is correct.[7] If, as a child, you have frequently heard "Because I told you so!" from a parent, this would be a sign of more conservative discipline. Hierarchical morality requires obedience based on the idea that the parent ranks higher than the child in size, strength, intelligence, and presumably wisdom.

Both types of families use "carrots" in a variety of ways. But the two disciplinary moralities differ most notably in their negative incentives. Hierarchical parents motivate child behavior mostly through fear; egalitarian parents do so mainly with guilt.

What is fear? Fear is the emotion people feel when they think about a specific threat that they seek to avoid (such as being spanked for breaking a rule). The maladaptive sibling of fear is anxiety, which can take on a generalized form that pertains less to a certain object or time; a person in this state of anxiety often does not know how to respond.[8]

What is guilt? People feel guilty when they think they have acted immorally or have hurt someone and worry that they may lose that person's favor. Guilt motivates a child to behave better toward a parent for fear of losing the parent's love. Like fear, guilt also has a maladaptive sibling, whose name is depression. Depression occurs when a person attributes guilt to him or herself and perceives it as a permanent condition, as opposed to being situational and absolvable.

Interestingly, anxiety and depression are the two most common mental-health problems worldwide.[9] Another of the most prevalent disorders involves the abuse of substances, which people often use to self-medicate these first two conditions. For example, in spite of its negative health impacts and externalities, alcohol is a great drug for reducing anxiety in many people. Later in this chapter, we'll explore a connection between alcoholism and extreme hierarchical parenting.

None of this is to suggest that fear and guilt lack adaptive functions. A hunter who does not fear large game could be killed in the hunt. A hunter who feels no guilt for hoarding away the kill from the rest of the band might be ostracized or worse. It's just that parents manipulate these emotions, often unconsciously, to discipline and control their children. The relatively high levels of maladaptive anxiety and depression around the world suggest that these mental afflictions may have some connection to family conflict. The origin of these mood disorders could be either biographical (for example, having an abusive parent) or genetic (inheriting an imbalanced personality trait that might have served ancestors better in a more egalitarian or hierarchical environment).

To summarize, family disciplinary strategies come in two flavors: hierarchical and egalitarian. It's clear that a preference for the former is one of the strongest predictors of political conservatism, while liberals prefer egalitarian morality. These attitudes toward authority reflect variation in people's tolerance of inequality.

What's *not* nearly as clear is whether actual biographical events from one's childhood and socialization affect one's adult political orientation, as the Berkeley authors of the *Authoritarian Personality* asserted. Altemeyer avoided this quandary by focusing exclusively on a subject's present-moment *attitudes* toward family discipline. By doing so, he showed that his RWA scale could work even in the absence of psychoanalytic premises of childhood traumas. In a sense, Altemeyer's simplification was an important advancement that made the RWA test more scientific and more widely accepted. But a fascinating question remained unanswered: does one's childhood environment influence adult political views?

There are many problems in trying to determine how an individual was actually disciplined as a child. A person's introspective reports about childhood discipline (or even those of their parents) might not be accurate. Memories can be subjective or forgotten or sanitized. They are difficult to confirm or standardize. Most importantly, family discipline is hard to measure, because doing so would involve intruding on the relationship between parents and young children—and then waiting a couple decades to look for a potential political impact.

But at least one scientific study has done exactly this. Psychologist R. Chris Fraley, at the University of Illinois, Urbana-Champaign, and his colleagues studied seven hundred children from a longitudinal study funded by the National Institute of Child Health and Human Development. Between the ages of six and fifty-four months, mother-child interactions were videotaped while the children tried to perform tasks just beyond their abilities. Observers then rated the mothers on their sensitivity in their interventions. Those children who had more sensitive mothers scored lower in a test of conservative ideology at the age of eighteen years. This relationship was very weak but still significant.

But something else did have a larger impact. When the children were one month old, researchers measured their parents' attitudes toward discipline. The questions were quite similar to the RWA items listed at the beginning of this chapter. Parents defined as having authoritarian attitudes agreed with statements like, "*Children should always obey their parents*," and "*The most important thing to teach children is absolute obedience to whoever is in authority*." Eighteen years later, their children were more likely to be conservative. Egalitarian parents, who thought that "*children should be allowed to disagree with their parents if they feel their own ideas are better*," and that "*a child's ideas should be seriously considered in making family decisions*," were more likely to have liberal children eighteen years later.[10]

The correlation between childhood discipline and adult ideology does not explain whether nature or nurture is more responsible for political ideologies. Exposure to maternal sensitivity could have caused liberalism (nurture), or a mother could have an inherently more nurturant personality that she genetically passed down to her child, indirectly impacting the child's political disposition (nature). Whatever the case, this relationship is real and interesting. And for now, it's as far as we can go scientifically.

Nonetheless, some have ventured even further: anthropologists have drawn connections between a group's political tendencies and its child-rearing culture;

and psycho-biographers have made similar connections between the political careers and childhoods of notable leaders. These approaches are less than scientific, and they inevitably face harsh criticism. But it's entertaining and informative to peruse a few of the political consequences commonly attributed to egalitarian and authoritarian families.

Egalitarian Families

Some cultures have, on average, a more egalitarian approach than others to childrearing. These are often the same cultures that have greater gender equality, for the reasons discussed in chapter 12. One such culture mentioned then was the Machiguenga tribe of the Peruvian Amazon. Psychiatric anthropologist Allen Johnson remarked that the Machiguenga "place the highest value on equality." Parents allow their children to question their authority and even to rebel, to some degree. Johnson reports that Machiguenga parents "tolerate great emotional expressiveness in small children, including a protracted temper-tantrum phase."

Eventually, the children grow into young adults, and Machiguenga culture socializes them to behave respectfully toward other family members and households. Their culture emphasizes "peace and cooperation in a local network of friends and relations."

Just as Machiguenga childrearing allows young children to question parental authority, their society tolerates only a high degree of equality *between* families. If a family feels that another party has not respected it as an equal, it may break off from its hamlet to live alone in the jungle for long periods of time.[11]

The Machiguengas' hunter-gatherer, swidden-agriculture society couldn't be more different from 1960s America. Nonetheless, both cultures share some similar root causes of protests against societal inequalities. In 1967, psychology professor Kenneth Keniston at Yale Medical School contemplated a paradox: the university students most likely to protest the Vietnam draft were the group *least* likely to be drafted; in fact, this protesting demographic was the most likely to receive student deferments on account of their graduate studies.

What, then, *could* predict which US college students were likely to engage in left-wing, antiwar protests during the late 1960s? The most protest-prone students had been raised in families whose "dominant ethos" was "unusually equalitarian [and] permissive." In these egalitarian families, "children talk back to their parents at the dinner table," Keniston observed. Additionally, he noted:

> As a group, [left-wing peace] activists seem to possess an unusual *capacity for nurturant identification*—that is, for empathy and sympathy with the underdog, the oppressed and the needy.[12]

In other words, the leftist students had the *opposite* of a belief in a just world. They had a belief in an *un*just world. Collectively, this sentiment arguably had historical repercussions. Why did it erupt when it did?

Keniston and other social scientists attributed the egalitarianism of student demonstrators in the 1960s to the empowerment of women in the American family. Left-leaning activists were more likely than nonactivists to have "career-type" mothers.[13] Indeed, figure 35 in chapter 12 shows that the percentage of US women with college degrees was quickly rising in the 1960s. The percentage of women contributing to the labor force, in fact, had also been on the increase since at least the 1940s. These women, who were extracting a greater proportion of resources from the economic environment, depended less on men; greater independence also afforded them easier exit strategies. Therefore, gender equality rose. And more parents used increasingly egalitarian childrearing practices. Their post-War "baby boomer" children came of age en masse and reacted to a constellation of social, economic, and geopolitical inequalities that struck their moral sensibilities as unjust.

Hierarchical Families

The film *The White Ribbon, a German Children's Story* won the Palme d'Or award at the Cannes Film Festival in 2009. The Austrian-German movie also received nominations for Best Foreign Language Film and Best Cinematography at the 2009 Academy Awards. The drama's script opens with the following narration:

> I don't know if the story that I want to tell you reflects the truth in every detail. Much of it I only know by hearsay, and a lot of it remains obscure to me even today, and I must leave it in darkness. Many of these questions remain without answer. But I believe I must tell of the strange events that occurred in our village, because they may cast a new light on some of the goings-on in this country.

The story that unfolds depicts the family and social life of a fictitious German village called Eichwald between 1913 and 1914, the years leading up to World War I. Throughout the film, many types of stronger individuals impose severe punishments on weaker ones. Most prominently, parents use harsh discipline on their children, often crossing the line to abuse. The stronger children, in turn, behave extremely cruelly toward weaker ones, and they secretly exact violent revenge on the strict adults in their lives.

The opening narration, then, suggests that the extreme authoritarian discipline received by German children caused many of them to become Nazis when they came of age amid the economic depression and political turmoil of the 1930s. *The White Ribbon*'s writer and director, Michael Haneke, told the Austrian newspaper *Kurier* that his film is about "the origin of every type of terrorism, be it of political or religious nature."[14] Haneke explained to a different magazine that his movie was not only about German fascism; rather, "it's about a social climate that allows for [political] radicalism."[15] In other words, the film's central argument is that authoritarian discipline in the family engenders political and religious authoritarianism in society.

Beyond fictionalized portrayals such as *The White Ribbon*, social scientists have also frequently attributed the strength of the fascist movement in early twentieth-century Germany to disciplinary patterns in the German family. According to political psychologist Neil Kressel:

> Germans were more authoritarian than other Europeans, displaying a greater respect for leaders, a stronger sense of duty, and a heightened willingness to obey orders. This pattern of psychological authoritarianism may have resulted from the structure of the German family life and its rigid, obedience-oriented, child-rearing practices.[16]

Others have also called attention to punitive child-rearing practices that were commonplace in German families and schools.[17]

Beyond these kinds of qualitative observations, we do have some quantitative data that substantiate a link between authoritarian discipline and German fascism. The Allies, who eventually defeated the Nazis, had to "De-Nazify" post-War Germany. To do so, they commissioned studies to determine who could have very easily become a Nazi but instead became an anti-Nazi dissident. Psychiatrist David Levy, a medical officer in the US military, carried out one such study. Levy discovered that the comparatively left-leaning anti-Nazis were more likely than a Nazi control group to have had a father who did not use corporal punishment, or who had died during the subject's childhood. A dissident's father was also more prone to talk "freely to his children and [to have] an easy relationship with them."[18]

The British also conducted similar studies to select Germans to serve as leaders in the post-War democratic government. The agency mandated with this task was the German Personnel Research Branch. This institution developed surveys aimed at sorting out "those [Germans] of democratic from those of fascist temperament." Roger Money-Kyrle, a British psychologist who served in this de-Nazification process, concluded that "authoritarians, with an almost monotonous regularity, spoke of the strict patriarchal nature of their early environment to which they gratefully attributed their own regard for discipline."[19]

Future Nazis who had authoritarian fathers may have been emotionally prone to accept Hitler as a father figure. Photographs and stylized images of Hitler on Nazi propaganda posters show the leader of *das Vaterland* (the fatherland) with stern, harsh, and forbidding expressions. These emotions greatly contrast with the faces of political leaders further to the left of the spectrum, including in their own propaganda.

In America, hierarchical discipline, often reaching physical abuse, surfaces in the family histories of those people who espouse a far-right-wing worldview. The young Glenn Beck, for example, had to contend with a physically abusive man who dated his mother. Later on in his career as a Tea Party–backed political commentator, Beck would fantasize about violently punishing his liberal adversaries in Congress.[20]

On the extreme right end of the US political spectrum, the members of white hate groups often have abusive, alcoholic fathers. Dave Holland, the founder of the Southern White Knights of the Ku Klux Klan, grew up near Atlanta under the roof of one, according to ethnographer Raphael Ezekiel. As soon as his father died, Holland joined the Klan. The hierarchical worldview of this hate group may have replaced the hierarchical discipline that Holland received at home. If the Berkeley authors of *The Authoritarian Personality* were correct, then people like Holland displace hatred for abusive parents onto scapegoat out-groups. But this psychoanalytic logic is difficult to prove.

Ezekiel has found a similar family pattern among members of a Detroit-based neo-Nazi organization. The leader of the group, a pseudonymous "Paul," had a father accustomed to inebriating himself with beer and throwing the bottles at his wife. The beatings were so severe and frequent that other parents in the neighborhood prevented their children from playing with Paul. Ninety percent of the twenty Detroit neo-Nazis that Ezekiel studied had lost a parent when young. Their mothers' subsequent stepfathers and boyfriends "typically were cold, rough, and abusive." The future hate-group members usually lacked extended family members or other community institutions to shelter them from this mistreatment.[21]

Since the mothers of the Detroit neo-Nazis lived in poor neighborhoods and had minimal family-support systems, they likely depended on the stepfathers and lacked exit options. These conditions would maximize gender inequality and increase the chance that the children would bear the brunt of an extremely hierarchical relationship with a stepfather. Indeed, studies suggest that women with high levels of economic dependence on their husbands are the most likely to suffer violent physical domestic abuse.[22]

Does Hierarchy or Egalitarianism Have a Gender?

In hierarchical families, discipline comes more powerfully and lopsidedly from the father. In egalitarian families, the mother tends to have relatively more authority—as in the case of the economically empowered professional mothers of the 1960s peace protestors. Perhaps this is why Chris Matthews, the host of MSNBC's political talk show *Hardball*, once called the liberal Democrats the "Mommy Party" and the conservative Republicans the "Daddy Party."[23]

In all earnestness, though, George Lakoff believes that these disciplinary styles are the fundamental differences that distinguish American conservatives from their liberal counterparts. The left and right, he argues, think about political issues through contradictory "nation as a family" metaphors. Conservatism, he asserts, "is based on a Strict Father model, while liberalism is centered around a Nurturant Parent model."[24]

But there's an asymmetry here between strict *fathers* and nurturant *parents*. What happened to the mothers? Harvard psychologist Steven Pinker has taken Lakoff's political work to task for this issue (and numerous others[a]). Pinker argues that the term "Nurturant Parent" reflects Lakoff's pro-liberal partiality:

> The metaphors in our language imply that the nurturing parent should be a mother, beginning with "nurture" itself, which comes from the same root as "to nurse." Just think of the difference in meaning between "to mother a child" and "to father a child"! . . . Dictionaries list "caring" as one of the senses of "maternal" but not of "parental," to say nothing of "paternalistic," which means something else altogether. But it would be embarrassing if progressivism seemed to endorse the stereotype that women are more suited to nurturing children than men are. . . . So political correctness trumps linguistics, and [Lakoff's] counterpart to the strict father is an androgynous "nurturant parent."[25]

Pinker's response implies that liberals like Lakoff are averse to stereotyping the genders—especially the physically weaker one. Pinker is correct: political correctness has become so egalitarian that it rejects a worldview in which either gender is intrinsically more or less of any value (such as "nurturant") than the other.

Some segments of the far left believe that a hyperegalitarian form of political morality once existed in a distant, *female-dominated* past. The myth of matriarchal origins stretches back at least to the nineteenth century. In 1861, Swiss anthropologist Johann Jakob Bachofen published a book titled *Mother Right: an Investigation of the Religious and Juridical Character of Matriarchy in the Ancient World*. In his book, Bachofen argued that humanity's religious, social, and moral institutions stemmed from an earlier "Mother Right" phase of development. Friedrich Engels, who coauthored the *Communist Manifesto* with Marx, drew on Bachofen's ideas. In doing so, Engels linked patriarchal oppression to the structure of the monogamous family and to the privatization of property. "Matriarchal Studies" later became a popular field advanced by radical leftist feminists in the 1970s. The historian Christopher Lasch has called the critique of patriarchal authority "one of the most enduring ideological expressions of [the far left's] revolt."[26]

There's only one problem: scientific anthropologists all agree that patriarchy is a human universal; matriarchy has never existed outside of politically motivated myth. The impetus for this myth comes from a hyperegalitarian political ideology. But the enduring believability of ancient matriarchy, even in the absence of all evidence, seems to have deeper footholds in our psyche. According to psychologists Janine Chasseguet-Smirgel and Alan Dundes, we can "observe projected on to the history of civilizations the individual adventure of development in men and women." Or, in the more transparent words of sociologist Steven Goldberg, "every infant does indeed live in a matriarchy."[27]

The key point here is this: males, on average, are stronger, and they are traditionally dominant over females. So conservatives are more supportive of male authority because of their higher BJW. Liberals are more egalitarian, while the hyperegalitarianism of the far left favors female authority.

Liberty, Equality, Fraternity?

Apart from gender, another important inequality in families exists between generations. Some political theorists have pointed out that conservatives and liberals differ over which generation of the family they idealize. According to this analysis, the right identifies with the "strict father" of the older generation, while leftists identify with the younger generation of siblings. Twentieth-century political theorist Harold Lasswell wrote: "Some individuals cherish a fraternal and some a paternal ideal. The anarchist, socialist, and democrat talk the language of equality among a family of brothers; the monarchists preserve a father."[28] The English psychologist John Carl Flügel thought in a similar vein:

> The [political] leader represents the father, the group represents the sons, and the slogan "Liberty, Equality, Fraternity" of [the French Revolution] does but express more explicitly than the usual the ideals dimly apprehended by rebels of all time.[29]

The conservative counterpart to this French slogan was "The Nation, The Law, The King." This politicization of generations follows the same underlying logic that we've applied to other inequalities in the family and in society: the right identifies with the powerful parties and hierarchical relations (in this case, the strict father or the king); the left favors weaker parties and more egalitarian relations (in this case, siblings or "the people").

As a case in point, the term *fatherland* (from the German word *Vaterland*) was heavily used in Nazi propaganda (and then in anti-Nazi propaganda) during World War II. The term became so imbued with connotations of German fascism that post-War English has generally dropped the use of "fatherland" in non-Nazi contexts.[b]

According to linguist Gabriella Klein, other right-wing governments have glorified their "parent" civilization, in opposition to a notion of the contemporary culture as an inferior child. Schools in fascist Italy described Latin as the language that "had the virtue of expressing man and humanity both in sovereign dignity and in lucid and harmonious spirituality." Fascist ideology thus admired the strength of the older, Roman Empire, but feared the "degeneracy" of the vernacular daughter languages.[30]

Leftists show the opposite attitude toward power inequalities within the family. Chairman Mao once told an American visitor: "There were two 'parties' in [my] family. One was my father, the Ruling Power. The Opposition was made up of myself, my mother, my brother and sometimes even the laborer."[31]

The leftist perspective described by Mao imputes an egalitarian ideal to the solidarity between siblings and others against a dominant father. The slogan of the French Revolution also pairs "Equality" with "Fraternity." The desire of the far left to equally redistribute private property to "the people" recalls the socialized nature of childhood property among siblings. Conversely, the political right's strong support for private property evokes respect for a father's possessions, which remain "off limits" to children.[32]

Despite the association of "Equality" with "Fraternity," a closer look at siblings reveals another microcosm fraught with major *inequalities*.

BIRTH ORDER AND SIBLING RIVALRY

The treatment between siblings, in fact, is often far from egalitarian. According to the Judeo-Christian Bible, the idea of sibling rivalry dates back to the first brothers: Cain kills Abel and defiantly asks God, "Am I my brother's keeper?" (Genesis 4:9)

Subsequent biblical siblings also fight with one another. When Rebekah was pregnant with the twins Jacob and Esau, the brothers fought within her womb.

"Jacob" means "heal-grabber," because Jacob was born seizing Esau's heal, trying (unsuccessfully) to hold back his older brother Esau from being born first (Genesis 25:26). While Moses was on Mount Sinai advancing monotheism, his older brother Aaron was competing against him by fashioning a golden-calf idol (Exodus 32:4). And Joseph's brothers sold him into slavery (Genesis 37:28).

Sigmund Freud recognized the power struggles between siblings, and credited them as an origin of the political personality. In his book *The Ego and the Id* (1923), Freud wrote: "Even to-day the [moral and political] social feelings arise in the individual as a superstructure built upon impulses of jealous rivalry against his brothers and sisters."[33]

One of Freud's colleagues, the Austrian psychiatrist Alfred Adler, is better remembered as an early theorist of sibling rivalry. Adler used a vivid political metaphor to explain what he believed to be the origin of sibling conflict:

> For a while [the first born] is an only child and sometime later he is "dethroned." . . . Sometimes a child who has lost his power, the small kingdom he ruled, understands better than others the importance of power and authority. When he grows up, he likes to take part in the exercise of authority and exaggerates the importance of rules and laws. Everything should be done by rule, and no rule should ever be changed, power should always be preserved in the hands of those entitled to it.[34]

Adler devised a "position psychology" that attributed formative importance to an individual's birth order within the family.[35]

As Adler pointed out, a power discrepancy exists between siblings. Even after the first-born's "ousting" from sole power over the "child" niche of the parental environment, older siblings retain advantages over later-borns. During childhood, a first-born sibling is normally bigger, stronger, smarter, and often more bonded to parents than a later-born.

Today, Frank Sulloway is a world expert on the psychology of birth order, and the author of the book *Born to Rebel: Birth Order, Family Dynamics, and Creative Lives* (1996). Sulloway explains that, before the birth of a later-born, a first-born receives undivided parental resources; but "when the second [child] comes along, the oldest still gets half of all that [attention], so younger siblings never have a chance to catch up."[36] The undiluted parental investment initially received by first-borns may explain some studies suggesting that they have slightly (but significantly) higher IQs than their later-born siblings.[37]

The Success of First-Borns at Achieving High Rank

Perhaps Adler was correct that first-born children maintain a particular hunger for power; because inequalities between first-born and later-born siblings often do continue past the childhood years. In 1874, English polymath and statistician Sir Francis Galton (the half-cousin of Charles Darwin) noticed the effect of birth order in the sciences. First-borns and only children, Galton discovered, constituted a disproportionate number of eminent scientists.[38]

Subsequent studies over the years have found an overrepresentation of first-borns among corporate CEOs, college professors, Nobel laureates, world political leaders, US presidents, Supreme Court Justices, and legislators in the House of Representatives.[39] It can't be overemphasized, though, that birth-order phenomena such as these are not black-and-white certainties; rather, they're statistical trends.

But if you had to bet money on the birth order of a United Nations secretary general, you'd be well advised to guess "first-born." During the history of the UN, from its inception in 1946 through 2016, seven of the eight secretaries general have been first-born children. Only Dag Hammarskjöld was a later-born (the fourth of four children).[40] Kofi Annan was born first, along with a twin.[41]

The Rebelliousness and Adventurousness of Later-Borns

In comparison to their eldest siblings, later-borns have a greater streak of rebelliousness and adventurousness in their character. Political assassins in the United States, for instance, have been later-borns, often with older brothers. This birth-order fact holds true for every one of the successful assassins of US presidents (i.e., of Lincoln, Garfield, McKinley, and John F. Kennedy).[42] John W. Hinckley Jr., the failed assassin of President Reagan, was the youngest of three children (including an older brother). The birth orders of other failed assassins are difficult to find.

Frank Sulloway published one of his most creative studies in 2010. He found the birth orders of seven hundred brothers who had played Major League baseball, going back to the year 1876. Then Sulloway determined the statistics for stealing bases, which is considered a high-risk, antiauthority action. He discovered that the younger brothers stole or attempted to steal 10.6 times as many bases as their older brothers! Moreover, the younger siblings were 3.2 times more successful at doing so than first-borns.[43]

Sulloway has also discovered another telling indicator of risk-taking among younger siblings. Compared with the firstborns in his sample of historical figures, last-borns were 3.0 times more likely to circumnavigate the globe.[44]

Traveling around the world is a form of novelty-seeking behavior. This adventurous personality trait, in turn, relates to xenophilia. As we learned in chapter 11, xenophilia can motivate dispersal and lead to outbreeding.

Left-Right Political Orientation and Birth Order

Based on the very specific groups above, it seems as though birth order impacts personality. But let's take a look at more representative samples of siblings to see if their personalities vary—especially their *political* personalities.

The "Big Five" Personality Traits

One of the most conventional measurements of basic personality traits is the "Big Five" inventory (introduced in chapter 3). Frank Sulloway has done the most extensive analysis of the Big Five personality traits in relation to birth order. In his latest meta-analysis, Sulloway evaluated ten studies of over seven thousand brothers and sisters who rated themselves and/or their siblings on the Big Five dimensions. When Sulloway controlled for family size and socioeconomic status, he discovered that birth order had quite weak, but extremely significant, correlations with these personality traits.

Later-borns scored higher on "Openness." This means they tested higher in levels of novelty seeking, rebelliousness, and nonconformity. The first-borns with low Openness were more conforming, traditional, and closely identified with their parents.

The second Big Five trait most associated with political attitudes is "Conscientiousness." The meta-analysis found first-borns to be more conscientious than their younger siblings. Conscientiousness entails high levels of organization, achievement orientation, reliability, responsibility, self-discipline, and scholastic achievement.[45]

As shown in figure 6 (chapter 3), Openness correlates most strongly with the political left, while Conscientiousness correlates with the right. The Big Five factor that is by far the most correlated with birth order is Conscientiousness.[c]

Primogeniture

First-borns tend to be more conservative than their younger siblings because of their personalities. But Sulloway has suggested that, in the past, economic considerations might have also contributed to the conservativeness of first-borns: "Historically the practice of primogeniture was very common in Europe.

So firstborns had every reason to preserve the status quo and be on good terms with their parents."[46]

By the practice of primogeniture, the first-born (usually male) child would inherit all of his parents' property (and political titles)—to the exclusion of younger siblings. In areas with high population density or resource scarcity, this custom prevented land from splintering into smaller and smaller parcels with each generation.

But Sulloway's logic may be upside down. Parents in primogeniture-practicing cultures throughout the ages probably willed property to first-borns *because* of the typical personality of eldest children. Given the choice to bestow the family fortune on one of several children, it makes more sense for custom to select the conscientious, organized, responsible, self-disciplined first-born child over the rebellious, nonconforming later-born who identifies less with his parents. In other words, first-borns were not conscientious because they inherited the whole estate in some cultures; rather, they inherited everything because they were more conscientious to begin with.

Consequently, eldest children retain the same conscientious personality traits even when the rules of primogeniture no longer apply. A study carried out in the late 1990s among Japanese Americans illustrates this point. Compared with their younger siblings, first-borns were more likely to live in Japanese neighborhoods and identify with Japanese values. In addition, they were more likely to espouse their parents' Shinto or Buddhist faith, and with a higher degree of religiosity than later-borns.[47] The same principle applies to other cultures as well: in general, parents' social attitudes are nearly twice as strongly correlated with those of their first-born children as compared with those of their later-borns.[48]

These facts may account for ancient records testifying to how early civilizations discriminated between children according to birth order. The Judeo-Christian Bible forbids depriving the first-born of his birthright (Deuteronomy 21:15–17). When God *really* wanted to punish the Egyptians for refusing to liberate the Israelites from slavery, the last and most horrible of his ten plagues was the death of the first-born (Exodus 11:4–6).

Other ancient cultures also bequeathed more to first-borns. Such customs were recorded in the Middle Assyrian Laws (on Tablet B, no. 1), around 1100 BCE. The Nuzi-Akkadian Laws (no. 3) of Mesopotamia also describe the practice in the second millennium BCE.[49]

Overtly Political Behavior and Traits

With so many inequalities in the personalities and historical treatment of siblings, it's no wonder that even full-blooded brothers and sisters raised by the same two parents in a similar disciplinary environment often develop quite different political attitudes. In fact, behavioral geneticist Robert Plomin discovered in the 1980s that despite the close similarities in the physical characteristics and intelligence of siblings, their personalities are only slightly less different than children taken at random from the population; siblings have similar personalities to one another only about 20 percent of the time.[50]

In the subsections below, we explore the ways in which birth order predicts overtly political attitudes.

Scientific Revolutionaries through History

In Sulloway's *Born to Rebel*, the author describes an extraordinarily impressive historical database that he has compiled. This catalogue contains the names of 3,890 scientists who took different sides with respect to twenty-eight different scientific innovations and revolutions (such as the Copernican, Newtonian, Darwinian, and Freudian ones).

According to Sulloway's calculations, first-borns were 1.9 times more likely than later-borns to support conservative theories (such as eugenics, or doctrines that support the "Wisdom and Power of God" or "faith in Divine Providence"). The later-borns were 3.1 times more prone than firstborns to support liberal theories. But later-borns' backing of liberal theories was 7.3 times greater when their own first-born sibling opposed the theory.

Evolutionary theory in particular especially polarized eighteenth- and nineteenth-century scientists according to their birth order. During the century and a half before Darwin published his *Origin of Species* (1859), later-borns backed theories of evolution 9.7 times more frequently than first-borns.[51]

The US Supreme Court

The Supreme Court of the United States is the highest court in the country. It has ultimate appellate jurisdiction over all other state and federal courts. Of the eight thousand or so cases that percolate all the way up to this institution each year, the Supreme Court decides to hear only about 1 percent. To be selected, cases often involve questions surrounding the interpretation of law—especially when federal law may disagree with the verdict of a state-level court. Therefore, the Supreme

Court's decisions frequently depend on the moral philosophy of the justices with respect to conflicts between more and less powerful parties. In addition to the technical elements of law, then, Supreme Court decisions are often highly political.

To add to the politicization of the Supreme Court, the justices are appointed by the president. The selection of justices is especially crucial because they serve life terms. From the beginning of the modern party system in 1860, through 1996, Republican presidents have nominated first-born justices about twice as often as Democratic presidents have; and Democratic presidents have nominated later-born justices approximately twice as frequently as their Republican counterparts.[52]

Between 1789 and 1989, the Supreme Court deliberated over 459 "landmark" cases (as defined by *Congressional Quarterly*). These momentous cases involved upholding or striking down laws, establishing the Court's authority in new areas, or finding that some areas lie outside its competence. Only 129 of the landmark cases resulted in a unanimous ruling, while 330 had divided decisions. In these divided decisions, first-borns were significantly more likely to vote with the majority. Later-borns were more prone to rebel against the powerful majority.[53]

Contemporary Political Issues

Politically controversial issues in recent times have also divided siblings in different proportions according to birth order. Psychologist Richard Zweigenhaft has studied the use of marijuana among high school students. The investigation took place in the Rocky Mountain region of the United States between 1969 and 1981. During these years, the possession and use of marijuana was illegal throughout the United States. Involvement with the plant was considered a form of "rebelliousness against the status quo."

During this twelve-year period, researchers collected information from 364 students about their relationship with marijuana. Birth order was the only factor of those investigated (which included gender, father's education, and family size) that consistently predicted the answers to their questions; specifically, the last-born children were significantly more likely than first-borns and middle children to have ever used marijuana, to have kept a personal supply, and to have used the drug at a frequency of twice a week, when available. And middle-born students were more exposed to marijuana than first-borns. At graduation, 48 percent of the last-borns, 38 percent of middle-borns, and only 33 percent of first-born children had used marijuana during their senior year.[54]

Zweigenhaft has also found birth order involved in the participation of US college students engaging in civil disobedience on behalf of a labor union. For over two years in the late 1990s, Kmart employees in Greensboro, North

Carolina, protested against what they claimed were unacceptable working conditions, leave time, and benefits. They also made allegations of harassment and discrimination. After company-union negotiations failed, hundreds of people participated in demonstrations, a boycott, and civil disobedience. In the first three months of 1996, police made over 150 arrests. The protestors arrested included Guilford College students.

Zweigenhaft gathered a sample of seventy-three students, many of whom had participated in the civil disobedience. He discovered that birth order was significantly associated with whether and how many times students were arrested. Of the students *not* arrested, only 43 percent were later-borns. Of the twelve students arrested once, 50 percent were later-borns. And all five (100 percent) of the students arrested more than once were later-borns.[55]

Is it possible to show that birth order influences left-right political orientation itself? The study of Japanese Americans mentioned above found that first-borns were 1.4 times more likely to vote for conservative political candidates than their younger siblings.[56] And a 1993 study of over 193,000 college freshmen in the United States revealed that later-borns were more liberal than first-borns. Compared with their older siblings, the younger ones were more prone to be pro-choice on abortion, to favor the legalization of homosexual relationships, and to approve of casual sex. First-borns, in contrast, attended church on a regular basis more frequently than later-borns.[57]

Birth order sways political attitudes. From his analyses of numerous studies on birth order over the years, Sulloway has estimated that later-borns are between 20 and 43 percent more likely than first-borns to support a liberal political position, to back a liberal candidate, or to campaign for a liberal social cause.[58]

Siblings are often conscious of political differences between them. Sometimes the differences are subtle. Other times they vote for different parties. Once in a while, full-blown political rivalry erupts between siblings at the highest level of a country's political elite. In Ecuador in 2010, a public ideological clash broke out between the head of state and his older brother. Ecuadorian president Rafael Correa led a left-wing political party. While in office, he had allied his country with the socialist Bolivarian Alliance for the Peoples of Our America (ALBA), founded by Hugo Chavez. Rafael Correa was also known for his anti-"imperialist" rhetoric and his rebelliousness against government contracts signed with foreign oil companies and creditors.

Rafael's older brother, Fabricio, was notably more conservative. Fabricio threatened to run against his "little brother" in the 2012 presidential election. He attacked Rafael for his choice of collaborators; the older brother claimed that the president's cabinet was filled with homosexuals (whom he referred to as "the pink circle"). Fabricio also disapproved of Rafael's association with "communists" and "guerrilla sympathizers." If elected, Fabricio promised to crack down on crime. The first-born son would have also strengthened ties with Europe and the United States. Fabricio's prominent position against societal rule-breakers, coupled with his position in favor of more powerful foreign countries, suggests a much stronger belief in a just world.

Fabricio's campaigning against his brother could be explained only as sibling conflict, because public-opinion statistics show it was completely irrational. At the time, Rafael had an approval rating of over 50 percent; Fabricio polled at only 2.8 percent. Graffiti on the streets of the capital called Fabricio "Cain," in reference to the Bible's first-ever first-born, who murdered his younger brother Able.[59]

In part III, we've learned the following:

- Attitudes toward inequality and authority in society differ remarkably among the members of a group. Hierarchical, "Belief in a Just World" values correspond to political conservatism; an egalitarian worldview, on the other hand, is associated with the political left.
- Moral stances toward inequality in society do *not* correspond well or at all to self-interest. Yet these positions *do* relate to a parallel set of opposing attitudes concerning inequalities within the family. Conservatives prefer a hierarchical disciplinary strategy; liberals believe in egalitarian child-rearing practices.

  Thus, it's not by chance that both parents and diplomats speak of "carrots" and "sticks." Neither is it coincidental that some parents and politicians are predisposed to believe that incentives or punishments are more effective than the other—regardless of the circumstances.
- Birth order can create additional family disparities, which influence attitudes toward inequality and authority in general. First-born children tend to be more politically conservative than their later-born siblings.

Where do these power imbalances within the nuclear family ultimately come from? Why do these very close relatives have such different interests, to the point where parents need disciplinary strategies and rivalry between siblings is commonplace?

Next, part IV unearths the evolutionary roots of these tensions between the closest relatives. We'll learn why conflicts between parents and their children, and also between siblings, are universal facts of human existence.

PART IV.
THE BIOLOGY OF FAMILY CONFLICT

Part IV investigates the evolutionary logic of family conflict. We will learn how the biology of "parent-offspring conflict" creates tension between generations and among the siblings of a nuclear family. Specifically, part IV explains *why* divergences of interest arise between family members during various stages of the human lifecycle. As we shall see, the ways in which family conflicts unfold shape people's moral worldviews and influence their political orientation.

16.
WHY SIBLING CONFLICT OCCURS AND POLARIZES POLITICAL PERSONALITIES

Ecuadorian President Rafael Correa is a leftist. His eldest brother Fabricio is a conservative. The two men share 50 percent of their genes, yet Fabricio wanted to replace his little brother and take away 100 percent of his political power. If they share such a high percentage of their DNA, why are their personalities so different? Why don't siblings in general act as kindly toward one another as their parents normally do toward their children? After all, parents also share half of their genes with their offspring. Where does the additional antipathy between siblings come from?

THE ORIGIN OF SIBLING RIVALRY

In 1972, evolutionary theorist Robert Trivers published his famous paper on parental investment. This theory explains why human females are the highly investing (and choosier) sex: because women's investments in gestating, nursing, and rearing their children decrease mothers' ability to invest in other offspring far more than fathers' lesser investments do.[1]

Trivers's brilliant paper reverberated across multiple disciplines and was cited in more than eight thousand publications over the next forty years. But something was missing. His theory had correctly reasoned that parents invest in their young in such a way as to maximize the number of their surviving offspring; but it had assumed, in his own words, that "offspring are . . . passive vessels into which parents pour the appropriate care."

Trivers quickly corrected this problematic premise by recognizing that offspring are active agents that struggle to maximize their own reproductive success from the very beginning of the lifecycle. Thus, the theory of "Parent-Offspring Conflict" was born in 1974.

The core of this theory is the following: An offspring actively seeks to maximize its own fitness. Having a high fitness means reproducing as many viable offspring of its own when it reaches maturity. For the best chance of doing so, an offspring seeks to extract a higher amount of investment from its parent(s) than the parent(s) are selected to give.

A parent's own interest is to withhold more resources than an offspring would like in order to equally invest the resources in other offspring. The parent also retains resources to protect its own health and future reproductive fitness. Although an individual offspring wants more than its "fair" share of parental investment, the offspring's extractive efforts *do* have limits; it must not overly endanger the health of a parent, which could reduce or end the flow of resources and jeopardize the offspring's own survival.[2]

From an offspring's perspective, having more siblings can clearly be a disadvantage because brothers and sisters redirect parental investment *away* from itself, potentially reducing its fitness or even threatening its survival. On the other hand, Trivers pointed out that having additional siblings that share the original individual's DNA increases the individual's inclusive fitness. This means that having a brother is like having a son; both relatives share half of an individual's genes and can reproduce a quarter of them in their own progeny. So siblings can increase fitness just as well as children can.

Conflict nevertheless arises between siblings because to reproduce, a sibling must live until sexual maturity. In the case of humans, surviving until this phase of life in premodern environments is about as likely as winning a coin toss. During the most precarious years of infancy and early childhood, a vulnerable offspring depends on receiving as much food, attention, and healthcare as possible from its parents—*in competition with its siblings*.

Here is where a genetic asymmetry arises: an offspring shares only 50 percent of its genes with a full-blooded sibling—and just 25 percent with a half-sibling; yet the offspring shares 100 percent of its genes with itself. So an offspring's genes want it to receive a greater share of parental resources than its siblings receive. A parent, in contrast, shares 50 percent of its genes with *each one* of its offspring. Therefore, a fitness-maximizing parent wants to distribute food equally among its children.[a]

On account of their divergent interests, Trivers predicted that parents and their offspring would disagree over:

- The length of the period of parental investment;
- The amount of parental investment that offspring receive; and
- How altruistically or selfishly an offspring behaves toward relatives (such as siblings).

These inherent conflicts between parents and their offspring are what ultimately make discipline a universal aspect of childrearing. As discussed in the previous chapter, the two disciplinary strategies leverage different emotions (fear or guilt) for their coercive purposes.

In the context of this inter-generational conflict, Trivers expected that offspring would "employ psychological weapons in order to compete with their parents." For example, an offspring could deceptively withhold body language signaling satiation until receiving more food than a fitness-maximizing parent would otherwise feed it. Trivers emphasized that deceptive communicative tactics play an especially important role in species whose young are born dependent on parents and take many years to slowly lose this vulnerability. In human beings, for instance, Trivers pointed out the phenomenon that psychologists call "regression." When a child regresses, he or she reverts to an earlier stage of less mature behavior in an attempt to receive more investment than the parent would otherwise give.[3]

SIBLINGS COMPETE TO SURVIVE

In a hypothetical world where all siblings throughout human history survived until sexual maturity, receiving a bit more or less parental investment than one's brothers or sisters would be far less crucial. However, infant mortality has traditionally been a major hurdle to overcome. Among contemporary hunter-gatherer societies (such as the !Kung, Hadza, Hiwi, Aché, and Tsimane), as well as among eighteenth-century Swedes, the chance of living past the age of five is only between 43 and 50 percent.[4] In many developing countries today, where access to modern medicine and sanitation is limited, almost half of all children die before their fifth birthday.[5] This high child-mortality rate explains why life expectancy *at birth* remains much lower in some countries than in others.

Under these traditional conditions, where child mortality claims one in two children's lives, the greatest threat to a young child's survival is for a younger sibling to be born before the first one reaches its fourth or fifth birthday.[6] If a baby among the !Kung hunter-gatherers of southern Africa receives a younger sibling at the age of four, then the older child has a 90 percent chance of surviving childhood. However, if a !Kung mother has an additional child when she already has a two-year-old baby, then the likelihood that the two-year-old will die during childhood is over 70 percent.[7]

In developed countries today, the great majority of healthy children survive until the age of five. This dramatic reduction in child mortality has occurred thanks to improved nutrition, potable water, sanitation systems, and vaccinations. The lengthening of inter-birth intervals (due to the demographic transition) has also lowered childhood mortality.

Our early human ancestors, however, lived in an environment where conditions resembled those of the !Kung much more than those of a developed country. Their high rates of infant mortality constituted a strong selection pressure for the evolution of competition between siblings. Consequently, the human personality has adapted to contend with sibling rivalry—even though one's genes cannot know one's birth order.

How does a young offspring prevent its siblings from threatening its prospects for survival? In some species, the stronger sibling kills the weaker one.

Siblicide in Resource-Scarce Environments

Siblicide is the killing of an individual by his or her sibling, usually during infancy. Biologists have recorded this phenomenon across a range of animals, including insects, fish, birds, and mammals.

Siblicide is most common among birds that eat large prey, such as predators (like eagles) and seabirds (like pelicans and boobies).[8] Among Verreaux's eagles, chicks are killed by their siblings as a matter of course; that is, they practice what biologists call "obligate siblicide." Even though the parents can only hunt enough food to raise one bird, the mother eagle lays an average clutch of two eggs. Normally, the chick born slightly earlier pecks its younger sibling to death within only days of hatching. The second egg serves as "insurance" for the parents in case the first one is infertile, defective, or hatches an ill or weak chick.

In other species of birds, siblicide only occurs under conditions of resource scarcity. Blue-footed boobies, for example, subsist on a diet of fish and squid. Marine resources such as anchovies can fluctuate substantially from season to season. During a good year, booby parents have enough food to raise three chicks. In a bad year, their chicks' body weight drops due to lack of food. When an older chick's weight falls to 80 percent of normal body mass, the stronger chick (which is typically the first-born) pecks its younger sibling(s) to death. When this happens, booby parents do not interfere because doing so would reduce their fitness. It's better for the parents to have one strong chick instead of multiple weak or dead ones.[9]

The lack of parental intervention in cases of booby siblicide suggests that parent-offspring conflict does *not* occur. In some mammals, however, biologists *have* recorded notable conflict over sibling aggression.

Fur seals live in an adjacent habitat to the boobies in the Galapagos Islands. A young fur seal's risk of pup mortality increases by 60 percent upon the birth of a younger sibling.[10] The elder-born seal uses its greater size to bite its smaller sibling and distance it from their mother and her milk. This type of sibling conflict is especially common during El Niño years, when the sea temperature rises and the food supply drops. Mother fur seals intervene aggressively in defense of the younger sibling.[11]

Parent-offspring conflict occurs in this case because both siblings might survive. If they do, then the fitness of the mother (which shares half of her genes with both offspring) increases. The older pup, though, prefers not to risk malnutrition, illness, or death. It protects its own survival at its sibling's expense. After all, the larger pup shares only a half of its genes with its sibling (or a quarter of its genes with its half-sibling, as the case may very likely be), but it shares 100 percent of its genes with itself.

The human "clutch size" is normally only one, due in part to the optimal body size at birth. Like birds of prey, humans also occupy the top of their food-chain pyramid; so it takes large amounts of high-value resources over a long period of time to raise a child to self-sufficiency. In addition, hunter-gatherer women have difficulty in simultaneously carrying around two babies on their backs while provisioning staple foods to their band.[12]

Even though the birth of a younger sibling substantially bumps up the risk of child mortality for a child under the age of five in many environments, anthropologists have not found infantile siblicide to regularly occur among humans. The reason why infant humans do not kill each other is not necessarily because it wouldn't be adaptive for the older one. Rather, human infants are born precociously and highly dependent on their parents for survival. They simply don't have the means to commit siblicide (the fictional idea of infants that *would* have the strength to kill, however, has inspired many horror movies). By the time a child develops the ability to kill a younger sibling, he or she is much less vulnerable to child mortality. Thus, the motive disappears—countered instead by the benefit of having a sibling who shares a high percentage of genes.

Historians, however, *have* found a remarkable instance of systematic siblicide in a human culture. Between the fifteenth and the seventeenth centuries, fratricide regularly occurred among the ruling family of the Ottoman Empire. For the family of the sultan, economic resources abounded in enormous measure; but the resource of political power was extremely scarce—only one son could inherit his father's title.

In much of Christian Europe, the law of primogeniture determined that the first-born son would be coronated. This custom reduced, although not entirely, the potential for siblicidal conflict. The Ottomans, however, followed Islamic law. And sharia had abolished primogeniture for both political succession and the inheritance of property.[b]

Without primogeniture, succession was chaotic and costly for the early Ottomans. At the turn of the fifteenth century, a Turco-Mongol warlord invaded the Ottoman Empire and deposed the sultan. In his place, the warlord instated the defeated Sultan's son, Mehmed Çelebi, in 1402. Mehmed's brothers, however, refused to accept the legitimacy of his rule. A bloody civil war ensued between the brothers for thirteen years before Mehmed I crowned himself as sultan.

This "Ottoman Interregnum" nearly destroyed the empire. To prevent history from repeating itself, the descendants of Mehmed I developed a culture of fratricide. When an Ottoman prince became sultan, he waited until the birth of his first male heir and then had all his brothers and half-brothers strangled with a silk cord. His brothers' cadavers were placed into weighted bags (along with the bodies of their wives and pregnant concubines), and then tossed into the Bosporus Strait. Mehmed III, who reigned from 1595 to 1603, notoriously murdered nineteen brothers to secure the throne.[13]

Siblicide regularly occurred among the Ottoman ruling family for a couple centuries. Ottoman siblicide compares to that of nonhuman animals in some respects: the killings took place at the top of the food chain and in an environment of scarcity. For fur seals, scarcity is economic; for the Ottoman royals, it was political. On the other hand, Ottoman siblicide also differed from that found among animals: the sultan's children killed each other as adults and not as infants.

Since human infants are incapable of siblicide, they have evolved to use less violent means of limiting competition from siblings: infants engage in behaviors designed to *prevent* their parents from producing younger siblings too soon.

Weaning Conflicts

In the absence of modern contraceptive methods, one of the most effective ways of preventing a woman from becoming pregnant is for her to regularly nurse an infant.[14] For about six months after birth, 99 percent of continuously breastfeeding mothers will not ovulate. The birth-control effect of this "lactational amenorrhea" can endure, with lesser efficacy, for two years or even longer.

For infants in traditional environments, contracepting their mothers by nursing for as long as possible serves as a potential life-saving strategy. Preventing the birth of a younger sibling within the first years of life eliminates one of the most serious threats to the older child's life.[15]

If a mother weans her infant off of nursing too early because of the arrival of a new baby, the older child can suffer from gross protein deficiency. The medical name for this condition is *kwashiorkor*. The word comes from the Ga language of Ghana, and means "the disease of the displaced one," or "the sickness the baby gets when the new baby comes." *Kwashiorkor* causes swollen feet, a distended abdomen, enlarged liver, hair thinning, tooth loss, skin depigmentation, and dermatitis (see figure 39).

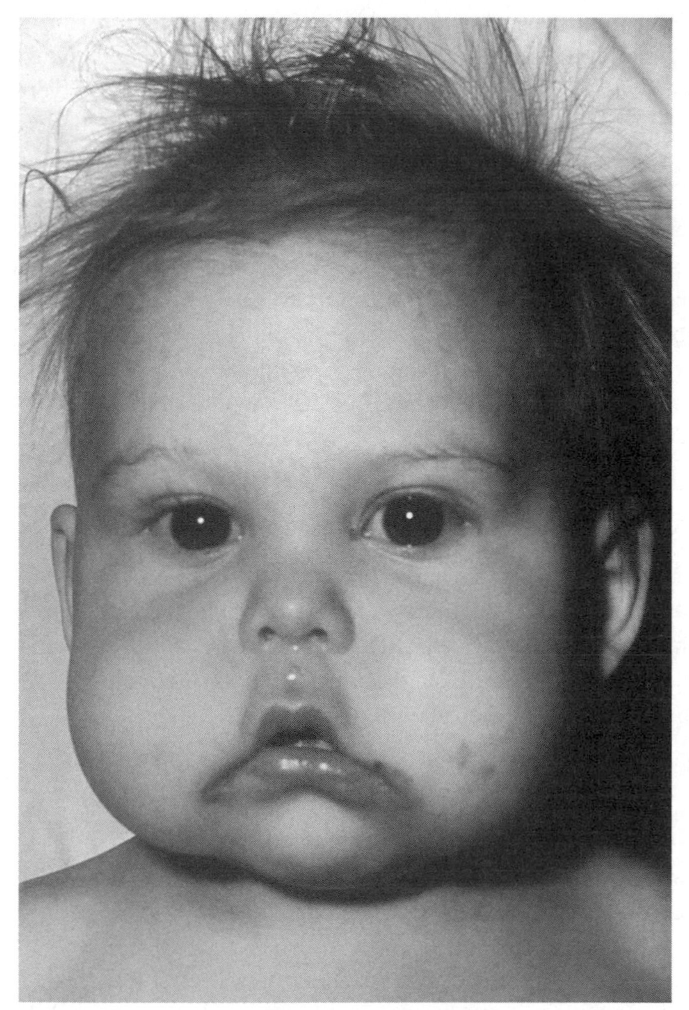

Figure 39. Child Suffering from *Kwashiorkor*.

To prevent the birth of a younger sibling and the loss of parental investment, it's evolutionarily adaptive for human babies to derive pleasure from suckling. Babies are simply wired to do so. Indeed, nipple substitutes such as thumbs and *pacifiers* can appease this instinct so a baby won't cry.[16]

At a certain point, however, the baby's interests diverge from its mother's. The infant would benefit by continuing contraceptive breastfeeding; but the mother incurs the opportunity cost of not being able to have another baby. At this point, the nursing infant's fitness rises at the expense of its mother's. Consequently, parent-offspring conflict occurs over weaning.

Humans aren't the only mammals in which parents and offspring disagree over the timeframe of weaning. Researchers have also documented weaning conflicts in baboons, langurs, rhesus and other macaques, caribou, vervets, dogs, cats, and rats. As predicted by Trivers's theory, weaning conflict increases in length and intensity in species where an offspring can "expect" its future siblings to be half-siblings.

Among baboons, for instance, unrelated males frequently sire offspring with the same female. From the perspective of a nursing baboon infant, a future sibling that would divert resources from its mother would probably have a different biological father. So the younger half-sibling would share only 25 percent of its older sibling's genes. But the mother would share half her DNA with each of her offspring. Indeed, Trivers pointed out that baboon weaning conflicts can last for months. Baboon infants undergoing weaning cry loudly, even though these primates normally keep silent to avoid predation by lions, leopards, and hyenas.[17]

Rhesus monkeys also reproduce quite promiscuously. Consequently, primatologists have observed notable conflicts among these animals over weaning and parental investment in general. According to Frans de Waal, a mother monkey has to loosen her infant from her belly when it refuses to let go. Just as in humans, these conflicts of interest necessitate parental discipline. The rhesus disciplinary method that De Waal describes is on the hierarchical end of human approaches. Rhesus mothers, he writes,

> use aggressive behavior to punish and inhibit "unacceptable" behavior patterns in young monkeys. They tend to use threats for this purpose, but can do so only after threats have become effective through previous associations with physical punishment. Thus, in the rhesus monkey, it is not unusual to see a mother bite her own offspring.[18]

Although rhesus discipline is "hierarchical," they lack the monkey equivalent of human "strict father" morality; males play virtually no role in childrearing.

Infantis Interruptus

After the age of two, a human infant can no longer hope to contracept its mother by resisting weaning from breastfeeding. Nonetheless, it's still in the infant's best interest to do everything possible to prevent any younger siblings from being born for another two to three years.

Between the ages of two and five, in fact, children engage in behaviors designed to sabotage their parents' attempts to procreate. In aboriginal Australia, families traditionally sleep together out in the open of the desert. Young children are so wary of their parents' nocturnal activities that the parents do not often have intercourse at night; rather, they have other adults distract their children during the day, when they can sneak off into the bush.[19]

Psychiatric anthropologist Nicholas Blurton Jones has researched similar behavior in industrialized societies. Here, 20 percent of children between the ages of one and three wake up at night and cry enough to get the attention of their parents. Dr. Blurton Jones has suggested that toddler night waking is adaptive for crying children because this behavior delays the birth of the next sibling.[20]

Later on, young children commonly argue with their parents over their bedtime. Invariably, the children are the party who want to stay up later, while the adults want them to go to bed earlier. By staying awake later, the children reduce the chance that the parents will have sexual intercourse. Through this *infantis interruptus* strategy, the children enhance their own fitness at the expense of their parents' fitness.

Folklorist Alan Dundes once traced the roots of parent-offspring conflict over infantis interruptus all the way back to the ancient Near East. In the Babylonian flood myth, the gods decide to destroy humankind because humanity made so much noise at night that it prevented the gods from sleeping. Dundes explains: "If one reads 'parents' for 'gods' and 'infants' for 'humankind,' one can see that the absurdity is not absurd at all. Rather we have another version of the child's wish to disturb or separate the parents in bed."[21]

Indeed, efforts by young offspring to prevent their parents from mating are probably even older than humanity itself. According to Frans de Waal, chimpanzee children try to interrupt mating among adults:

> If we consider that the [chimpanzee] children harass half of all copulating couples and that a good quarter of their interventions result in copulation being broken off, it is not surprising that the males often half-playfully chase the children away before they make advances to an estrus female. But the children are like tiresome flies: they come back again and again. The little ones seem to be magnetically attracted to sexual contacts among their elders.

De Waal concludes that the little chimpanzees are "jealous" of the mother's resources, which they could lose to a younger sibling. One of the chimp children discussed by de Waal, named Fons, even sank his teeth into an adult male's scrotum. The adult males are tolerant so as to not lose the favor of the females they're trying to copulate with. However, male aggression increases and female protection decreases when the chimp children reach the age of about four.[22]

For a human child, reaching the age of five means that he or she has overcome the substantial risks of child mortality. Children of this age in a premodern environment no longer have as high a chance of dying from malnutrition or certain infectious diseases. Since they become less vulnerable, children begin to play about unattended more often—even though they are far from self-sufficient. At the ages of five and six, the fear of snakes peaks among children (according to a study of almost twelve thousand British youths). This instinctual fear may help keep children away from new dangers as they venture outside the environment of their parents' vigilance.[23]

While the births of younger siblings once posed a hazard, having additional siblings *raises* the inclusive fitness of children over five. That is, these older children benefit by having more relatives who share copies of their genes.

At this point, stronger, older, and more autonomous children can begin to care for and babysit their younger siblings. Biologists call this behavior "alloparenting." Alloparenting increases the fitness of the young caretakers, the parents, and the younger children alike. Psychologists believe that alloparenting makes older siblings more responsible and helps them identify with parental authority. Thus, first-borns score higher in the Big Five dimension of "Conscientiousness" than later-borns (who are generally *not* the main authority among siblings). Conscientiousness, in turn, is associated with political conservatism.

SURVIVAL PRESSURES MOLD SIBLING PERSONALITIES

Politically relevant personality differences between siblings (such as Conscientiousness) originate from each individual's adaptation to different niches within the family. As we've seen, the differentiation of personalities occurs amid an often-conflicted love triangle consisting of parents, offspring, and siblings. Each individual struggles to maximize his or her own fitness. Sometimes interests coincide with one another, as in alloparenting; sometimes, interests collide.

The important point is that a nuclear family does not function as a monolithic environment that makes siblings psychologically similar to one another. To the contrary; as birth-order researchers have learned, siblings differ in personality far more than they differ with one another in intelligence or physical characteristics. The force that shapes siblings' psychological traits so distinctly is competition for their parents' attention, love, and investment. Competition drives siblings to occupy different personality niches. Why don't they simply compete for the same niche?

Birth-order expert and MacArthur fellow Frank Sulloway has compared the differentiation of sibling personalities to the evolutionary concept of "divergence." Darwin theorized that, all else being equal, natural selection favors species that face the least amount of competition from other organisms. Thus, the principle of divergence explains why both species and siblings become more and more dissimilar over time: they avoid direct competition by specializing in disparate ecological or personality niches. In the case of siblings, personality differences "are Darwinian adaptations for enhancing parental love and attention," Sulloway explains.[24]

For instance, suppose that a first-born excels in academics. This eldest child might have a leg up on a younger sibling in intelligence, experience, and identification with a particular subject area, such as math, by several years. The later-born might feel as though she cannot compete with her older sister's math skills, so she decides, at some level, to specialize in a different niche, such as the arts.[25]

Older siblings often identify with the parent most capable of providing affection and investment, or with the parents' strongest values and aspirations. That is, a first-born normally occupies a preferential niche by adapting to the central components of the caretakers' egos. A later-born, in contrast, may specialize in identifying with the other parent more closely to avoid competition on an uneven playing field.[26] Later-borns also tend to diverge by claiming *numerous* of the parents' more peripheral or secondary interests. So younger siblings generally have a greater number of diversified interests than first-borns.

In the mid-1960s, psychologists at the University of Florida studied the personalities of first-born and second-born sons from two-child Protestant homes. The researchers discovered that older siblings took an approach to life "dictated by conscience and a striving for achievement." These first-borns apparently identified with the authority of their parents, and had absorbed (or "introjected") their parents' greatest aspirations for their children's success. The psychologists characterized the younger siblings, in contrast, as "having a wide variety of interests."[27]

Later-borns, as we learned in chapter 15, tend to be more politically liberal. And political liberals, like later-borns, also tend to have more diversified interests than conservatives. Scoring to the left on an ideological self-placement test correlates with having many (as opposed to few) CDs in one's bedroom. The same applies to genres of music and to the quantity and subject matters of books. Liberals tend to have more of these objects and also more varieties of them in their rooms than conservatives.[28]

First-borns have the easiest opportunity to specialize in their parents' core interests. Second-borns, such as those from the University of Florida study, diverge into a greater number of peripheral interest niches. In families with a large number of children, it becomes harder and harder for the youngest siblings to occupy novel niches within the environment of parental love and attention. Parents therefore tend to be more attached to their older children. Moreover, the older a child becomes, the less likely it is that he or she will succumb to the causes of child mortality. These two facts explain why parents sometimes bond more strongly with first-borns—and why younger siblings are less and less likely to receive vaccinations. Vaccination rates fall by 20 to 30 percent for each increasing birth rank (independently of socio-economic class).[29] Having or lacking a vaccination can be a matter of life and death. Research conducted in Latin America during the 1970s found that birth order significantly impacted the chance of surviving childhood. According to Sulloway, "children of fifth and higher birth ranks experienced two-to-three times the usual rates of infant mortality."[30]

The evolutionary struggle between siblings for parental attention clearly molds personalities in divergent directions. As we learned in the previous chapter, first-borns are more likely to identify with parental authority, to be achievement oriented, and to score highly on Conscientiousness. Later-borns, who diversify interests to compete and survive, score higher on Openness and are more likely to take risks and to rebel. These traits shift political orientations further to the right or the left, respectively. Knowing someone's birth order is even more useful than knowing their socio-economic status in predicting whether they believe in a just world.

The tendency for siblings' politically relevant personality traits to diverge has evolved to increase their chances of surviving the perilous life stage of childhood. Once personalities take form, however, they usually endure for many years. So even as adults, Rafael and Fabricio Correa feud with each other from across the ideological spectrum; each man seeks (to retain) the scarce resource of political power. Like many brothers, they accord one another less kindness than most parents show toward their children—even though each individual in these relationships shares half of the other's genes.

In general, siblings fight with one another more intensely than suits their parents' fitness interests. This conflict has created the need for parents to control their children. Parents therefore use different approaches to discipline unruly offspring. The hierarchical disciplinary strategy leverages fear, and the egalitarian strategy leverages guilt. Quarreling siblings may also use these same moralities against one another. A larger first-born child can harness fear more effectively, while a smaller, later-born has more hope in wielding guilt to manipulate an older sibling. A preference for the hierarchical approach is associated with political conservatism, while liberals favor the guilt-based egalitarian style.

PART V.

ARE PEOPLE BY NATURE COOPERATIVE OR COMPETITIVE?

The last of the three personality clusters that underlie political orientations concerns perceptions of human nature. These perceptions span across a spectrum: on one end, people view human nature as cooperative; on the other, human nature is viewed as competitive.

Debate over the nature of human nature has always been central to political psychology. In fact, arguments that people are essentially "cooperative" or "competitive" stretch back across many centuries of political philosophy. These two opposing philosophies of human nature underlie a great deal of the concepts discussed earlier, in parts I and III; thus, in figure 12 the white "human nature" rectangle overlaps with the grey "tribalism" rectangle, and also with the black "disciplinary" rectangle.

To make these overlapping concepts more concrete, imagine two men—one named Jean, and the other named Tom. Jean believes in a cooperative human nature. Therefore, he's more likely to form business and family alliances with members of out-groups. Tom, on the other hand, sees people as fundamentally competitive, so he's prone to mistrusting members of out-groups even more than those of his in-group. Thus, each man's conception of human nature also influences his level of xenophilia and ethnocentrism.

Likewise, beliefs that cooperative or competitive tendencies prevail within other minds relate to which form of discipline someone approves of. Tom, who thinks people are self-interested, is more likely than Jean to favor a hierarchical coercive strategy when dealing with conflictive children, siblings, or political adversaries. Jean prefers a more egalitarian approach, since he believes others are cooperative by nature.

Chapter 17 briefly reviews notable philosophical debate about human nature throughout the ages. Then we learn how, over the past century, political psychology has consistently associated particular perceptions of human nature with different segments of the political spectrum. Next, chapter 18 inquires whether people's perceptions of human nature change in any predictable way over the course of their lifetime.

17.
SAGES THROUGH THE AGES

POLITICAL PHILOSOPHERS WRESTLE WITH HUMAN NATURE

Ever since the ancients first recorded their contemplations about politics, the topic of human nature has invariably arisen. Indeed, all political theories have a theory of human nature.[1]

Just as inevitably, the debate over human nature has polarized the great political theorists. From East to West, and from millennium to millennium, one set of thinkers has believed that people are in essence cooperative; according to this view, people place the interests of others and groups ahead of themselves, just as often as not. The other set of political philosophers has seen human nature as fundamentally competitive. From this perspective, people first and foremost prioritize their own interests.

The "People Are Cooperative" School of Thought

In the fourth century BCE, the Greek philosopher Aristotle composed his *Politics*. In this seminal work on political philosophy, he compared the relationships of power within the household to those of the city-state. Aristotle's treatise famously declared that "man is by nature a political animal." By this statement, he meant that people are social creatures that "desire to live together," even in cases "when men have no need of assistance from each other." Just as the nature of men and women is to couple, reproduce, and form household units, the nature of households is to form villages. According to this Aristotelian line of thought, "the city-state is the end of the other [lower-level] partnerships," such that "every city-state exists by nature." The ultimate reason why society exists is for its members to collectively and individually attain "the good life."[2]

Aristotle didn't entirely deny the existence of self-interest, yet he believed that socio-political relationships are mutually beneficial for all parties involved, despite the inequalities they entail. For example, he found common interests being fulfilled in paternal authority over a wife and children—and even in the authority that a master exercises over a slave. So long as an authority was virtuous, Aristotle believed that unequal relationships benefitted the subjugated, since they could have their lives structured by correct reason (which he thought women, children, and "natural slaves" lacked in a full-fledged capacity). Because individuals are not self-sufficient, Aristotle argued, people are inclined toward cooperation, which he considered to be an essential quality of human nature.[3]

The Good-Slate/Blank-Slate Hypothesis

Political philosophers who have argued that people are fundamentally cooperative have had to confront the problem of why so much selfishness nonetheless exists in the world. Most of these thinkers have argued that selfishness comes *not* from the original nature of the individual (which they maintain is either good-natured or empty), but rather from an unfavorable societal environment.[a]

A key Chinese philosopher who emphasized the importance of societal environment was Mencius (孟子; Mèng Zǐ), who was born in 372 BCE—about a dozen years after Aristotle. Mencius is believed to have studied with Confucius's grandson, and then to have spent decades living as an itinerant sage, wandering around China advising rulers on the Confucian theory of good government. By the seventh century CE, scholars began to consider Mencius to be the most legitimate transmitter of Confucius's teachings. Mencius's own philosophy eventually became canonized. Today, he is deemed the second most important Confucian thinker after Confucius himself.

Since Confucius did not elaborate in any great detail on the issue of human nature, Mencius's view on the subject became the orthodox Confucian position. Mencius believed in the innate goodness of the individual—that people are born with a sense of commiseration, a sense of shame, a reverential attitude toward others (especially for older family members), and a sense of right and wrong. Mencius called these natural dispositions the "four sprouts."

By this analogy, he implied that healthy plants come from sprouts but also depend on proper cultivation. Likewise, human beings depend on a positive ethical environment to fulfill their natural tendency to morally flourish. In addition to warning of potential societal pitfalls, Mencius emphasized the importance of a proper family upbringing in the moral cultivation of an individual.

In the Confucian tradition, Mencius asserted that good government depended heavily on the correct moral development of the ruler. Without a solid ethical foundation, the leader would lose the good will of public opinion. And without the hearts and minds of the people, a ruler risked relinquishing the "mandate of heaven" (天命; *tiānmìng*), or political legitimacy.[4]

In the seventeenth and eighteenth centuries, Europe witnessed the Age of Enlightenment. During this period, intellectual elites championed the idea that reason could transform the political structure of society by limiting the power of church and state. The British political philosopher John Locke (1632–1704) was an Enlightenment thinker of paramount importance, often referred to as the Father of Liberalism. Locke's ideas deeply influenced the American revolutionaries as well as their Declaration of Independence.

Regarding human nature, Locke believed that people are born with a mind that is a *tabula rasa* (a blank slate). Since people had a mind as void of innate ideas as "white paper," he claimed that human knowledge originates entirely from sensing our environment and reflecting on these perceptions.[5]

Locke's theory of blank-slate minds had egalitarian implications. In his *Two Treatises of Government* (1689), he argued that, since God had "furnished [people] with like faculties, sharing all in one community of nature, there cannot be supposed any subordination among us, that may authorize us to destroy one another, as if we were made for one another's uses, as the inferior ranks of creatures are for ours." Locke's declaration that people are born free and equal refuted the traditional view that God had granted monarchs dominion over lesser subjects.

Instead, Locke asserted that humans in "the state of nature" (that is, the hypothetical condition of humanity before the foundation of the state) were equally endowed with "natural rights." Specifically, these pre-state humans had the rights to life, liberty, health, and property.[b] By rooting these rights in "natural law," Locke claimed that these basic entitlements run deeper than the man-made law of any particular society (today's discourse on human rights follows in the same tradition).

According to Locke, people living in the state of nature normally exercised enough reason and tolerance to respect each other's rights to life, liberty, health, and possessions. However, conflict could arise on occasion when someone unjustly violated the rights of others. To better protect their natural rights, Locke conjectured that the cooperative primordial people agreed to conditionally transfer some of their power to a government.

This "social contract" theory assumed that people willfully consented to being governed because it was in the best mutual interests of all those involved. If a government broke this contract by trampling on the natural rights of its citizens, then the people could justly oppose—and even overthrow—the illegitimate government.[6]

Within ten years of Locke's death, another major political philosopher was born on the European continent: the Genevan thinker Jean Jacques Rousseau (1712–1778). The central concept in Rousseau's worldview was that human nature is originally compassionate where self-preservation allows, but that people are corrupted by society. By this line of reasoning, he famously proclaimed that "Man is born free, but is everywhere in chains." How could this happen?

Rousseau envisioned a hypothetical state of nature in which humans were innately empathetic. Yet primitive people, he imagined, lived independent lives. They subsisted with minimal needs and scarcely relied on each other. Natural humans differed little from animals, except for their free will and perfectibility.

When increasing complexity eventually arose in society, Rousseau believed, it corroded human psychology. Rising population density, he argued, led to greater sexual rivalry for mates. As new technology and economic specialization came about, interdependence and competition also increased. Rousseau thought that the development of hierarchy produced pride and vanity and cost people their original independence, freedom, and authenticity. Thus, selfishness appeared as an emergent property of society (in spite of Rousseau's belief that society was made up of naturally cooperative individuals).

Through his political writings, Rousseau sought the redemption of human freedom. His solution to preserving freedom under the authority of the modern state involved a social contract, whereby a state's legitimacy was based on the "general will" of equal and sovereign citizens. Rousseau also wrote on the topic of childrearing. *Émile, or On Education* (1762), which he considered his most important book, prescribed a "child-centered" form of parenting that focused on preventing the deterioration of a child's natural goodness. This egalitarian disciplinary strategy emphasized free, experiential learning—as opposed to subjection to the dominant will of an authority figure.[7]

Rousseau's political philosophy greatly influenced the French Revolution, which began ten years after his death. This momentous event marked the beginning of a new era, in which republics, democracy, and secularism would supplant monarchies and reduce the power of the church.

Few thinkers, however, have motivated as many political revolutions as the German philosopher Karl Marx (1818–1883) and his followers. Marx's *Communist Manifesto* is considered one of the most influential political documents of human history. What did Marx believe about human nature?

According to Marx's materialist theory of history, human nature is empty: "The human essence is no abstraction inherent in each single individual," he wrote. Instead, the nature of people comes from "the ensemble of the social relations." From this perspective, humans are not born competitive; rather, particular social and economic structures create selfishness and exploitation. Replacing capitalist modes of production with socialist ones, according to Marx, would therefore change human behavior—and eventually lead to a Communist utopia.[8]

Marx's theory, then, insisted that people could become cooperative. An ideological agenda to show cooperation in nature arose within the sciences under the Soviet Union. The Ukrainian agronomist Trofim Lysenko became one of the highest-ranking scientists under Stalin's regime, where he directed the Institute of Genetics within the USSR's Academy of Sciences. Lysenko believed that members of the same species didn't compete with one another in nature; rather, animals only preyed on members of other species. Intraspecific competition was "unnatural" in comparison with cooperation.

Lysenko also subscribed to the scientifically refuted theory that organisms directly inherit traits from their environment (as opposed to by Darwin's theory of natural selection, which entails the inheritance of genes). Lysenko's radical emphasis on cooperation and the environment—and his opposition to competition and genetic heredity—clearly stemmed from political ideals.[9] Human (and even animal and plant) nature could remain blank or good, and any evil in the world could originate in the outside environment or the structure of society.

Today, "Lysenkoism" refers to the distortion of the scientific process to reach an ideologically convenient conclusion. More specifically, Lysenkoism connotes an excessively environmentalist bias.

The "People Are Competitive" School of Thought

For nearly every political philosopher who believed in a human nature that was cooperative (or corrupted only by society), a contemporary rival reached exactly the opposite conclusion: that the individual is inherently competitive.

For example, Aristotle's cooperative view of human nature aimed to counter the earlier ideas of pre-Socratic philosophers like the Sophists. In the fifth century BCE, Sophists like Thrasymachus had apparently developed a hedonistic model of human nature (according to Plato's portrayal of him as a character in his *Republic*). As their name suggests, the hedonists believed that individuals naturally seek pleasure for themselves. Cooperation only arose as a learned custom by which society restrained people's inborn selfish tendencies.[10]

In Book I of Plato's *Republic*, Thrasymachus cynically declares that "Justice is nothing other than the advantage of the stronger." He also asserts that "injustice . . . is stronger, freer, and more masterly than justice." According to Thrasymachus, the interests of the elite determine justice in a city-state, and their rules often run against the natural self-interests of other citizens.[11] Unlike the blank-slate theorists who thought that society corrupts the individuals who cooperated to form it, Sophists like Thrasymachus argued that society forces cooperation from selfish citizens.

In the fourth century BCE, South Asia saw the rise of the region's pioneer political philosopher. Chānakya (also known as Kauṭilīya) advised the ruler Chandragupta. Under Chānakya's guidance, Chandragupta founded the Maurya Empire and became the first unifier of the Indian subcontinent.

Chānakya's thought emphasized the competitive element of human nature. One quote widely attributed to him (although perhaps apocryphal?) states that "there is some self-interest behind every friendship. There is no friendship without self-interests. This is a bitter truth." With this theory of human nature, Chānakya advocated the use of duplicity in statecraft and in life. For instance, he wrote: "Do not be very upright in your dealings for you would see by going to the forest that straight trees are cut down while crooked ones are left standing."[12] Consequently, historians have often referred to Chānakya as the Indian Machiavelli (even though Chānakya lived 1,700 years before this Italian philosopher). Today, Delhi's main diplomatic neighborhood bears Chānakya's name.

During the third century BCE in China, one of the most influential political philosophers of the Warring States period was the Confucian scholar Xunzi (荀子; Xún Zǐ). Xunzi's took on his contemporary, the elder Mencius, as an archrival. In direct opposition to Mencius's position on human nature, Xunzi believed that "human nature is evil" and inclined toward "waywardness."[13]

Xunzi argued that desires for bodily satisfaction and social prestige are innate and ineradicable. He believed that selfishness also belied the ideal of filial piety: "for a son to relieve his father of work or a younger brother to relieve his elder brother," wrote Xunzi, "acts such as these are all contrary to man's nature and run counter to his emotions."[14] Xunzi argued that Mencius erred in his belief that the instinctual aspect of human nature (the *xing*) was good; if people truly were good by nature, then they would cause no harm or disorder in the absence of prohibitions, laws, and government. The necessity of these institutions, according to Xunzi, proved the innate waywardness of people.

In Xunzi's philosophy, a moral, social, and political order required that people occupy hierarchical roles according to their "rituals and duties." Society's rules helped to ensure this order (and avert chaos) by preventing people from indulging their selfish desires. The custodians of the social order were to be gentlemen scholars who were superior to commoners in their ability to control their desires. A proper education, Xunzi believed, made moral rectitude possible.

Before the establishment of moral rule, ancient people's selfishness had led to conflict, disorder, and poverty, Xunzi imagined. To prevent this chaos, primordial sage kings would have instituted the "rituals and duties" that held society together. In this sense, Xunzi's thought foreshadowed the concepts of "state of nature" and the "social contract" that European philosophers debated nearly two millennia later.[15]

In seventeenth-century England, for example, the philosopher Thomas Hobbes devised a similar political philosophy, and he was the first thinker to coin these two terms. According to Hobbes, people's innately competitive traits made his "state of nature" a dismal place, where people were terrified of being killed by their fellow humans in a "war . . . of every man against every man." Before civilization, Hobbes wrote, there was "continual fear and danger of violent death." And "the life of man," he famously declared, was "solitary, poor, nasty, brutish and short."[16]

Through Hobbes's "social contract," people gave up their rights to kill each other in favor of a governing authority. Only an omnipotent and centralized sovereign power (that is, an authoritarian "Leviathan") could keep the peace.[c]

Even after the consolidation of Leviathan states, however, their "civilized" governments had no higher power to report to. Without an authority to hold the international system together, the leviathans would reproduce the state of nature that had tyrannized primordial individuals. Human nature was rotten, and so was society.

Younger philosophers, such as Rousseau, reacted strongly against Hobbes's view of people and his Leviathan government; yet even as they attacked these philosophical problems from the opposite ideological direction, they retained Hobbesian concepts such as the "state of nature" and the "social contract" when contemplating human nature.

In addition to Hobbes, history has no shortage of other political philosophers who believed that self-interest dominated human behavior. Across the Atlantic, the fourth president of the United States, James Madison, was also a political theorist. Before state legislatures ratified the US Constitution, Madison wrote the *Federalist Papers* to explain—and sell—the proposed constitution to the thirteen states. In his most famous essay, *Federalist* No. 10, titled "The Same Subject Continued: The Utility of the Union as a Safeguard Against Domestic Faction and Insurrection," Madison shared with his readership his theory of human nature:

> The latent causes of faction are thus sown in the nature of man . . . human passions have in turn divided mankind into parties, inflamed them with mutual animosity and rendered them much more disposed to vex and oppress each

> other than to cooperate for their common good. So strong is this propensity of mankind to fall into mutual animosities that where no substantial occasion presents itself the most frivolous and fanciful distinctions have been sufficient to kindle their unfriendly passions and excite their most violent conflicts.[17]

Madison saw people's tendency to form political factions as a potentially divisive threat to the political health of his young country. The tyranny of a majority party, for instance, could endanger the rights of the individual. To protect the individual and the nation's cohesion, Madison favored a federalist system that divided power between federal and state governments. In addition, citizens would vote for elite representatives to make decisions on the public's behalf (as opposed to one giant direct democracy without delegates or safeguards).

POLITICAL PSYCHOLOGY LINKS PERCEPTIONS OF HUMAN NATURE TO POLITICAL ORIENTATION

The question of human nature reliably polarized political philosophers across many centuries and several oceans. One group of these thinkers viewed people as innately cooperative or potentially compassionate; another group argued for an inherently competitive human nature. This division begs the question of whether the split corresponds to a difference in left-right political orientation. In some cases, such as that of Marx, there is little doubt about where to place the thinker on the political spectrum. In other cases, however, identifying the ideological leanings of historical figures is a task better left for professional historians.

In any case, history's great political philosophers are not the only people who disagree over the nature of human nature; the human-nature question is a perennial problem that also divides contemporary politicians and ordinary citizens. Below we'll explore what modern political psychology has discovered about this ancient philosophical puzzle. Statistical tools and laboratory experiments can determine precisely how an individual's perceptions of human nature can predict his or her political orientation.

But first, a very brief tour of more recent political leaders and movements reveals a notable trend: *conservatives tend to view human nature as competitive, while liberals are more prone to perceiving human nature as cooperative.*

In 1964, the Republican Party nominated Arizona senator Barry Goldwater as its candidate for the US presidential race. Although Goldwater lost the race to Lyndon Johnson, he set the ideological tone for the resurgence of right-wing politics in the 1960s, which earned him the nickname "Mr. Conservative." The book that launched Goldwater to national prominence was his *Conscience of a*

Conservative (1960). This widely read booklet laid out the senator's views on numerous controversial political issues of the day. It resonated with millions of conservatives across the country.

On the topic of human nature, Goldwater's book addressed "the corrupting influence of power": "the natural tendency of men who possess *some* power," he wrote, is "to take unto themselves *more* power. The tendency leads eventually to the acquisition of *all* power."[18] In Goldwater's worldview, man's competitive nature had no limit.

In 1980, when the conservative politician Ronald Reagan asked Americans for their votes at the end of his presidential campaign, he said: "As you go to the polls next Tuesday and make your choice for President, ask yourself these questions: Are you better off today than you were four years ago? Is it easier for you to go and buy things in the store than it was four years ago?" David Sears, a political scientist at UCLA, has pointed out how right-wing politicians like Reagan tend to make more appeals to the public based on the assumption of a self-interested audience.

Sears has contrasted Reagan's speech to that of the Democratic president John F. Kennedy. In 1961, Kennedy famously entreated his "fellow Americans [to] ask not what your country can do for you; ask what you can do for your country." In Kennedy's speech, the liberal president invoked a cooperative human nature.[19]

In the late 1960s, left-wing peace activists in US colleges were demonstrating against their country's war in Vietnam. Psychologists who analyzed the ideological themes of their protests noted "a strong antipathy to self-interested behavior." When these liberal students took psychometric tests, they measured significantly higher than nonactivists on humanitarianism, which included a strong "desire to help others" and a valuing of "compassion and sympathy." The researcher noted that the most radical left-wing activists also had unrealistic expectations about how cooperative others would be in supporting them when they graduated; this group of students planned to continue to "work full-time in the 'movement' or . . . to become free-lance writers, artists, [or] intellectuals."[20]

Even though our leaders' perceptions of human nature are normally less skewed, their biases can nonetheless have wide-reaching policy repercussions. In 2009, the conservative politician George W. Bush had finished eight years as president of the United States. Barack Obama then assumed office, bringing a liberal administration to the White House. Obama apparently believed that his predecessor's conservative view of human nature hindered US relations with the Muslim world by focusing too much on military interventions, counterterrorism measures, and coercive interrogation techniques. More than right-wing Americans, Obama assumed that human nature is cooperative—even across cultures. Thus, reaching out to Islamic countries was Obama's top foreign-policy priority.

On the very first day of his presidency, Obama called the leaders of the Palestinian Authority, Jordan, and Egypt. He also called Israeli prime minister Ehud Olmert to request that Israel cooperatively open its borders with the Gaza Strip (even though Gaza was under the administration of Hamas, which both the Israeli and US governments considered a terrorist organization). Obama then announced the appointment of a special envoy to promote a US-brokered peace process in the Middle East (a move that President Bush had resisted).[21]

Obama granted his first interview as president to the Arab cable TV network Al Arabiya. One of his first foreign trips was to Turkey and Egypt. At Cairo University, in the heart of the Arab world, the newly elected liberal president reached out to Muslims, offering them "a new beginning" based on "mutual interest and mutual respect." During his presidency, Obama would prohibit torture and ban the phrase "war on terror" from official government discourse.[22] Underlying these policy shifts was an assumption that Muslim societies had a predominantly cooperative nature—as long as the United States shifted its approach to them in a more egalitarian direction.

Differing perceptions of human nature may divide the left from the right on economic issues as well. Evolutionary economist Paul H. Rubin of Emory University has suggested that "preferences regarding altruism" translate into different fiscal policies.[23] Rubin means that liberals (who perceive human nature as more cooperative) favor greater income redistribution than conservatives (who seek to reduce taxes).

To the extent that people identify free-market capitalism with self-interest, capitalism has polarized the political spectrum. The far left has decried self-interested capitalism as the root of all evil, and accused the right of celebrating self-interest by worshiping the god of free markets. The far right, on the other hand, has denounced socialist control economies for impeding the pursuit of competition and sapping away motivation.[24]

Belief in a Dangerous World

If, as conservatives tend to believe, human nature is fundamentally competitive and self-interest prevails, then people live in a dangerous world. The "dangerous world" metaphor has long been associated with right-wing ideological views. In the last couple of centuries, though, this metaphor has taken the form of folk-Darwinism. University of Michigan philosopher Peter Railton has dubbed this worldview "your great-grandfather's Social Darwinism," in which "all creatures great and small [are] pitted against one another in a life-or-death struggle to survive and reproduce."[25]

In fact, folk-Darwinism's ruthless "survival of the fittest" concept is a one-sided (and frequently distorted) view of the fuller scientific picture of evolution that has developed over the second half of the twentieth century. Since the 1960s, biologists have made major advances in understanding how evolution motivates various kinds of altruistic cooperation in nature—in addition to self-interest (which we'll learn about in part VI). Nonetheless, public opinion's idea of folk-Darwinism, which situates people in a dangerous jungle world, has generally been evoked to support a right-wing moral philosophy.[26]

Numerous political psychologists have commented on the right's "Darwinian" dangerous-world metaphor. The *Authoritarian Personality* group at UC Berkeley remarked how highly ethnocentric subjects had "a conception of a dangerous and hostile world" that resembled an "oversimplified survival-of-the-fittest idea." One conservative subject recalled the discipline that he used to receive from his father: "I always accused him of being harsh. . . . And apparently this all falls in with Darwin's theory too."[27] Others who have linked folk-Darwinism's dangerous-world motif to conservatism include the British psychiatrist Roger Money-Kyrle (1951), Princeton political psychologist Fred Greenstein (1975), and Berkeley metaphor theorist George Lakoff (2002).[28]

The social-Darwinist survival-of-the-fittest idea appears most obviously and prevalently in the discourse of the extreme right. Adolf Hitler saw life as a zero-sum struggle between the races, in which one group would always seek to dominate the other.[29] In a 1928 speech that Hitler gave in Kulmbach, Bavaria, he envisioned a conflict between races in pseudo-Darwinian terms:

> The idea of struggle is as old as life itself, for life is only preserved because other living things perish through struggle . . . in the struggle, the stronger, more able, win, while the less able, the weak, lose . . . it is not by the principles of humanity that man lives or is able to preserve himself above the animal world, but solely by the means of the most brutal struggle.[30]

Hitler erred by confusing strength and animal brutality with fitness. He overlooked how cooperative behavior in human and nonhuman animals plays a major role in contributing to fitness, including the struggle for survival. Moreover, the inhuman acts committed by humans in the name of Hitlerism greatly surpassed the brutality of any other known animal. Nonetheless, Hitler viewed the world as extremely dangerous, and he attributed the danger to a misconstrued social Darwinism.

At the turn of the twenty-first century, ethnographer Raphael Ezekiel discovered the same, naturalistic dangerous-world metaphors among extreme right-wing groups in the United States. Richard Butler, the leader of the Aryan Nations, used animals to describe his political convictions:

> Of course, we know the jungle, that the lion will eat the rabbit, and a rabbit doesn't have any right. He doesn't have any right to life, he has a right to use the endowments that nature and the genetic program have, to save his life. In other words, to run down the hole when he sees something [stronger] coming after him, a coyote or something.

Tom Metzger, the leader of White Aryan Resistance, expressed a perception of human nature in which competition is taken to the extreme. Metzger believed that little had changed since Hobbes's state of nature, since life remained a war pitting man against man: "either I am strong enough to defeat you or you will smash me. It's simple," he said.

Ezekiel uncovered a similarly dangerous worldview among a Detroit cell of neo-Nazis. What most impressed the ethnographer about the neo-Nazis after his months of fieldwork with them was the emotion of fear: "These were people," he explained, "who at a deep level felt terror that they were about to be extinguished. They felt that their lives may disappear at any moment."[31]

A palpable fear of a dangerous world figures prominently even in the mainstream American far right. Glenn Beck, the Tea Party–aligned public-opinion leader, feels that people are out to get him. Beck wears bulletproof vests when speaking in public. He also planned to build a six-foot barrier around his home in New Canaan, Connecticut, to barricade his residence against bullets (but the security wall conflicted with local zoning ordinances).[32]

Beyond these individual cases of political leaders, is there any hard proof that right-wingers *in general* fear a dangerous world any more than run-of-the-mill liberals do? Beyond conventional questionnaires, innovative lab research has shed important light on this human-nature problem. University of New Mexico psychologist Jacob Vigil used computers to alter photographs of human faces; he simplified the photographs into "sketches," and then blurred them to create emotionally ambiguous expressions. Next, Vigil had 740 adults interpret the emotions on the faces. Republicans were significantly more likely than Democrats to see threatening or dominating emotions in the hazy faces (other factors such as gender, age, and employment status, however, did not affect their perceptions).[33]

Thanks to independent experiments later conducted at the University of Nebraska, we now know that as people process these facial emotions, there are also differences in brain activity between liberals and conservatives. Compared with liberals, conservatives process dominant emotions (like anger and disgust) more quickly. Karl Evan Giuseffi and John Hibbing discovered this curious physiological trend by using electroencephalograms to measure electrical activity in the brains of people as they viewed dominant and neutral facial expressions. Because the brain's electrical response to these faces occurs within a third of a second, these rapid, ideologically important reactions appear to take place at an unconscious level of awareness.[34]

Is there any other way to objectively determine whether conservatives truly fear a dangerous world more than liberals? Political scientists and psychologists from Rice University and the Universities of Nebraska and Illinois came up with a creative experiment to answer this question beyond a reasonable doubt.

The researchers showed a series of thirty-three images to a group of adults with strong political beliefs. Three of the pictures depicted threatening conditions: the first had a very large spider on the face of a frightened person; a second image showed a dazed individual with a bloody face; and the third one showed an open wound with maggots in it. In a control condition, the psychologists replaced the three startling pictures with nonthreatening ones (a bunny, a bowl of fruit, and a happy child).

While the subjects viewed the images, the scientists monitored their skin conductance to measure fear (arousal causes tiny amounts of perspiration, which alter how well electricity flows across the body's surface).

The research team discovered that the individuals who had a higher physiological response to the threatening images (i.e., the people who were more startled and therefore sweat more) were significantly more likely to have conservative attitudes. For instance, they tended to support capital punishment, patriotism, and the Iraq War. Those who were less startled by the threatening images generally supported pacifism, foreign aid, and liberal immigration policies. Physiological responses to bunny rabbits and bowls of fruit, of course, did *not* predict political orientation.[35]

This dangerous-world experiment is particularly valuable because it is quite clear what causes what. If an MRI scan reveals differences between the brains of liberals and conservatives, we cannot be sure whether nature or nurture is ultimately responsible. Brain differences could be innate or environmentally acquired or both. But the physiological fear responses in this "spider" experiment depend on sweat; and the sweat glands are controlled by the sympathetic nervous system, which is *involuntary*. It's highly unlikely that sweating in response to a spider depends on the environment in which a subject was nurtured (especially if the subjects are from the same location) and that the conservatives happened to have more traumatic spider experiences than the liberals.

Beyond introspective surveys and subjective ratings of facial emotions, then, it seems as though conservatives truly *do* perceive the world to be a more dangerous place than liberals do—even while asleep. Research on the dream lives of Americans found that Republicans reported nearly three times as many nightmares as Democrats. Conservatives were also more likely to initiate physical aggression in their dreams, and they were twice as likely as liberals to dream about male characters. Left-leaning dreamers, in contrast, reported more female characters in their dreams.[36]

Belief in a Degenerating World

In addition to believing in a competitive human nature and a dangerous world, right-wingers are also more likely to perceive that the world and its morality are increasingly *degenerating*. The term "conservatives" itself implies a desire to keep what is good and prevent it from deteriorating into something worse. The political left, in contrast, is more prone to thinking that human nature can evolve into something better. The term "progressives" implies this belief that the advancement of morality is possible and desirable.

The sensation of deteriorating social morality is not merely an artifact of the modern Western world; people in ancient Greece, Israel, China, Rome, and nineteenth-century Europe have expressed similar concerns, according to the research of social psychologist Richard Eibach. Likewise, many Americans commonly point to a perceived rise in teenage pregnancy as proof of moral decay. A 2003 poll revealed that 68 percent of adults thought teen pregnancy was on the rise—even though teen births had fallen by 31 percent over the last decade.

The likelihood of holding a "degenerating world" perspective, however, depends significantly on one's political orientation. The 2000 National Election Study asked Americans whether they thought that "*newer lifestyles are contributing to the breakdown of our society*." Fully 80 percent of conservatives agreed with this statement, compared with 59 percent of moderates, and 49 percent of liberals.[37]

The RWA test takes advantage of this relationship to measure political orientation. The scale's content within the "human nature" category heavily pertains to the "degenerating world" theme. Conservatives are more likely to agree with RWA items that include the following phrases:

- "*The way things are going in this country, it's going to take a lot of 'strong medicine' to straighten out the troublemakers, criminals, and perverts.*"
- "*. . . the rot that is poisoning our country from within.*"

- *"The facts on crime . . . and the recent public disorders all show we have to crack down harder on deviant groups and troublemakers if we're going to save our moral standards. . ."*
- *"In these troubled times . . ."*

Each of these phrases suggests a society in the process of moral decay. They express the sensation not only that human nature is bad, but also that it's growing even worse.

It's quite easy to find "reactionary" politicians on the right of the political spectrum who believe that the present and future are decadent eras, and who yearn to return to a past that they believe was more moral. Newt Gingrich is one of them. Gingrich served as a Republican congressman from the conservative US state of Georgia for twenty years. In March 2011, he announced his bid to run for president on the Republican ticket. In Gingrich's worldview, the moral fabric of America is increasingly deteriorating. He has especially expressed this "degenerating world" opinion during the administrations of liberal presidents.

In 1995, Gingrich had just become speaker of the House of Representatives, to lead his party in opposition to Bill Clinton. Gingrich claimed: "We had long periods in American history where people didn't get raped, people didn't get murdered, people weren't mugged routinely."[38] Fifteen years later, in 2010, Gingrich published a book titled *To Save America*. It argued that the values of the United States' founding fathers were quickly slipping away. America's morality was under dire threat, Gingrich argued, from President Obama's "secular-socialist machine."[39]

Further to the right, the "degenerating world" theme grows even stronger. Commentators have described Glenn Beck's television program as having an ethos of "extreme doom and pessimism." Even an anchor from Beck's own conservative outlet, the Fox News Channel, called Beck's show "the Fear Chamber."[40]

On the extreme right end of the spectrum, people's assessment and prognosis of human nature becomes the gloomiest. A Detroit neo-Nazi interviewed by Ezekiel offered his thoughts on the subject: "People are getting worse and worse and worse . . . and murder and torture and all this. Just for the sake of laughing and enjoying it."[41]

If conservatives fear that human morality is deteriorating, then the political left is well known for its hope that human nature will improve. Before he became a liberal president, Barack Obama gained his country's attention by publishing a book called *The Audacity of Hope* (2006). During his Nobel Peace Prize lecture at the end of 2009, the progressive president touched on his theory of human nature: "we do not have to think that human nature is perfect for us to still believe that the human condition can be perfected," Obama said. Not "improved," but "perfected"—the same term that Rousseau used.

On the far left, people have believed that, after capitalism, humanity would pass to the more moral stage of socialism, and eventually to the utopia of Communism. Communism entailed a classless society in which morality would be so fantastic that it would render government unnecessary.

So what is the truth about human nature? In part VI, we'll look beyond philosophy and the social sciences for answers. But first, one important question remains.

18.

DO PERCEPTIONS OF HUMAN NATURE CHANGE AS WE AGE?

Only about nine thousand people live in the little town of Praxedis, which stands on the southern side of the border between the United States and Mexico, thirty-five miles away from Ciudad Juarez. Yet Praxedis is located in what has become one of the most dangerous places on earth. In 2010, over two thousand five hundred homicides occurred in the Juarez Valley alone, as the Sinaloa and Juarez drug cartels fought to control smuggling routes into the United States. One of these contraband passages is the single highway that runs through Praxedis.

In addition to attacking each other, the rival mafias have killed civilians for resisting corruption, for working for an enemy cartel, or to intimidate the relatives of their victims. The drug cartels have especially targeted and slain hundreds of police officers and their family members. These assassinations often involve gruesome mutilations, which function as extreme coercive measures to enforce the cartels' interests.

In 2009, for example, the cartels tortured and then decapitated the police chief of Praxedis. No one wanted to replace the head of police for fear of meeting the same fate. The mayor asked nineteen people to take on the position, but one after another, each candidate refused. And so, the position remained vacant for over a year.

Then in October 2010, the twentieth person considered for the position was a twenty-year-old criminology student named Marisol Valles García. Although Ms. Valles hadn't yet finished her degree, she agreed to become her town's chief of police. International media outlets descended on Praxedis to interview the brand-new head policewoman. Ms. Valles told the press that she was "just tired of everyone being afraid."

Once appointed, Chief Valles's first move was to hire more policewomen onto her thirteen-member force. Although the officers were mostly female and unarmed, Valles told reporters that she planned to fight crime "with principles and values." In other words, her approach to law enforcement would be as extremely egalitarian as the cartels' enforcement was hierarchical. Valles intended "to foster greater cooperation among neighbors so they [could] form watch committees." To make progress, then, the police chief depended on a community whose nature was (or could become) cooperative—even in the face of grave danger.

The media dubbed Ms. Valles "THE BRAVEST WOMAN IN MEXICO." Many members of the public agreed, but others argued that she was far "too young and naïve" to succeed. One newsreader commented:

> Good luck to her. Her "Pollyanna" attitude will likely get her and her female officers raped, tortured, and killed. She clearly has no idea of what police work is like and certainly no concept of the viciousness and cruelty of drug gangs. Adios Señorita [Valles] Garcia.

After a few apparently uneventful months on the job, Marisol Valles disappeared. When she stopped showing up at work, the town had to dismiss her from her position.

In March 2011, Ms. Valles surfaced somewhere near El Paso, Texas, where she and her family were hiding while they sought asylum in the United States. What happened after the Valles family vanished?

What we know is what Ms. Valles told American media outlets on the northern side of the border. In one interview, she said that the cartels had threatened to kill her and her newborn baby boy if she refused to work with them. When Ms. Valles told American journalists of her ordeal, she cried and explained: "We were fighting for our people, our community. I never thought that things would turn so ugly. Maybe I was too naïve." The once-fearless former police chief also told the media: "I was afraid, I couldn't sleep anymore, and I was always thinking about when they would come for me. . . . The fear will never go away. What I experienced is a fear that will last a lifetime." Despite fleeing to the United States, Ms. Valles pleaded to her constituents across the border "not to let fear get the better of us."

Some readers of the Spanish-language press admired Ms. Valles for her efforts. Others considered her well meaning but inexperienced. Many of these readers, however, questioned whether the former police chief had truly put her community's interests ahead of her own. They accused her instead of seeking fame and having planned everything from the beginning to acquire US citizenship.[1]

In essence, the controversy boiled down to whether Ms. Valles had been exceptionally naïve or exceptionally cynical. What did she truly expect of human nature? Did the young woman really believe she could work with a cooperative community against the drug cartels? Or did she expect the ruthlessly competitive drug cartels to threaten her, after which she would whisk her family away to safety and launch a high-profile petition for asylum in a safer country?

In general, are young people any more or less cynical than adults? If perceptions of human nature do change with age, does political orientation shift as well?

PHILOSOPHICAL AND CULTURAL BELIEFS ABOUT CYNICISM AND POLITICAL ORIENTATION ACROSS THE LIFESPAN

Aristotle believed that views of human nature dramatically reverse as one ages. In the fourth century BCE, the Greek philosopher wrote in *Rhetoric* (his treatise on persuasion) that the young

> look at the good side rather than the bad, not having yet witnessed many instances of wickedness. They trust others readily, because they have not yet often been cheated. They are sanguine; nature warms their blood as though with excess of wine; and besides that, they have as yet met with few disappointments. . . . They are easily cheated. . . . Their hot tempers and hopeful dispositions make them more courageous than older men are; the hot temper prevents fear, and the hopeful disposition creates confidence. . . . They are ready to pity others, because they think every one an honest man, or anyhow better than he is: they judge their neighbor by their own harmless natures, and so cannot think he deserves to be treated in that way. (Book II, Ch. 12)

Aristotle argued that by the latter half of life people's rosy perception of others is completely turned on its head:

> Men who are past their prime . . . have often been taken in, and often made mistakes; and life on the whole is a bad business; they are cynical; that is, they tend to put the worse construction on everything. Further, their experience makes them distrustful and therefore suspicious of evil. . . . They are small-minded, because they have been humbled by life: their desires are set upon nothing more exalted or unusual than what will help them to keep alive . . . their experiences taught them how hard [money] is to get and how easy to lose. They are cowardly, and are always anticipating danger. . . . They live by memory rather than by hope. (Book II, Ch. 13)[2]

In the Far East, Chinese culture also expects people's view of human nature to change over the life cycle, and in the same manner. This assumption underlies the saying that people are helped by Confucianism when they are young and by Buddhism when they are old. What does this mean?

This expression assumes that young people are prone to an egalitarian ethic, and to rebel against authority. To temper these dispositions, Confucianism provides a hierarchical framework that emphasizes filial piety to parents, and also obedience to the proper societal authorities. That is, Confucianism is believed to temper the relative liberalism of youth with the philosophy's conservative worldview.

If the old tend to grow overly cynical, then Buddhism, according to the saying, would help balance them in the other direction. This religious philosophy's third Noble Truth promises that the path to eliminating suffering is to eliminate craving. Releasing attachment to desire—to what secular people consider their self-interest—is believed to advance one toward enlightenment.

Most schools of Buddhism practiced in China also do *not* believe in a competitive human nature; rather, Mahayana Buddhism insists that all beings have an innate, enlightened essence called Buddha-nature (佛性; fó xìng). Therefore, all beings have the potential to achieve liberation (however gradually). The original expression, then, implies that Buddhism acts as a liberalizing force to temper the conservatism of the elderly.

In early twentieth-century Europe, the Irish playwright George Bernard Shaw (1856–1950) also assumed that the young are—or at least should be, in his opinion—more idealistic than their elders. Shaw famously wrote: "Any person under the age of thirty, who, having any knowledge of the existing social order, is not a revolutionist, is an inferior."[a]

Winston Churchill, the British prime minister who led his country through World War II, is supposed to have once said: "If you're not a liberal when you're twenty-five, you have no heart. If you're not a conservative by the time you're thirty-five, you have no brain." Again, these words echo the same assumptions as the previous statements—that the young, led by idealistic hearts that overestimate human nature, lean left. And that older adults lean to the right because of their cynical minds.

Like so many other famous quotes, however, this one is apocryphal. The actual author of the quotation was the French statesman François Guizot (1787–1874), who said: "Not to be a republican at twenty is proof of want of heart; to be one at thirty is proof of want of head."[3] By "republican," Guizot meant a liberal who supported a republic (as opposed to a conservative who favored monarchy). Variations of the quote have been either borrowed or misattributed to a long string of nineteenth- and twentieth-century politicians from France, England, Germany, and the United States. In subsequent cases the word "republican" has turned into "socialist" or "liberal." So the original meaning has never changed: a sensible person is supposed to begin life on the left by virtue of emotions, and to migrate to the right with a mature mind.

"My God! I went to sleep a Democrat and I've awakened a Republican."

During the turbulent late 1960s in the United States, the leftist counterculture produced a warning that associated their political enemies with age: "those over thirty are not to be trusted," some leftists factions would say, "because of their increasing investment in the system as it exists."[4] This belief may have assumed that people over the age of thirty became more conservative because they were more likely to own a house, to earn a higher salary, and to have too much at stake to back a revolutionary call to destroy the existing social, economic, and political orders.[b]

Even though philosophers and cultural beliefs do not agree on the actual essence of human nature, at least one consensus has emerged: people's level of cynicism increases as they mature, pushing them further to the right on the political spectrum than where they began. But is there actually any scientific proof that people grow more conservative with age?

AGE AND THE BIG-FIVE PERSONALITY DIMENSIONS

One quick way to see whether political orientation shifts with age is to check the enormous literature on personality studies. As we learned in chapter 3, a couple of the major personality traits that psychologists conventionally measure are well associated with political attitudes; of the Big Five dimensions, higher Openness correlates with left-leaning voting, and higher Conscientiousness predicts approval of right-wing parties.

Studies in the United States, Germany, Italy, Portugal, Croatia, and South Korea have all drawn the same conclusion: as subjects in each country live through their twenties, Openness significantly drops and Conscientiousness significantly rises. Indeed, most large personality changes take place before the age of thirty. Perhaps François Guizot and the 1960s counterculture were on to something: if an individual's political personality hasn't changed by the time of his or her thirtieth birthday, it's not likely to differ all too much at forty or fifty. This tendency doesn't mean that no change can take place after thirty—only that normal personality change in middle age occurs at a substantially slower pace. For all practical purposes, the personality stabilizes after thirty.[5]

These findings coincide with an even larger study, which surveyed thirty-six cultures in seven ethno-linguistic families across Africa, Europe, and Central, South, and East Asia. All over the world, the average Big Five personality traits of human beings between the ages of eighteen and twenty-two systematically differed from those of older adults.[6]

How can we be sure that these lifecycle personality shifts *really* correspond to greater political conservatism? And what role might perception of human nature play?

AGE AND RWA SCORES

In contrast with the Big Five studies, RWA is a much more direct measure of political orientation. But the drawback is that fewer RWA studies exist, and they have measured much smaller samples of people.

Robert Altemeyer began one such study in 1974, when he took the RWA scores of a group of eighteen-year-old college freshmen (see figure 40). Four years later, he retested the same group of students, who had become graduating seniors at the age of twenty-two. Finally, twelve years after the first test, the psychology professor succeeded in tracking down seventy of the original students. By this time, the cohort was thirty years old.

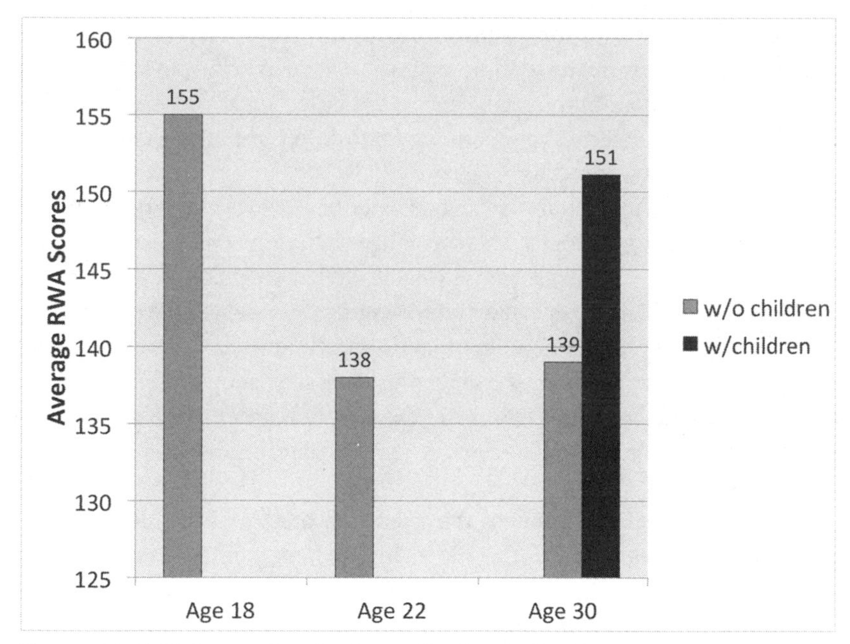

Figure 40. RWA Scores Tracked over Twelve Critical Years.

Between the first two tests, the impact of college made the cohort's average RWA score drop by 11 percent. Why did Altemeyer's students become more liberal, even as they aged four years? The main reason is probably sample bias: the students were a self-selected group that had chosen, in the course of a liberal-arts education, to study psychology. And social scientists, according to a 2010 study published in the *New York Times*, are over twice as liberal as the average worker.[7] So four years of taking classes from liberal professors could have made psychology students more liberal. This theory coincides with Altemeyer's own data: between their freshman and their senior year, the RWA scores of initially conservative students (in the top quartile of RWA scores) *dropped* 2.4 times more than the scores of the liberal freshmen.

Between the ages of twenty-two and thirty, the average RWA score grew more conservative by 5.4 percent. A huge difference emerged among the cohort at the age of thirty, however, when Altemeyer separated the students who had had children from those who hadn't. His twenty-three former students without children had virtually the same average RWA score as they did as college seniors; their score had increased by less than 1 percent. But in the case of the forty-eight former students who had children, their RWA scores became 9.4 percent more conservative.[8]

Of course, we still don't know for sure whether having children makes people more conservative, or whether people who are conservative in the first place are more likely to have children. At least in Altemeyer's longitudinal study, the former possibility seems to be the case, since the freshman RWA scores of those who later had children hardly differed from the freshman scores of the childless thirty-year-olds.

Why would having children push one's political views to the right? Richard Eibach, a political psychologist who specializes in judgmental biases, offers one possibility. For new parents, Eibach explains, "the responsibility of taking care of a vulnerable child requires a new type of vigilance, in which the parent is alert to sources of danger that a nonparent could safely ignore."[9] The result can be an "illusory increase in external dangers." Eibach has shown that new parents tend to believe that crime rates have increased with respect to the period before their transition to parenthood—even when actual crime rates have dropped dramatically.[10]

With parenthood, then, comes the need to protect a young child who has minimal defenses against the threats in his or her environment. As parents become attuned to these "new" threats, their attention focuses more and more on the dangerous elements of the world.[c] Since "dangerous world" thinking is associated with a more self-interested view of human nature, parenthood increases conservatism.

Perhaps the left-wing members of the 1960s counterculture admonished each other not to trust anyone over thirty because older adults are more likely to be parents. Along with their conservative focus on a dangerous world, parents are people who have become the prototypical authority figure. So parents are more likely to identify with societal authorities. For their part, political candidates spend no small amount of energy in persuading public opinion that they're capable family authorities who can transfer these skills to managing government. For the same reason that the head of a family represents authority, however, the radical left explicitly opposed the conventional family and considered marriage "unhip."

We should also keep an open mind to the possibility that factors other than parenthood can raise the conservatism of aging humans. After all, certain personality traits linked to conservatism seem to slowly increase even in later adulthood. Experts on aging believe that people's personalities may be hard-wired to shift over time; these changes would better adapt people to different phases of the lifecycle.

Robert McCrae, for instance, has speculated that "Openness" (in addition to the Big Five trait of "Extraversion") is higher during youth because being outgoing helps young adults to find mates. Perhaps the sanguine optimism and "warm blood" that Aristotle attributed to them serves this function as well. An increase in "Conscientiousness" in the late twenties, on the other hand, would better suit parents to the task of raising a family; being a good parent requires a great deal of responsibility.

If becoming a parent doesn't alter these personality traits, then what would? McCrae proposes that age-related changes in gene expression may shift an individual's personality traits, including Openness and Conscientiousness. Several facts lend credibility to this possibility: (1) personality traits themselves have a high level of heritability; (2) numerous processes known to be under genetic regulation develop over time; and (3) "personality changes from late adolescence to early adulthood are themselves modestly to moderately heritable."[11]

The big picture is quite clear: perceptions of human nature shift over the lifespan, as do political orientations. In late adolescence and much of the twenties, the political personality is particularly flexible. By the age of thirty, however, personalities tend to stabilize. Even though normal variation exists in these traits at any age, they do tend to shift from their starting points. Some time during the mid-to-late twenties, people grow more cynical and conservative. Parenthood apparently accelerates this conservatism. In addition, it's possible that genetically regulated personality changes help people better adapt to each stage of life. We'll unravel this mystery much further in chapter 21.

From any angle one looks at the question of human nature, it is solidly linked to our constitution as political animals. So what can science reveal about our true human nature?

PART VI.
ILLUMINATING OUR TRUE HUMAN NATURE

The characters in Sergio Leone's spaghetti Western *The Good, the Bad, and the Ugly* (1966) personify, if somewhat exaggeratedly, the political philosophies of human nature discussed in part V. Some thinkers have viewed people as fundamentally cooperative, like the Good character, Blondie (played by Clint Eastwood); this honest, lone gunman saves the lives of wounded soldiers, even at his own risk. Other philosophers have believed human nature to be essentially competitive, like the Bad character, Angel Eyes (Lee Van Cleef); this ruthless, sociopathic mercenary assassinates men for money—including his own clients.

But who is correct? Is human nature intrinsically cooperative, as argued by Aristotle, Mencius, and Rousseau? Or is our core nature competitive, as claimed by the Sophists, Xunzi, and Hobbes? The question is crucial because this perceptual bias impacts our political orientation.

Today, numerous scientific disciplines have clarified various aspects of human nature, encroaching on an area of inquiry that once belonged exclusively to philosophy. Instead of allowing subjective political biases to influence the debate, these modern fields have objectively defined and measured altruism and self-interest.

As our understanding of human nature comes into higher resolution, we find the way in which philosophers once posed the original human-nature question to be increasingly problematic. Evolutionary economist Paul Rubin has remarked: "the true original state of humans is so different from the hypothesized state of nature that arguments proceeding from this basis generally cause confusion rather than enlightenment."[1] Political scientist Glendon Schubert has also rejected "the presumption that our political theory as a species began 2,500 years ago in Athens" or that a social contract originated "as described in 'naturalistic' fables (whether optimistic like that of Rousseau or pessimistic like that of Hobbes)." Instead, Schubert insisted in 1976 that

> the roots of political behavior go back not thousands but millions of years. . . . The implications of contemporary research in physical anthropology . . . and related sciences are going to jack political philosophy off its classical assumptions—once political scientists become better educated in, and start facing up to the facts of biological life including their own life history as a species.[2]

Unlike the philosophical and folk-psychology perspectives that view human nature as basically "good" or "bad," evolutionary research finds that human altruism and self-interest are quite . . . complex.

Perhaps the science of the complex is better embodied by the third protagonist of Leone's Western film, the "Ugly" bandit named Tuco Benedicto Pacífico Juan María Ramírez (played by Eli Wallach). Tuco shows greater loyalty to his family than to non-kin. Yet he warily reciprocates trust with friends, and makes alliances when it's convenient. If Tuco feels betrayed, he attempts to exact vengeance on his enemies.

Unlike the "Good" and the "Bad" characters, Tuco is the only well-developed and realistic character. He has a past and a story—unlike Angel Eyes or "Blondie," who is "the man with no name." The Good and the Bad are merely philosophical foils, invented as literary devices to highlight the moral forces tugging from different directions at Tuco's lifelike conscience.

Instead of jumping to philosophical conclusions on human nature, then, part VI argues that it's time for political science to approach human nature from a scientific perspective. It's time to integrate the field, whose paradigms have shifted every twenty years, into an evolutionary framework, whose permanence extends beyond the foreseeable future.

From the scientific perspective, a human nature *does* exist. As the great evolutionary biologist William D. Hamilton once remarked, "the *tabula* of human nature was never *rasa*," and it "is now being read."[3] How do we decipher our slate? What does it say?

Part VI shows that the biology of altruism is the best guide to human nature. And studying altruism reveals the most accurate portrait of wo/man the "political animal." We'll learn about altruism from studies on nonhuman animals—and also from experiments carried out among diverse human cultures. We have gained much of this knowledge about ourselves through "economic games," which are often played with real money; these experiments help quantify altruism, trust, self-interest, and betrayal.

The following chapters explore a rich typology of altruisms, beginning with the simplest forms and progressing to the most complex. Chapter 19 explains kin-selection altruism. Chapter 20 covers reciprocal altruism, including game theory, indirect reciprocity, costly punishment, and cheating.

Next, chapter 21 considers how neurological changes that occur over the human lifespan appear to change perceptions of self-interest. That is, the maturation of the brain may alter an individual's assessment of how altruistic he truly is, and how altruistic others are.

Chapter 22 describes a concept that we'll call "self-deceptive altruism." Here, we learn why it is evolutionarily adaptive for humans to have a distorted perception of reality, such that people believe themselves to be more altruistic than they actually are. This theory of "self-deceptive altruism" helps explain the substantial confusion over the nature of human nature. It also sheds light on why political scandals perpetually shock both the public and the perpetrators themselves. Finally, we'll see why self-deception over human self-interest explains the similarities in authoritarian governments on both the extreme right and the extreme left of the spectrum.

The last chapter reflects on the phenomenon of heroic rescuers who incur enormous risk to their own lives to save unrelated strangers. This behavior is difficult to account for, and it may lie beyond the explanatory power of current scientific knowledge.

19.
THE CONSERVATIVE ALTRUISM
KIN SELECTION

WHAT IS ALTRUISM?

In part V, we surveyed views of human nature from philosophical and psychological perspectives. Distinguishing "cooperative" from "competitive" seemed to be a fairly straightforward exercise. However, as we dig deeper beneath the surface to find the evolutionary roots of our moral nature, we will concentrate less on subjective perceptions and more on objective actions. This approach demands greater scientific precision in defining concepts and measuring behavior.

The notion of "cooperation" becomes too vague. What does it really mean to cooperate? A traveler could cooperate with a highway robber instead of fighting. Or two merchants from different regions could cooperate by bartering corn for potatoes. A lifeguard could rescue a drowning baby from a swimming pool—or an ocean-going snorkeler from sharks. In each case, the potential costs and benefits to the cooperator are extremely different.

This is why evolutionary biology focuses instead on a concept called "altruism." Although people commonly confuse altruism with cooperation, it's important to understand that altruism is a very specific type of cooperation. An act is altruistic if it benefits some individual, but at a cost to the altruist who willingly performs the act. Moreover, we should be able to define costs and benefits in terms of Darwinian fitness (that is, the ability to survive and reproduce).[1]

In practice, it's usually difficult to prove that an altruistic act directly enhances the survival or reproductive success of the beneficiary at a cost in fitness to the altruist. Suppose that an individual gives an amount of food (or money) to another in an environment of economic abundance. A single, small donation might not actually affect the health of the individuals concerned or the quantity of offspring that each one produces in the course of their lifetimes. Theoretically, however, giving away resources linked to survival and reproduction has some bearing on the fitness of the individuals on both sides of the transfer. The impact might be very little. But if we can measure an evolutionarily costly behavior that benefits another, then we consider altruism to have occurred. In the coming chapters, we'll explore a number of noteworthy experiments that have devised different ways to measure altruism.

Within a population, the tendency for individuals to perform altruistic acts varies. Simply put, some people are more generous than others. And experiments have uncovered a substantial genetic component to altruism. We can even estimate the proportion of altruism that is heritable by studying twins. One study asked over 1,100 twins to report on a number of altruistic acts, such as donating blood. By calculating the correlations between the altruistic behaviors of fraternal twins, and then comparing them with identical twins, the researchers determined that the heritability of altruism is 56 percent.[2] This figure means that a slight majority of the variation in altruistic behavior between individuals occurs due to genetics, while the environment determines the rest.

Why would an individual willingly reduce his or her own fitness to increase that of another? Wouldn't evolution reward selfish individuals by multiplying their genes the most? The following chapters explore the different reasons why altruism *does* exist—as well as the evolutionary logic that makes altruism adaptive in the long run.

Here, we begin with the most elemental type of all, which is called kin selection. We find this form of behavior in many nonhuman animals, and it also plays a prominent role in the daily life and political nature of our own species.

KIN SELECTION

Imagine that you are a female Belding's ground squirrel, scurrying along an alpine meadow in the Californian Sierra Nevada. Suddenly, you sense a hawk swooping down to snatch up a meal. Instinctively, you seek cover and let out a high-pitched whistle to warn your fellow squirrels.

Is this cooperative alarm call an act of altruism? Statistics suggest that it is *not*; as all squirrels in hearing range flee for shelter, chaos ensues, and the sentinel actually increases its own chances of surviving, even as it helps others.

Suppose, however, that a weasel or a badger approaches the colony hoping to eat a squirrel. If you spotted one of these land animals, or a coyote, you could alert your colony by belting out a second type of alarm call. To warn of a terrestrial threat, you would trill half a dozen five-kilohertz shrieks per second. But there's a dilemma. Yelling out the trill alarm for ground predators raises your chance of being eaten by announcing your location. Yet the warning also increases the odds that other squirrels can safely escape. What's a squirrel to do?

If she is surrounded in the meadow by unrelated individuals, a squirrel will keep silent when she senses a terrestrial threat. But if she has relatives nearby, she will start to trill. We know this thanks to Cornell biologist Paul Sherman who, along with ten research assistants, spent over three thousand hours in the 1970s painstakingly tagging ears, dying rodent hair, and mapping out the squirrels' family trees.[3]

In addition to Sherman's squirrels, we now know that a range of other mammals and birds also share preferential alarm calls; in these species, individuals are more likely to warn close relatives about approaching predators despite a personal risk to the signaler.[4] But would a squirrel—or a vervet monkey—imperil its own safety to benefit a distant cousin? How exactly are they wired to deal with these life-or-death situations?

Hamilton's Rule of Kin Selection

In 1964, British biologist William D. Hamilton developed a rigorous theory that had the power to explain exactly when an animal should make the altruistic alarm call. His work, which became known as Hamilton's Rule of Kin Selection, revolutionized evolutionary theory. In less than fifty years, more than ten thousand publications have cited his seminal paper.

Instead of thinking in terms of the reproductive success of an individual, Hamilton considered the fitness of a gene. From the gene's standpoint, there were two ways to reproduce: (1) the gene could replicate itself directly, through sexual reproduction; or (2) the gene could cause its "host" to behave in ways that would help multiply other organisms likely to contain identical copies of the same gene. In other words, a gene could induce altruistic acts toward kin. Since family members are related by common descent, they are more likely than non-kin to share identical copies of a given gene. Richard Dawkins's book *The Selfish Gene* (1976) vastly popularized this concept.

An elegant and easy-to-grasp equation lies at the core of his Rule of Kin Selection:

$$C < B \cdot r$$

In plain English, this means that a gene for altruism can evolve in a population, given the following condition: the gene would cause an individual to behave altruistically if the **C**ost in fitness to the altruist is less than the **B**enefit in fitness to the kin, times the genetic **R**elatedness of this beneficiary (relatedness means the proportion of genes that the two individuals share in common).[5]

Let's make this more concrete. According to the rule, the squirrel should warn others of the approaching coyote if two littermates (i.e., full siblings) and a cousin are in the area. Since each full sibling shares half of the alarmed squirrel's genes, and the distant cousin shares an eighth, the altruistic alarm call could benefit more copies of the gene than it would endanger. In the worst-case scenario, the warning call could cost the life of the altruist squirrel (so the cost, *C*, equals 1), but the alarm could save two siblings ($2 \cdot 1/2$ the genes shared in common), plus one cousin ($1 \cdot 1/8$ shared genes), which equals 1.125. Since the potential cost of the sentinel's genes is less than the benefit to the shared genes of relatives ($1 < 1.125$), the squirrel should warn its three relatives.

With his theory of kin selection, Hamilton formalized an intuition held by his predecessors, the importance of which had not yet been fully recognized. The British geneticist and evolutionary biologist J. B. S. Haldane (1892–1964) was once asked if he would risk his life to save a drowning brother. Haldane responded: "No, but I would to save two brothers or eight cousins."[6] At least he would have broken even.

Suppose that Haldane did jump into a dangerous river to save two brothers at great risk to himself. If he were 100 percent certain that he would die and save both of them, it wouldn't really be worth it ($1 = 2 \cdot 1/2$). But if Haldane believed that there was some chance that all three brothers could come out alive, it would make sense for him to jeopardize his own life.

If Haldane came out of the river alive, he would be able to reproduce his own offspring, which would directly increase copies of his genes. Having more direct descendants would increase Haldane's *classical fitness*. Having saved his two brothers, they could both reproduce as well. Every two children that the brothers produced would replicate Haldane's genes as if he had an additional son of his own. By acting altruistically toward his brothers, Haldane would increase his *inclusive fitness*. This term encompasses both of the ways that the selfish gene replicates (through sexual reproduction and by aiding other organisms likely to host the same gene).

Hamilton's Rule explains not only altruism, but also selfishness. According to the Hamiltonian equation, it doesn't make sense for someone to sacrifice his life for only one brother. Robert Trivers has pointed out how Hamilton's Rule applies to parent-offspring conflict over sibling rivalry. According to this model, a sibling should only refrain from selfishness when the benefit to his brother is more than two times greater than the cost to himself. For a shameless sibling, then, the ideal distribution of resources (say, candy) would be "one for you, two for me" (sound familiar?). After all, the selfish sibling shares 100 percent of his genes with himself, but only half of his genes with his full-blooded brother.

Parents, however, share half of their genes with each child, so parents prefer to distribute resources equally among their offspring. Investing equally in each offspring maximizes the parents' inclusive fitness (provided each offspring has the same reproductive potential). It suits the parents' interests for the offspring to commit altruistic acts toward one another any time that the benefit to one sibling is greater than the cost to the other. Since a sibling only wants to do so when the benefit is more than twice as great as the cost, conflict arises between parents and offspring, Trivers predicts, over how well siblings treat one another.[7] Sibling rivalry is particularly intense during childhood, when the evolutionary stakes are high.

A theory is only as good as its ability to explain empirical facts. If we put the theory of kin selection to the test, how well does this form of altruism predict real-life human behavior?

Experiments for Testing Kin-Selection Altruism

At the University of Oslo, a group of biologists set out in 1998 to see how well Hamilton's theory could account for the behavior of their own graduate students. The best time of year to measure altruism, the professors decided, was during the season of giving. Just after Christmas, the researchers sampled a group of fifty Christian students. None of them had a salaried job, so purchasing gifts may have signaled a bit more generosity for the students than for those with greater disposable incomes.

The experiment was simple. The biologists determined how much each individual spent on presents, then compared the amount of money with the proportion of genes the giver shared with the recipient. The students invested roughly twice as much on parents, siblings, and children (with each of whom the givers shared 50 percent of their genes) as they did on their grandparents, nieces, and nephews (who shared only 25 percent of their genes).

The percentage of students who bought presents for relatives generally declined in relation to shared genes. All of the students gave gifts to relatives who shared half of their genes (parents, siblings, and children). Eighty-four percent bought presents for their grandparents (a quarter shared genes), 20 percent gave to their first cousins (an eighth shared genes), and none of them bought presents for second or third cousins.[8]

However, the biologists discovered a couple noteworthy exceptions to this trend. First, the gift givers invested the greatest amount of resources in their romantic partners. Why did they spend the most on the least-related individuals? The graduate students, who ranged in age from twenty-three to thirty-one, were just beginning their reproductive careers. Sacrificing a small amount of one's own fitness for a (potential) reproductive partner makes evolutionary sense: the transfer is an honest signal of commitment; moreover, a (future) Norwegian parent is, according to this small study, 100 percent likely to give Christmas gifts to his or her children. So giving to a stranger who would highly invest in your own children makes sense.

The second exception to Hamilton's Rule was a bit trickier. The amount of money that the students spent on aunts and uncles (who shared 25 percent of their genes) was as low as what they spent on their first cousins, with whom they only shared 12.5 percent of genes. Why? From the point of view of an individual trying to maximize inclusive fitness, it makes more sense to invest disproportionately in relatives with a higher reproductive potential. Cousins are younger than aunts and uncles; therefore, transferring resources to them could have a larger impact on reproducing family genes.

Finally, the students spent much more on their grandparents than on aunts and uncles, even though they shared a quarter of their genes with each of them. They probably did this because their grandparents belonged to the same direct, vertical lineage; so a student could have expected a greater inheritance from grandparents than from their aunts and uncles (who would tend to will their resources to their own children and grandchildren). It would be worth analyzing wills to determine whether this is true—and whether kin-selection altruism can explain inheritance patterns.

Various social scientists have done exactly that. A study of one thousand British Columbian wills revealed that the deceased benefactors bequeathed the greatest proportion of their estates to their closest genetic relatives, a lesser amount to more distant genetic relatives, and the least amount to non-kin.[9] A second study of fifteen hundred Californian inheritances between 1890 and 1984 also substantiated Hamilton's Rule.[10] As in Norwegian Christmas giving, however, the major exception in both the inheritance cases involved the distinction between relatives in a direct lineage (direct progenitors and descendants) and peripheral relatives (genetic relations who aren't direct progenitors or descendants, such as aunts, uncles, nieces, nephews, cousins, and siblings). Even though kin-selection theory predicts the same amount of altruism toward a grandson and a nephew (since both share 25 percent of one's genes), inheritance studies find that people favor their direct relatives (grandchildren) over peripheral ones (nieces and nephews).

In fact, people will their children almost five times as much of their property as they do to their siblings. The explanation may involve sibling rivalry, or perhaps the greater reproductive potential of the young.[11]

Europeans and European Americans, however, tend to have different family structures than those of other cultures. In many parts of the world, people have greater contact—and closer social ties—with their extended family members. Does Hamilton's Rule still apply in such societies, or do people treat their extended relatives more like close ones? To answer this question, British psychologist Elainie Madsen of University College London (UCL) and her colleagues devised an experiment to compare kin-selection altruism among students in the United Kingdom with two South African Zulu populations.

Madsen's team wondered how much pain individuals in each location would endure to win a prize for themselves, compared with a prize for family members of varying genetic relatedness, or for non-kin (friends or a charity). In the United Kingdom, the psychologists awarded the UCL students £1.50 for every twenty seconds they could withstand an isometric ski-training exercise. The pose involved sitting in a chair position with one's back against the wall, with calves and thighs forming a 90° angle. Within the first two minutes, holding the position becomes more painful—increasingly quickly.

When told that the prize would be for them (100 percent shared genes), the UCL students endured the pain significantly longer than when the prize would go to a parent or sibling (50 percent shared genes). By the same token, the students held the pose significantly longer to benefit their nuclear family members than they did to award the prize to uncles, aunts, nieces, or nephews (25 percent shared genes). However, there was no difference in the amount of pain the students would withstand for uncles/aunts/nieces/nephews compared with first cousins (12.5 percent)—or even between cousins and a charity.

The first Zulu population that the researchers tested was the remote community of Emmaus, located deep in the Drakensberg Mountains. The people of this relatively isolated village lived in traditional extended-family homesteads, and their impoverished economy revolved around subsistence farming. Due to "security and ethical reasons," the psychologists awarded the rural Zulus basic food staples instead of cash. For every twenty seconds an individual held the stress pose, the more oil, sugar, and tinned fish someone would receive.

Just like the British students, the Zulus from Emmaus endured significantly more pain for themselves than for a nuclear family member. But there was no difference in the amount of time the individuals held the pose for their "nuclear" family members or for genetic aunts, uncles, nieces, and nephews. And yet they held the pose longer for aunts/uncles/nieces/nephews (25 percent) than for first cousins (12.5 percent), and longer for the cousins than for charity (0 percent related).

Why did the British students discriminate between nuclear family members and uncles/etc., but the Zulus treated 50-percent relatedness the same as 25-percent? Madsen and her colleagues speculated that the extended-family residential system caused a closer childhood association with this larger kin group, making them like a larger nuclear family. The psychological effect of growing up in the extended-family compounds would be similar to the Westermarck effect (see chapter 11), in which childhood association de-eroticizes relationships—and likely imprints an individual with a template for identifying close kin. So being raised in close proximity to uncles and aunts made them more like mothers and fathers. But after the 25-percent-shared-genes boundary, an individual's supply of altruism apparently begins to run down.

The second Zulu community where the psychologists ran the stress tests was called Hluhluwe, located a couple hundred miles northeast of Emmaus. Although rural, Hluhluwe was much more integrated into a modern wage economy than Emmaus, since the town stands near a major trade route between South Africa and Mozambique. The Zulu population of Hluhluwe also lives in extended-family compounds, but a high proportion of men emigrate from the town in search of economic opportunities. According to Madsen, this emigration combines with polygamy and a limited availability of contraceptives such that establishing paternity—and therefore genetic relatedness—is difficult. Although these conditions apply to both Emmaus and Hluhluwe, male emigration and infidelity are likely higher in Hluhluwe because of its close location to the trade-route highway (the HIV prevalence was 53 percent higher in Hluhluwe, reaching 41.2 percent).

How did the residents of Hluhluwe perform on the altruistic stress test? Like the other two populations, individuals trying to win the prize for themselves withstood pain significantly longer than when bearing pain for siblings or parents. But in Hluhluwe, there was no difference in pain endurance for nuclear family members versus uncles/aunts/nieces/nephews. Neither did the townspeople hold the stressful position any longer for uncles vs. cousins (they did, however, try harder for cousins than when donating the award to the local school).[12] Perhaps the greater economic migration in Hluhluwe increased the townspeople's uncertainty about genetic relatedness, which in turn disrupted the extended gradient of kin-selection altruism that existed in other communities.

The Christmas-present, inheritance, and pain-endurance experiments show how people are generally willing to give more to relatives who share a larger proportion of genes in common with the altruist. But does *trust* in others also depend on genetic relatedness?

Psychologist Lisa DeBruine of McMaster University adapted a "trust game" to explore this question. The game involved two people, a proposer and a responder. The proposer begins with an amount of real money and has to make the first move. He can either (1) split the money equally to share one half with the responder, or (2) decide to pass all the money to the responder. In the second case, where the proposer trusts the responder, the amount of money multiplies when the responder takes it. The catch, however, is that the responder can either (1) split the larger pot two ways, or (2) take more than half for himself.

DeBruine told her players to play an online version of the game with sixteen different partners, who were supposedly located at universities in other locations. In reality, the human beings played the trust game against a computer. The variable that changed was the photograph of the "other player," whom the humans believed to be their partner. Most of the photographs showed strangers of the same sex as the player. But DeBruine also manipulated photographs by digitally morphing each player's own face with a photograph of a stranger. Although a morphed face contained 40 to 50 percent of a subject's face, people could not consciously realize that the photograph resembled them or had been altered.

What happened? The proposers who saw a photograph of their "responder" that had been morphed with their own face were significantly more likely to trust their partner by giving him or her the money to distribute. When put in the responder position, though, the human players were just as unselfish in distributing the larger pot of money to strangers as to the morphed photographs of themselves.[13]

THE POLITICAL CONSEQUENCES OF KIN SELECTION

If people place greater trust in those whose faces more closely resemble their own, then kin-selection altruism could have political consequences. After all, part of a voter's decision-making depends on which candidate they trust the most (or perhaps on which candidate they distrust the least!). Of course most people vote along party lines. But then again, "most people" aren't necessarily the ones who determine the outcome of an election. Often times, swing voters decide who wins. This swing-vote minority comprises moderates, the undecided, the unknowledgeable, people uninterested in politics, and those who have a low ideological coherence. In short, many swing voters *could* be influenced by appearance.

Stanford University communications professors Jeremy Bailenson and Shanto Iyengar have actually tested this hypothesis. A week before the 2004 US presidential election, they gathered a representative sample of hundreds of American voters. Then the researchers used computer software to morph the subjects' faces with photographs of the two candidates—the Republican incumbent George W. Bush, and his Democrat challenger John Kerry.

As shown in figure 41, the manipulated photos blended 40 percent of the subjects' facial features with 60 percent of the politicians' looks. Only 3 percent of the participants thought that the doctored photos might have been retouched (since the spliced photos tend to look more airbrushed). But not a single subject realized that their own image had been spliced into the politicians' photos.

Figure 41. Presidential Candidates, and Subliminally Morphed Photographs of Subjects and Candidates.

The Stanford professors then divided the sample into three groups and showed them pictures of the two candidates: a third of the voters viewed natural photographs of George W. Bush and Senator John Kerry; another third had their faces morphed with that of President Bush; and the last third had their faces blended with Kerry's countenance.

The group of voters that saw the two real photographs favored Bush over Kerry by a two-point margin (a week later Bush actually *did* beat Kerry, with 50.7 percent of the popular vote versus Kerry's 48.3 percent). But the group that saw their faces morphed with that of Bush gave the incumbent Republican a thirteen-point victory in the simulated vote. The voters who had been morphed with Kerry "elected" him by a seven-point win.

Bailenson and Iyengar emphasized that the morphed photographs had no effect on people with a strong partisan identity. So if you really hated or supported one of the candidates, even having 80 percent of your face blended into his photograph (a manipulation that remains consciously undetectable) still wouldn't make you change your voting behavior. Yet for independents, weak partisans, and people with a low interest in politics, there was a decisive impact when a candidate looked subliminally more like the voter.[14] In other words, it's perfectly plausible that the proportion of genes that swing voters share with presidential candidates could change the outcome of an election.

The political implications of Hamilton's Rule are truly far-reaching. As explained in chapter 10, kin-selection altruism exerts part of the centripetal force of ethnocentrism. And ethnocentrism, of course, is associated with cultural conservatism. This is why the political right extols the altruistic sacrifice of individual interests for the benefit of the tribe.

Yet regardless of political orientation, close family members tend both to behave more altruistically toward each other than toward distant kin and to treat distant kin better than non-kin. When we translate Hamilton's Rule from biology into the language of politics, kin-selection altruism toward close kin becomes "nepotism" and the preferential treatment of distant kin becomes "tribalism."[15]

MECHANISMS THAT COULD MODERATE VARIATION IN KIN-SELECTION ALTRUISM WITHIN A GROUP

In any population, some individuals are simply more ethnocentric, while others are more xenophilic. Scientists have recently discovered several biological mechanisms that likely account for this variation in altruism toward members of one's in-group.

One candidate is oxytocin, which has been called "the hormone of love." This chemical, which is also a neurotransmitter, is released by the hypothalamus into the brain and bloodstream. Traditionally, oxytocin is best known for its role in the female body during childbirth and breastfeeding. In addition to mediating contraction of the uterus and mammary glands, the hormone stimulates empathy, trust, cooperation, and generosity.[16]

As scientists learn more and more about oxytocin, they have discovered that the hormone is also involved in social bonding beyond the mother-child relationship. But the social altruism that oxytocin promotes is kin-selective; oxytocin enhances altruistic behavior toward an individual's in-group but not toward members of out-groups.

Research conducted by University of Amsterdam psychologist Carsten K. W. De Dreu demonstrates how the hormone of mother-child bonding doubles as a hormone of ethnocentrism. Dr. De Dreu studied a group of Dutch college students to see how they felt about two out-groups: Muslims and Germans. He chose these minorities because previous public-opinion polls had revealed that 51 percent of Dutch citizens had unfavorable views of Muslims, and that a substantial proportion of the Dutch considered Germans to be "aggressive, arrogant, and cold."

Of course, a social-desirability bias could have easily prevented the Dutch students from expressing negative views toward minority groups—whether they held such attitudes or not. Therefore, Dr. De Dreu used an implicit association test. Described in chapter 6, this type of test typically has people pair positive and negative words (like "good" and "bad") with photographs of people of different races. Social scientists find this test useful because the more ethnocentric a test-taker is, the longer it takes them to pair a positive word (and the more quickly they can pair a negative word) with the photograph of a member of an out-group.

In Dr. Dreu's experiment, though, instead of photographs he used names. For example, Peter is a common Dutch name. Names like Helmut connote German ethnicity, while "Ahmad" and "Youssef" are associated with Muslims. Compared with a control group, the Dutch students who had squirted oxytocin into their noses prior to the implicit association test received significantly more ethnocentric scores.

In another experiment, De Dreu presented his students with a classic moral dilemma: he had them imagine five people (anonymous with respect to ethnicity) about to be killed by an oncoming train. The only way to save their lives was to throw an innocent bystander onto the tracks. In some cases the victim had a Dutch name; in other scenarios, the bystander had a Muslim name. De Dreu discovered that the students who had sniffed oxytocin were much more prone to sacrificing a bystander named Muhammad than one named Maarten.[17]

Were the oxytocin sniffers primarily eager to save the Dutchman, or was their sacrifice of the Muslim aggressive? Another one of De Dreu's experiments clarified this question. In this setup, he had people play an economic game where they had three choices: (1) to keep money for themselves, (2) to distribute money to members of their in-group, or (3) to give the money to their in-group while subtracting money from an out-group. In a control group, Dr. De Dreu found that 52 percent of people kept the cash for themselves, 20 percent gave it to the members of their ethnic group, and 28 percent chose the option that hurt the out-group. Under the influence of oxytocin, however, only 17 percent of the players kept the money for themselves; 58 percent behaved altruistically toward their distant kin; and 25 percent reduced the resources of the out-group. So oxytocin's effect on kin-selection altruism was very significant in this game, but the hormone didn't appear to increase aggression toward non-kin.[18]

If oxytocin induces ethnocentric altruism but not xenophobic aggression, what could cause the latter? Implicit association tests conducted at Stanford University have found that xenophobia correlates with the response of the amygdala region of the brain. In one study, European Americans with the most negative implicit attitudes toward African Americans also had the strongest amygdala reaction to unfamiliar African American faces. White peoples' amygdalas also habituate faster to unknown white faces, in comparison with black faces. Black people's amygdalas do the opposite.[19]

What can we make of the amygdala, which is highly implicated in processing emotions? In this context, neuroscientists believe that amygdala activity may show sensitivity to perceived threat.[20] Other studies conducted by neuroscientists from Harvard and Yale have found the same link between implicit racial bias and amygdala response to photos of other-race faces. The amygdala effect is strongest when they showed an other-race face for only thirty milliseconds (a time so short that the viewer can only make out a flashing "mask"). But when the researchers showed participants black-and-white faces for 525 milliseconds (which gave them enough time to consciously make out the faces), the neuroscientists measured a greater response in other regions like the anterior cingulate. The anterior cingulate is associated with inhibition, control, and empathy. Dartmouth neuroscientist Jennifer Richeson has suggested that the anterior cingulate's activity attempts to control unwanted prejudicial responses to other-race faces.[21]

As we learned in chapter 2, conservative individuals have a significantly larger right amygdala. Greater liberalism, on the other hand, is associated with having a significantly larger anterior cingulate.[22] Although it's not clear whether the volume of these brain regions depends on nature or nurture, there is evidence that at least some amygdala activity can be traced to genetics.[23]

Do we have any concrete evidence, then, that an individual's DNA can explain variations in kin-selection altruism? There's a hormone and neurotransmitter found in most mammals, including humans, whose chemical structure is very similar to that of oxytocin; it's called arginine vasopressin, or AVP. Compared with oxytocin, AVP differs by only a couple amino acids. And the genes that code for both hormones are normally located quite close together on the same chromosome. Among other regions, the binding sites for this neurotransmitter are concentrated in the amygdala.

The human brain has three major receptors for AVPs. One of these (receptor 1a) came to the attention of biologists because individual differences in this receptor notably alter the social behavior of prairie voles. Male voles that have a long allele for this AVP receptor are more likely to trust unknown juveniles, to be more sexually monogamous, and to invest more time rearing pups. Their counterparts with short alleles for the receptor are more selfish.

A group of Israeli neuroscientists, led by Hebrew University's Richard Ebstein, wondered whether differences in altruistic behavior between individual human beings also had a "hardwired" component. After all, many aspects of the AVP systems are quite similar between species. The logical thing to do was to recruit a couple hundred university students to play an altruism game. Then, the researchers could compare variations in altruism with AVP genes.

Ebstein and his colleagues had the students play an online version of the "dictator game" mentioned above. The first player is the "dictator." He or she makes a unilateral decision about how to share an amount of real money with the second player (the "recipient"). Since the recipient is completely powerless, the dictator is free to give any amount of money she feels like—or even to keep all of it and give away nothing. The simple game is supposed to measure pure altruism. Even though economics would predict that a "rational," self-interested actor in the dictator's position would keep all the money, in reality 80 percent of dictators donate some money to the recipient. Twenty percent of dictators even split the money evenly with the powerless recipient (revealing strong egalitarian values).

How well did altruism correspond to alleles for the AVP receptor? The "dictators" with long copies of the gene gave away significantly more shekels to their recipients than did dictators with short versions of the allele. That is, the same gene that promoted altruistic behavior in prairie voles functioned similarly in college students! Thus, Ebstein and his colleagues found an evolutionary continuity between altruism in nonhuman animals and altruism in our own species. In the student dictators, a difference of only two to thirty-five base pairs in different versions of the receptor gene made all the difference.[24]

Why do individuals within a population differ in how altruistically they behave toward their fellow vole, or their fellow classmate? The following chapter helps explain why some people are more generous than others.

20.
THE LIBERAL ALTRUISM
Reciprocity

If you've ever had a friend who wasn't from your family or your ethnic group, then you probably realize that there are other forms of altruism besides kin selection. In fact, altruistic relationships can even develop between individuals of different species.

At first glance, it doesn't seem like a zebra would ever trust a hippopotamus. Hippos, which weigh up to three tons, are the world's third-largest land mammal, and they are among the most aggressive animals known to man. Even crocodiles have to watch out for them. Hippos are equipped with self-sharpening teeth and a jaw that can crush a small boat. But in March 2010, a zebra at Zurich Zoo intentionally stuck its striped head right into the enormous, trap-like jaws of a hippopotamus. For fifteen minutes, the hippo held its fearsome mouth open as the zebra cleaned its ivory teeth. When the zebra finished eating its meal, it safely went on its way.[1]

Even though the zebra didn't belong to the hippo's pod, the two animals' relationship was mutually beneficial; the dentistry increased the hippo's fitness, and the nutrients dislodged from the hippo's mouth fed the zebra. To make this exchange possible, the hippo refrained from biting the zebra's head off and the zebra had to trust the hippo.

Even closer friendships than this one are possible between much more distantly related creatures. In 1990, a Costa Rican fisherman named Chito found a wounded crocodile on the side of a road near the Parismina River. A cattle farmer had shot the crocodile in the left eye after he discovered the reptile preying on his cows. Chito took the crocodile home and nursed it back to health. Chito, who named the crocodile "Pocho," recalled: "He was very skinny, weighing only around 150 pounds, so I gave him chicken and fish and medicine for six months to help him recover." But when Chito released Pocho back into the river, the crocodile refused to swim away.

Figure 42. Pocho and Chito. Adam C. Smith Photography.

Twenty years later, the human and the seventeen-foot, 980-pound crocodile remained best friends. Pocho would come like a dog when Chito called his name. The massive crocodile, which had forgone an easy human meal for two decades, instead helped Chito earn a living by playfully rolling around with the fisherman in the water, and "wrestling" with him on land, for tourists (see figure 42). The fisherman has explained his unusual relationship with Pocho to bewildered journalists: "I just wanted him to feel that someone loved him, that not all humans are bad. . . . It meant a lot of sacrifice. I had to be there every day."[2] Why do such unrelated creatures put their own safety at great risk to help each other?

THE EVOLUTION OF RECIPROCAL ALTRUISM

The story of Chito and Pocho is certainly uncommon. Yet the friendship between the fisherman and the crocodile—two potentially dangerous creatures—is merely an extreme case of a prevalent phenomenon called "reciprocal altruism."

In 1971, evolutionary biologist Robert Trivers formally developed a theory on the evolution of reciprocal altruism. His paper has since been cited over six thousand times in scholarly publications. In simple terms, reciprocal altruism occurs when one organism reduces its fitness in order to increase the fitness of another individual that is not a close genetic relation, with the expectation that the beneficiary will reciprocate with altruistic behavior at some point in the future.

In his article Trivers predicted that reciprocal altruism would be most likely to evolve among long-lived social organisms with a high degree of interdependence. For example, a low dispersal rate during a substantial portion of an organism's lifespan would ensure long-term interaction with a set of familiar individuals. The need for defense against predators could promote interdependence. So many primates live in ideal societies for the evolution of reciprocal altruism.

As the interdependence requirement suggests, reciprocal altruism needs a "political environment" that offers some minimal level of egalitarianism. Male baboons don't live in one; rather, they have an extremely hierarchical social structure, where the highest-status baboon can simply take whatever he wants without too much cooperation or consent from others.[3] These notoriously hierarchical dominance relations make it difficult for reciprocal altruism to take place between males in baboon society. Nevertheless, female baboons maintain long-term, relatively more egalitarian bonds with other unrelated females. Joan Silk, an anthropologist at UCLA, has shown that reciprocal altruism enables these females to increase their own longevity, and also the survival of their offspring, even more effectively than rising to a higher social rank.[4]

In addition to conditions of sufficient egalitarianism and interdependence, Trivers stipulated that reciprocal altruism entails a benefit to the recipient that is much greater than the cost to the altruist. This imbalance between cost and benefit makes sense because this specific form of cooperation occurs between unrelated individuals. Trivers suggested that reciprocal altruism among human beings would include the following types of behavior:

- Helping in times of danger (e.g., accidents, predation, intraspecific aggression);
- Sharing food;
- Helping the sick, the wounded, or the very young and old;
- Sharing implements; and
- Sharing knowledge.[5]

In the case of each of these examples, altruists might reasonably fear that, during some point in their lifetime, they might find themselves in a position of need where the beneficiary could later reciprocate. In this sense, reciprocal altruism is like a rudimentary insurance system: pay only a little bit when you're safe to support those in great need, and receive a great amount from others when you find yourself in dire circumstances.

By definition, then, the cost in fitness to the altruist is far lower than the increase in fitness to the beneficiary. This characteristic limits how much of a sacrifice a reciprocal altruist can make. William Hamilton noted that the sacrifice borne by a kin-selection altruist can be far greater than that incurred by a reciprocal altruist: "reciprocal altruism can never be suicidal," he pointed out, "whereas suicidal nepotistic altruism [an extreme form of kin selection] can and has evolved . . . in worker sacrifices in the social insects."[6]

Hamilton's observation applies to human beings as well. Suicidal altruism generally occurs only when soldiers or terrorists perceive that their death will benefit close kin or their ethnic group. An ethnically heterogeneous political cause, in contrast, is unlikely to have the equivalent of Kamikaze pilots (suicide aviators in Japan's Pacific campaign during World War II) or "Black Tigers" (Tamil suicide attackers in Sri Lanka).

Palestinian suicide bombers during the Second Intifada knew that then-Iraqi dictator Saddam Hussein would pay $25,000 to their immediate families.[7] This large financial benefit to close kin suggests that at least some of the bombers' self-detonations could be considered acts of "suicidal nepotistic altruism." In addition to this earthly incentive, sermons aired on Palestinian television have promised that one of the rewards for the *shahid* (martyr) is to be "a heavenly advocate for seventy family members."[8] That is, suicide bombers would be able to help their kin get into heaven, so the effects of kin-selection altruism would continue into the afterlife.

Is Hamilton correct that "reciprocal altruism can never by suicidal"? For nonhuman animals, he is probably right. But humans can also be strongly motivated by ideas—including notions of the hereafter. On August 2, 2011, at the beginning of the Islamic holy month of Ramadan, the Palestinian Authority aired a television programed called "The Best Mothers." The episode featured the mother of Yusuf Shaker Al-Asi. Al-Asi, a young bomb-maker for Al-Aqsa Martyrs' Brigades (Fatah's military wing) was killed in a shootout when the Israeli army attempted to arrest him. His mother recalled:

> I said to him (my son), "I want to marry you off."
> He said, "Are you laughing at me? Just one wife?"
> I asked him, "What do you want? Four, according to tradition?"
> He said: "I won't rest until I have seventy [wives]."[9]

Al-Asi was referring to the "Seventy-two Dark-Eyed Virgins" who he believed would marry him in paradise, as often described by imams and depicted in music videos on both Fatah- and Hamas-controlled television channels.[10] If Islamist suicide attackers believe that Allah will reward them for their martyrdom with seventy-two virgins in paradise, then suicidal reciprocal altruism, in some sense, might exist. Regardless of whether or not the attackers are reciprocated with virgins, the earthly honor, reservations in heaven, and financial compensation benefit the direct kin of the suicide bomber.

The fact that suicidal altruism is kin-selective is underscored by the political agendas of the groups that use it. Suicide bombers serve nationalistic goals of claiming, attaining, or defending territory for an ethnic group, or to punish an in-group's perceived enemies. In short, where suicidal altruism is a cultural pattern, it exists in the context of extreme right-wing politics.

The political agendas of extreme left-wing organizations, in contrast, do not typically involve an ethnocentric cause. Consequently, when the extreme left uses terrorist tactics, their bombers are hardly ever willing to kill themselves in the act—especially not as a cultural pattern.

One could therefore consider kin-selection altruism as more of a conservative force, while reciprocal altruism has more liberal characteristics. For one, reciprocal altruism is not limited by tribalism; it can easily transcend the boundaries of ethnic groups, which gives it a comparatively more xenophilic property. In the extreme case, even a man and a crocodile can reciprocate acts of altruism. In addition, reciprocity requires relatively egalitarian social relations (unlike kin-selection altruism, which can easily function under the most hierarchical conditions). Finally, reciprocal exchanges are regulated largely by guilt, which is an instrumental emotion in egalitarian disciplinary strategies (see chapter 15).

When we translate reciprocal altruism from biology into the language of politics, it becomes "cronyism" (just as kin-selection altruism has the political cognates of nepotism and tribalism).

In Trivers's paper on reciprocal altruism, he showed mathematically how this cooperative behavior could evolve. He did so with the help of a classic problem in game theory called the "Prisoner's Dilemma."

What's the Dilemma? Imagine that you committed a crime with an associate. The police apprehend both of you and isolate you in separate cells where you cannot communicate with each other. Since the police lack enough evidence to make a conviction, they visit both you and your associate separately and offer each you the same deal. If you rat out your associate and he remains silent, then you go free and he receives eight years in jail (and vice versa). If both of you testify against each other, then both will be sentenced to five-year sentences. But if you both cooperate and keep quiet, then the police will only be able to keep you both incarcerated for three years on less serious charges. What should you do?

Given the risks, a "rational" prisoner who only wants to minimize his sentence is better off testifying against the other suspect. But most of life isn't really like the Prisoner's Dilemma. Rather, life among long-lived, interdependent social organisms is more like an *iterated* Prisoner's Dilemma, where two individuals can repeatedly play the same game—and without knowing how many rounds will take place over their lifespans. Moreover, reciprocal cooperation can lead to mutual *gains* in fitness over time—not merely shorter prison sentences.[11]

Under these life-like conditions, cooperative behavior could evolve. It could even evolve if individuals in an initial population had genes coding for selfish behavior in the iterated "Prisoner's Dilemma" game. At first, a genetic mutation for greater cooperation might arise. As pairs of individuals played millions of games over thousands of generations, they would be unaware of the long-term consequences of their strategies. Nonetheless, natural selection would keep tabs on increases and decreases in fitness resulting from the iterated game over long periods of time. The individuals with genotypes that caused more reciprocally altruistic behavior would, according to Trivers's theory, eventually replicate more than selfish individuals, leaving a greater number of offspring.[12]

In the 1980s, University of Michigan political scientist Robert Axelrod held a computer tournament in which he invited participants from around the world to submit programs to compete against each other in the iterated Prisoner's Dilemma. The programs could remember and react to the "behavior" of other programs. Programmers submitted a wide variety of algorithms, many of which were very complex. By using computers, Axelrod could easily simulate thousands of games repeated over many generations.

When he finished running all the programs against each other, Axelrod discovered that the selfish programs did quite poorly in racking up fitness points. The winning algorithm was actually the very simplest, and it came from a mathematician named Anatol Rapoport. Rapoport's strategy is known as "tit for tat." It begins the first round by cooperating, and then copies whatever move its opponent makes. In a small minority of cases, the "tit for tat" code will cooperate even after the opponent defects.[13] These computer simulations supported Trivers's theory: that reciprocal altruism could evolve as an adaptive strategy in certain biological environments.

Incidentally, Rapoport had a deep interest in connecting his work with the political sphere. In 1965, at the University of Michigan, he had led the first "teach-in" against the Vietnam War. Rapoport also applied game theory to reduce the threat of nuclear conflict, and he founded an NGO called Science for Peace.

As Rapoport's tit-for-tat strategy demonstrates, neither a pure reciprocal altruist nor a pure reciprocal defector would fare well in increasing its fitness relative to tit-for-tat players. The program's key features are that it punishes defections by other players, but these punishments are only strong enough to dissuade future cheating. Since punishment is expensive, forgiveness is crucial. Thus, social environments where reciprocal altruism has evolved are full of "wary cooperators"—to use political scientist John Alford's term. Alford believes that the conditional element of "tit for tat" helps the strategy successfully adapt to different environments, in which cooperation is more or less likely.[14]

The perennial problem with reciprocal altruism is that the system is particularly susceptible to cheating. Obviously agents can completely "defect" by not reciprocating at all; but the more insidious risk is that individuals can reciprocate but at a lower cost than that previously paid by a partner. Trivers proposed that numerous human emotions have evolved because they have proven adaptive for regulating reciprocal altruism. Among them, he includes:

- Friendship (liking and disliking);
- Moralistic aggression;
- Gratitude;
- Sympathy;
- Trust;
- Aspects of guilt (to motivate a cheater to compensate for defecting); and
- Some forms of dishonesty and hypocrisy.[15]

The human psychology surrounding reciprocal altruism likely grew increasingly complex, in an arms race of sorts, as our species became more and more intelligent. But we can still find the basic form of this type of cooperation among a few other social animals.

Primatologist Frans de Waal has spent many hours carefully observing chimpanzees at the Yerkes National Primate Research Center in Atlanta, Georgia. In one of his numerous studies, De Waal logged 4,653 chimpanzee interactions involving food, which involved over 2,300 exchanges between individuals. Did chimps that gave away food have their generosity reciprocated later on? Indeed, the frequency that a given chimpanzee (chimp A) shared with another (chimp B) moderately correlated with how often B shared food with A. Similar behavior occurs in the wild.[16]

Felix Warneken, a psychologist at the Max Planck Institute for Evolutionary Anthropology in Leipzig, Germany, has put chimpanzees into more novel situations than what they would face in nature to see if they would still reciprocate cooperative acts for one another. For example, Warneken had his chimps watch their group-mates as the latter tried to pass through a locked door. The only way that the group-mates could go through was if the observer unlocked the door for it from an adjoining cage. The researcher discovered that observer chimpanzees were more likely to open the door for the group-mate if the same group-mate had previously opened the door for it when the situation had been reversed.[17]

Were these chimpanzees simply hardwired to reciprocate, or did their reciprocity rest atop a deeper egalitarian ethic? Do primates less related to us have a sense of fairness? Cognitive scientist Sarah Brosnan has worked with Frans de Waal to see if brown capuchin monkeys respond to unfair exchanges with any indication of moral indignation.

Brosnan and De Waal gave to their monkey participants tokens that the capuchins could then exchange with the primatologists for food. In the first experiment, both monkeys received cucumber for their token. The monkeys felt fairly happy with this trade, judging by the 96 percent of them that accepted the transaction.

In the next test, however, one monkey received cucumber in exchange for a token, but the second monkey got a grape for the same price. To a capuchin monkey, cucumbers are all right, but grapes are delicious. When the second monkeys discovered that they could only purchase cucumbers (after seeing their partners buy grapes), 45 percent of the cucumber-receivers refused to participate. The indignant capuchins would hand back the token instead of taking the cucumber. Sometimes they would refuse to eat the less appealing food. Or they would angrily toss the token or the cucumber out of the test chamber!

But Brosnan didn't stop there. In another test, she gave away a *free* grape to a monkey and then tried to charge the second one a token for the cucumber. Fully 80 percent of the second monkeys refused to eat their cucumber or participate.[18] So even our quite distant primate relatives feel aversion to unequal exchanges. Without this moral emotion, reciprocal altruism could not exist among intelligent social animals.

Human beings, of course, are also reciprocal altruists. Remember the Norwegian Christmas-gift givers from the last chapter? The total number of presents they gave away to others was highly correlated with the number of presents they received.[19] So within their social networks, people reciprocated generosity for generosity, and stinginess for stinginess. Tit for tat. Many of the people with whom the students exchanged gifts, however, were close relatives, so this isn't really a pure measure of reciprocal altruism. Let's explore some of the interesting wrinkles in the ways that non-kin reciprocate.

INDIRECT RECIPROCITY (ALSO KNOWN AS COSTLY SIGNALING)

Among highly social and intelligent apes, the outcomes of social interactions between pairs of individuals do not remain in the dark. Third parties can directly observe, or hear secondhand, about who is generous and who cheats. So a given individual doesn't have to engage in an indefinite number of games with every other member of the population to know who to trust. Individuals develop reputations. And others take them into account when determining who to act altruistically toward, who to do business with, or even who to mate with.[20]

The only problem is that you can't see a reputation (much less genes associated with altruism). You can only see or hear about behavior. Therefore, the incentive arises for an individual to signal to others that he is generous and trustworthy. An individual might scratch another's back not only expecting direct reciprocation later on, but also hoping that others will observe and communicate about the backscratcher. According to Harvard mathematical biologist Martin Nowak, the logic becomes "I scratch your back and someone else will scratch mine." Individuals want to get into the backscratching business with known backscratchers—and not to scratch the back of a freeloader for naught.

This concept is called "indirect reciprocity," whereby individuals rely on "costly signaling" to build a reputation. Nowak's mathematical investigations prove, at least on paper, that evolution can favor indirect altruists. That is, signaling costly acts of altruism to uninvolved third parties can actually increase one's fitness in the long run, because others will be more eager to help those who help others.[21] Does this actually work in the field?

Another of Frans de Waal's studies involved a population of nine captive chimpanzees. The Dutch primatologist meticulously determined how many times each ape gave away food to others, per each hour of food possession. Once De Waal knew each chimp's rate of food distribution, he calculated how frequently other individuals in the nine-chimp social network aggressively rejected requests for food. The chimps that were the most generous in sharing food with the group were much less likely to be spurned when they had to beg for food from others.[22] Although we can't know for sure, it seems like the chimps developed reputations among their peers. The most generous chimp, who gave away food seventy-two times per hour of food possession, was *never* spurned when it had to beg.

Ethnographers have found similar patterns among hunter-gatherer populations. The Aché of Paraguay subsist on resources from foraging and horticulture. Individuals who share their produce more than average receive more food from a greater number of people when injured or sick. Aché who score below average in food sharing receive less aid from the group when in need.[23]

Numerous researchers have replicated similar findings in economic-game-type experiments in developed countries. One study has shown that players who are allowed to gain a reputation by giving are twice as likely to donate their resources.[24] Another experiment found that the players who publicly gave away the most money to a charitable organization (UNICEF) indirectly benefited by gaining a better political reputation; these exceptionally generous donors were more likely to be elected to represent the interests of their group after the game.[25]

Studies such as these may help explain why some people donate to blood banks, although their blood likely benefits non-kin, and although they can still receive the bank's blood without previous donations. Even if the bank doesn't directly reciprocate with a cookie or money afterward, the sticker or T-shirt with the words "I donated blood" could indirectly benefit the donor. If people could receive no recognition for giving to blood banks, then perhaps there would be many fewer donors.

STRONG RECIPROCITY (ALSO KNOWN AS COSTLY PUNISHMENT)

There's one more important piece to the reciprocal-altruism puzzle. Sometimes, when an individual fails to reciprocate without a valid reason, others punish the defector. This phenomenon is called "strong reciprocity" as well as "altruistic punishment" or "costly punishment." The punishment is altruistic because the punisher receives no material gain, and in fact incurs a cost: the risk that the defector will retaliate. Despite the cost, time, and effort taken by the altruistic punisher, he protects the reciprocal system from free riders.

According to Martin Nowak, costly punishment could evolve as an evolutionarily successful strategy. Game-theory models of the Prisoner's Dilemma, as well as individual-based simulations, yield the same result: a population can initially consist of entirely selfish individuals that always "defect." But if a couple altruistic punishers invaded a population (or even if a genetic mutation arose in a single individual), then strong reciprocity could survive as a stable strategy. The strong reciprocators would stop engaging with the selfish individuals after being exploited in the first round, and then they would begin many rounds of altruistic exchanges with each other.[26]

Social cooperation among larger populations might even *depend* on the evolution of altruistic punishment. Evolutionary anthropologists Robert Boyd and Peter Richerson created computer simulations that modeled the demographic conditions of the small-scale societies in which our early human ancestors probably lived. After simulating the passage of two thousand years, they discovered that without costly punishment, cooperation rapidly drops by the time a group reaches only sixteen people. With altruistic punishers, however, cooperation levels remain high until a population reaches sixty-four people, and then cooperation declines quite slowly.[27] Other simulations run by Austrian economist Ernst Fehr have replicated these findings. Fehr used his multi-person Prisoner's Dilemma program to simulate two thousand generations. Again, cooperation broke down without punishment in groups of sixteen individuals or more. But with altruistic punishers, cooperation rates can stay as high as 70 to 80 percent in populations of several hundred individuals.[28]

Since costly punishment is a fairly sophisticated behavior, one might assume that we'd find it only among human beings. Yet biologists have discovered altruistic punishers in a number of other creatures.[29] Certain species of ants attack and kill workers that try to "cheat" the system by laying their own eggs.[30] Macaques that find coconuts (one of their favorite foods) are supposed to call out to announce the news to their group-mates. Those that cheat others by keeping quiet bear the aggression of angry punishers. Consequently, female defectors actually eat *less* food than those who share with the group.[31]

Many human societies around the world depend on the cooperation of their members. By harnessing communal labor, a group can build and maintain public infrastructure or grow crops on shared land. For example, Shuar hunter-horticulturalists of the Ecuadorian Amazon carry out a collective-labor activity called a *minga* (a practice common across Andean indigenous groups as well, where it's known by the same name). One group of Shuar sugarcane cultivators convenes *mingas* to clear their fields of weeds. When they sell their crop, each member receives an equal share of the profits. As the game-theory computer simulations suggest, a medium-sized group of individuals that did not punish free riders would quickly find cooperation in decline. Just as expected, Shuar cultivators who miss a *minga* without a valid excuse have to pay a US$2 fine (the equivalent of a day's wage for working on a non-Shuar farm). Evolutionary psychologist Michael Price, who conducted this Amazonian fieldwork, discovered that the Shuar cultivators with the best attendance records had the most punitive attitudes toward the amount of money that slackers should be fined. The same dynamic occurs in industrialized societies.[32]

Across history, human societies have devised numerous ways to show that an individual has been punished for cheating rules. These signals range from scarlet letters, to shaved heads, to amputated fingers.[33] In developed countries, tax-compliance systems essentially work on the same costly punishment principle as Shuar *mingas*: fear of punishment keeps the cooperation rates high. Defectors pay fines or go to jail.[34]

But how far are altruistic punishers willing to go in the name of fairness? Will they sacrifice their own resources to make a moral point? If so, to what extent? There's a simple test called the "Ultimatum Game" that measures costly punishment. The game involves two players. Sometimes they are anonymous to each other and only know that they belong to the same community. The first player begins with a pot of money. Then this "proposer" offers a fraction of the pot to a recipient. Unlike in the dictator game, the recipient decides beforehand the minimum percentage that she would accept. If the proposer has offered a large enough portion, then the deal goes through; otherwise, neither player gets anything!

Evolutionary anthropologist Joseph Henrich has had people from all over the world play the Ultimatum Game. His data come from Africa (Ghana, Kenya, Tanzania, and Zimbabwe), Asia (Mongolia and Siberia), the Pacific (Fiji, Indonesia, and Papua New Guinea), and the Americas (Bolivia, Colombia, Ecuador, Paraguay, Peru, and the United States). Henrich went to the ends of the earth to play the Ultimatum Game because he wanted to represent the entire "breadth of human production systems." The ethnic groups that played the game were foragers, hunters, fishermen, horticulturalists, farmers, seminomadic pastoralists, wageworkers, and college students.

In each location, Henrich put the equivalent of about one day's wage into the money pot. After all the ultimatums had been played out, Henrich uncovered universal trends—but also telling differences between populations. In every group, the less proposers offered recipients the more frequently the latter would punish the former by voiding the deal. The anthropologist could find no group where extremely unequal offers went unpunished. Nonetheless, punishment behavior varied greatly. Some populations punished very little, and reluctantly. Others punished small inequalities quite freely. Henrich even found that 40 percent of the populations punished overly *generous* offers (greater than 50 percent of the pot).

The average offers in these diverse cultures ranged from a quarter of the pot to over half. The undergraduates at Emory offered about 40 percent.[35] Why did results vary so greatly?

One of Henrich's outliers was the Machiguenga tribe of the southeastern Peruvian Amazon. As in other locations, the Machiguenga players were anonymous to each other, but they knew that they belonged to the same community (which comprised seventy people). The money in the pot, in this case, was worth two-to-three days' of infrequently available wage labor. In other diverse locations throughout the world—from Asia, to Indonesia, to the United States—the most *common* offer (the mode) is typically 50 percent. But the modal Machiguenga offer was only 15 percent. Why is this tribe an outlier?

The Machiguenga have traditionally lived in single-family homes, scattered across the jungle. More distant relatives may live nearby, or not. As mentioned in earlier chapters, this Amazonian indigenous group is fiercely independent. If a family feels that another party has not respected it as an equal, its mobile unit may break off from its hamlet to live alone in the jungle for long periods of time.[36]

The Machiguenga can afford to be intransigently autonomous because each family is economically self-reliant. A household can easily subsist on hunting, fishing, foraging, and manioc-based gardens. According to Henrich, "cooperation above the family level is almost unknown, except perhaps for cooperative fish poisoning." It's possible that the Machiguengas' mega-rich ecological environment simply hasn't pressed them to develop a culture of greater cooperation. Therefore the Machiguenga recipients didn't consider their proposers to be selfish: "Rather than viewing themselves as being 'screwed' by the proposer, they seemed to feel it was just bad luck that they were responders, and not proposers," Henrich reports.[37]

Explanations of deviations in the Ultimatum Game in other parts of the world are apparently evasive. When Henrich tried to analyze the variation between populations and within them, no simple factor (or combination of factors) could explain it—not income, wealth, geography, sex, age, education, or household size.

But Amherst economist Herbert Gintis, who has also collaborated with Henrich on his cross-cultural Ultimatum Games, has a slightly different opinion. Gintis believes that two variables *can* explain up to two-thirds of the global variation among the ultimatum players. The more people buy and sell, or work for a wage (i.e., when there's more "market integration"), and the greater incentives they have to cooperate in their productive activities, the more they share in economic experiments such as the Ultimatum Game.

Gintis has also suggested idiosyncratic cultural and ecological explanations for the varied results of the game in different locations. For instance, the Au and the Gnau peoples of Papua New Guinea are horticulturalists and foragers; yet in the Ultimatum Game, proposers commonly offer *more* than half the money pot. And many recipients actually *reject* these "hyper-fair" offers. To understand why the Au and Gnau punish such "generosity," Gintis explains, one needs to know that gift giving in Melanesian culture buys political status. The recipients who turned away more than half of the money were punishing the "altruist" for seeking social superiority.

Among the Lamalera of Indonesia, almost all proposers offer half the money or more to the recipient (the average offer is 57 percent). But this is acceptable to them because the Lamalera traditionally hunt sperm whales with bamboo spears. The large catch is too big for one family (or even one community) to consume in time, so they're accustomed to distributing majority proportions of their resources to others.

The Hadza nomadic foragers of Tanzania, in contrast, make relatively low offers in the Ultimatum Game. And Hadza recipients frequently reject the deal. According to Gintis, this pattern of costly punishment makes sense in the context of the Hadza's small-scale foraging economy: even though hunters are supposed to share meat, in practice a great deal of conflict occurs over this scarce resource. So hunters often try to hide meat from the rest of the group.[38]

Perhaps the most important lesson from the Ultimatum Game studies, however, is not that substantial variation exists between cultures in their levels of altruism and costly punishment. Rather, the key finding is that every known society behaves contrary to the expectations of rational choice theory—which has been a dominant paradigm at the heart of microeconomics, and which has influenced many other disciplines, including political science. In fact, human behavior in the Ultimatum Game doesn't even come anywhere close to "rationality." If it did, then a self-interested individual would offer the lowest nonzero amount of money possible, and then the recipient would always accept. But this *never* happens.[39] It doesn't even happen if Indonesians are playing with a pot worth a quarter year's wages (which would be the equivalent of about $10,000 to the average American). In this high-stakes Indonesian game, almost all offers were between 40 and 50 percent—and over 10 percent of recipients *rejected* their offers.[40]

Henrich's global experiments have found that even third parties spend their own money to punish stingy proposers. About two-thirds of observers will use half a day's wage to punish a proposer who gives nothing to the recipient.[41]

NOT WIRED FOR UNCONDITIONAL MATERIAL GAIN

If we're not wired for unconditional, self-interested material gain, then what's going on? Neurobiologists who have actually looked at our wires report that our motives during these games are cooperation and revenge. James Rilling and his colleagues at Emory's psychiatry department have hooked people up to an imaging machine (fMRI) as they play the iterated Prisoner's Dilemma. During mutual cooperation, brain areas implicated in reward processing are activated (including the mesolimbic dopamine pathway). Conversely, when an individual cooperates and his opponent defects, this dopamine system responds negatively.[42]

Other neuroscientists, such as Dominique de Quervain, have searched for mechanisms that motivate costly punishment. De Quervain had subjects play a trust game similar to the one mentioned in the previous chapter. The trusting person could give real money to a recipient, to whom it arrived multiplied by a factor of four; if the recipient was trustworthy, he would split the increased pot with the first player. Otherwise, he defected and kept everything. When the opponent defected, the researchers gave their subjects a chance to punish the cheater. The punishment could be effective (in which case it decreased the defector's payoff) or merely symbolic.

As the subjects were cheated and determined how to punish the defector, De Quervain's team monitored blood flow in people's brains (with PET scans). When the subjects opted for the effective punishment, their caudate nucleus lit up. And the greater the activation of this brain region, the more the vengeful subjects were willing to pay from their own funds to punish the cheater.

What's the significance of the caudate nucleus? This structure is highly innervated by dopamine neurons. Neuroscientists have linked it to reward processing in rats, nonhuman primates, and humans. "Reinforcers" such as nicotine, cocaine, and monetary rewards activate the caudate nucleus. Since costly punishment also activated this structure, the scientists had to conclude that "revenge is sweet."[43]

What's the moral of the story? Are we agents of rational choice? Money does motivate people. But our *primary* wiring is not for unconditional, self-interested material gain. First and foremost, we've evolved as social animals. So successfully cooperating, and avenging cheaters, also stimulates our reward system. Among the Melanesian Au and Gnau peoples, where *giving away* resources increases status, being offered the majority of the stakes in the Ultimatum Game can motivate revenge. In other societies, receiving too little money instigates costly punishment. Money is only as good as the survival, status, or fitness advantages that it can buy.

21.

ALTRUISM ACROSS THE LIFESPAN

THE NEUROLOGICAL DEVELOPMENT OF CYNICISM

In our quest to identify the evolutionary underpinnings of the human-nature question, we have yet to return to the question of age. Was Marisol Valles, the twenty-year-old police chief from Praxedis, Mexico, naïve because of her youth? Did she overestimate the strength of her community's altruism while under threat from feuding drug cartels? Would she grow more cynical in later adulthood?

This chapter considers how perceptions of altruism and self-interest develop over the human lifespan. Several dynamic variables come into play as the phases of life progress; these include: (1) the types of altruism to which a person is most exposed, (2) the economics of dependence and support, and (3) the maturation of the brain. Each one of these shifting phenomena can alter an individual's assessment of altruism in the world. Therefore, these changes also impact political orientation.

TYPES OF ALTRUISM ACROSS THE LIFESPAN

The different forms of altruism that an individual encounters are not distributed evenly throughout each phase of life. Sometimes kin selection predominates; other times reciprocity prevails. These two types of altruism greatly contrast with one another in strength and in what we could call *genetic diameter* (how far their effects reach).

Kin selection is the stronger form of altruism, as evidenced by the fact that people will even sacrifice their lives for their families and ethnic group (just as some social insects will do for their nest). Whereas this nepotistic suicidal altruism exists, no suicidal counterpart exists for reciprocal altruism. To the contrary, the cost of a cooperative action to a reciprocal altruist is substantially *weaker* than the benefit to the recipient. So in terms of costs, kin selection is stronger than reciprocity. The "weaker" altruism (reciprocity) is also fraught by subtle cheating, defection, and retaliation.

With respect to "radius," however, kin selection is the lesser force; it only reaches as far as the percentage of genes shared with a recipient. The reach of reciprocity, on the other hand, can transcend even the barriers between species. For example, a human feeds a dog leftover food, and the dog defends the human from thieves in the night.

These two forms of altruism affect the social environment of people in different ways depending on their age. The human lifespan begins in a world encompassed by an extremely short genetic radius—within the closest reach of kin selection. Therefore, the strength of altruism that close kin extend to a child is very great. In fact, an infant lives almost entirely within a protective shell of kin selection (and parental investment, of course, which isn't altruistic when parents are increasing their reproductive success). Infants *depend* on the strongest of altruistic forces because they have no way of extracting resources from their environments on their own to survive; they must instead rely on the donation of food and protection from close genetic relations. Without kin selection and parental investment, they would soon die.

While infants and young children remain heavily dependent on extracting resources from parents and close kin, the personality best adapted to these circumstances is one that *expects* generosity and cooperation from others. Moreover, it suits these young humans to protest and rebel if doing so can acquire a greater amount of resources and parental investment. Receiving just a bit more food or care could afford the young child a survival advantage over siblings.

As children become adolescents, the genetic radius of their social environment expands. Although they come into contact with a wider range of individuals, a large proportion of their time continues to revolve around their family, so their world remains relatively buffered by kin selection. Therefore, a high percentage of the altruism in their lives is the stronger form. And adolescents still need family support, since they do not fare well in a state of complete independence (we'll soon see why). Based on what adolescents have learned from this limited environment—assuming it to be a healthy family and not an abusive one—they might assume that people in general are more cooperative than not. A short genetic radius of social experience, then, could artificially bias a young person's sample of the world, as well as his or her perception of the "goodness" of others.

Some degree of dependence on close kin typically lasts the greater part of two decades, and frequently into a third. By the age of thirty, however, most individuals have reached psychological and economic maturity and have also had a few years to adjust to this independence. Traditionally, one sex will have dispersed to marry into a different kin group. In some cultures both sexes disperse. This neolocal residence pattern is common among nomadic peoples, and also in many developed Western countries. The more time an individual spends away from close family members, the more the genetic radius of his interactions expands. The predominant form of altruism in his environment becomes reciprocal, which is of a lower strength and prone to defections. As Aristotle observed, older adults, who have had long-term exposure to reciprocal altruism, "have often been taken in . . . they are cynical."[1] Human nature appears more self-interested to them.

Finally, adults often have children of their own. Instead of rebelliously trying to extract greater amounts of resources from their parents than offered, as they did as children, parents have to prudently divide resources among their own children, who frequently behave competitively and selfishly with respect to their siblings. So it's more adaptive for the adult personality to be sensitive to the self-interests of children, whose expectations of great altruistic sacrifice from their parents may increase their chance of actually receiving it.

In their new role of authority figures, parents also become more aware of their own self-interest. On the one hand, parents find out how their own needs and desires sometimes encroach on the investment they could otherwise expend on their children. On the other hand, most parents realize how much they love their children and how "selfishly" they would behave if they had to protect them at the cost of others.

Due to these changing roles and exposure to different types of altruism, aging can increase perceptions of self-interest. This accounts for the rightward political shift that occurs in early adulthood, especially after having children. But let's look at this same phenomenon from a more precise economic point of view.

THE ECONOMICS OF CYNICISM

There is another, more concrete way to understand how self-interest shifts over the lifespan. Here, we can measure the economics of dependence and support among hunter-gatherer groups, thanks to the work of University of New Mexico anthropologist Hillard Kaplan. Kaplan has studied the Aché foragers of Paraguay to determine how many calories individuals produce and consume on a daily basis relative to their age.

Before crunching the numbers, it's important to note that young humans' extended period of economic reliance on their families markedly contrasts with "the standard mammalian pattern"—and even with the life histories of other nonhuman primates. Among these other closely related species, juveniles become economically independent as soon as their mothers wean them from breast milk. This independence means that, even before they reach reproductive maturity, other young mammals and primates can produce virtually all the food they must consume to survive.

Why, then, are humans a glaring exception to the rule? Why must human parents continue to provision their offspring with food for a protracted period after nursing ends? Because humans occupy high-level ecological niches that heavily depend on fully developed intelligence and advanced skills to extract food resources. Unlike us, Kaplan explains, nonhuman primates mostly subsist on leaves, fruits, and insects; hunted meats make up only a small proportion of the diets of some species, such as chimpanzees. By comparison, the nutrient-rich foods consumed by foraging groups of humans are extremely difficult to extract. The staple plant food for the Aché, for instance, is palm starch. To produce it entails cutting down a tree, opening the trunk, and processing the pulp and fiber. Aché women do not master these challenging skills until the age of thirty-five, when they reach their maximum productivity as palm-starch extractors.

Unlike nonhuman primates, most hunter-gatherer groups rely heavily on a diet of hunted game. To put meat on the table, traditional foraging peoples depend on their advantage in intelligence over prey that can run, fly, or swim much faster than people move. Kaplan elaborates:

> Human foragers have detailed knowledge of the reproductive, parenting, grouping, predator avoidance, and communication patterns of each prey species, and this too takes decades to learn. For example, in a test with wildlife biologists, an Aché man could identify the vocalizations of every bird species known to inhabit their region, and claims to know many more, which the biologists have yet to identify. . . . Following most hunts, the details of the hunt and the prey's behavior are discussed, and often recounted again in camp. Even the stomach and intestinal contents of the animal are examined to determine its recent diet to be used for future reference.[2]

Because becoming economically independent involves learning a great deal of knowledge, as well as a skillset that likely hinges on psychological maturity, Aché men do not produce as many meat calories as they consume per day until the beginning of their third decade (see figure 43, which describes an Aché population during a period of complete subsistence on hunting and gathering).

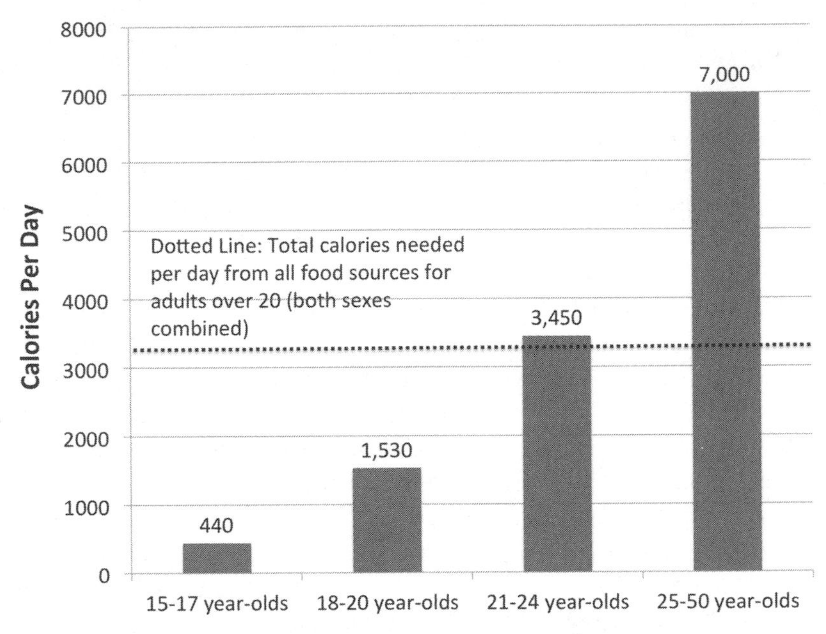

Figure 43. Meat Calories Per Day Acquired by Aché Males.

In environments of greater scarcity, hunter-gatherer peoples take even longer to reach economic independence. Hadza boys in Tanzania, for instance, spend long hours stalking game but bring home very little meat. Among the !Kung of the resource-poor Kalahari Desert, men younger than twenty-five "are known to be incompetent hunters."[3] If young !Kung men cannot provide for themselves, they obviously cannot provide the reliable meals that young children need. Consequently, the median age when most !Kung men first marry is 25.5 years (!Kung women, in contrast, first marry at a median age of 17.4 years—around the onset of sexual maturity).[4]

Before the age of twenty-five, then, a typical !Kung man depends on his family to meet his nutrient-consumption requirements. So the young !Kung's experience of the world is likely biased by an environment rich in kin-selection altruism. It would also suit him to *expect* cooperation from his close kin. After the age of twenty-five, however, when he becomes a net producer and is regularly asked for food from competitive dependents, he should grow more attuned to self-interest.

Americans are actually not so different from these hunter-gatherer groups when it comes to the economics of dependence and support. Like the Aché males who first become self-sufficient in their early twenties, Americans typically leave home for full-time employment (or college) at about the same time. This important life transition marks the beginning of psychological independence from the family environment. And being on one side or the other of this divide has notable effects on political orientation.

A study of over seven thousand American twins in the 1990s compared the political attitudes of identical twins with those of fraternal twins over the lifespan. From ages eleven through twenty, the correlations of political attitudes between fraternal twins were just as high as they were between identical twins. This suggests that their shared environment—especially their dependence on their parents' resources and subjection to parental authority—completely overwhelmed the genetic differences that influence personalities later in life.

In the first half of their twenties, however, the correlations of the political attitudes between fraternal twins dropped substantially, while those of the identical twins remained high. And the correlations remained at these substantially different levels past the age of seventy-five (the average adult correlations for fraternal and identical twins were 0.38 and 0.63, respectively). Once psychological and economic reliance on the original family environment ended, the strong effect of this shared environment fell away and genes exerted their influence on political orientation.[5] After leaving home, people become closer to who they really are.

THE NEUROSCIENCE OF CYNICISM

However, a deeper question remains: what exactly changes in the third decade of life that allows Americans to leave home, and that turns the Aché males from dependents into supporters? Why do net consumers become net producers so long after weaning, and even years after reaching sexual maturity? Perhaps the knowledge needed to make this economic transition is simply so complex that it takes twenty-plus years to acquire. On the other hand, maybe the human body must develop sufficiently to handle the task. Could changes in the brain during early adulthood lead to economic independence, or shift an individual's perceptions of self-interest?

The problem with the brain is that researchers have traditionally faced formidable challenges in observing it—let alone any changes that might occur inside. Dr. Jay Giedd, a psychiatrist at the US National Institute of Mental Health (NIMH), explains why this mysterious organ has long remained largely out of reach: "Wrapped in a tough leathery membrane, surrounded by a protective moat of fluid, and completely encased in bone, the brain is well protected from falls, attacks from predators, and the curiosity of neuroscientists."[6]

Yet the invention of imaging technology has revealed some of the living brain's secrets. Although early scanning devices such as X-rays had the drawback of exposing this sensitive organ to radiation, magnetic resonance imaging (MRI) has provided an alternative without ionizing radiation. Thanks to MRI, science can now scan the brain for basic research, even when not medically necessary. With this safer technology, longitudinal studies can image the same individuals' brains multiple times over a long period. Moreover, MRI shows the body's soft tissues (such as the brain) more clearly than X-rays.

Dr. Giedd has directed one of the largest and most important longitudinal studies of the human brain to explore whether changes arise as children grow up. This investigation, which began in 1991, planned to track the brains of thousands of children between the ages of three and sixteen. But when it came time to end the study, Giedd's researchers discovered that the sixteen-year-olds' brains were still undergoing major changes, even though their bodies had already grown substantially. This continued brain development surprised the neuroscientists, who had assumed that the brain stopped growing after puberty. Giedd decided to extend the NIMH study another two years, to follow their subjects through age eighteen. And then they had to extend the study until age twenty, and again until twenty-two. By the time the final results were in, the researchers concluded that the human brain does not reach full maturity until at least twenty-five.

One major change that Giedd's team observed during the maturation of the brain is called synaptic pruning. In this process the brain weeds out the synapses that it uses less, and reinforces the synaptic configurations that have become the most important pathways for the brain's owner. Although synaptic pruning reduces the number of neurons, the structural changes associated with this process make the brain's key circuits function more efficiently. So someone specializing in hunting skills for many years might become a more efficient hunter. The most dramatic pruning takes place by around the age of twenty-five.

The other central finding of the NIMH study concerns the late development of the prefrontal cortex. This region of the brain lies directly behind the forehead. It's responsible for regulating emotions, controlling impulses, and making complex cost-benefit judgments that weigh immediate incentives against future consequences. The proportion of the brain occupied by the prefrontal cortex is larger in humans than in any other species, reflecting our advanced ability to make long-term rational decisions. Unlike most regions of the brain, the prefrontal cortex continues to grow and develop well into the mid-twenties—and so do its cautionary functions.[7]

Long before the prefrontal cortex matures in the mid-twenties, however, a part of the brain called the limbic system rages into activity during puberty.

The limbic system is highly involved in producing emotions.[8] It plays a central role in sexual arousal, pleasure, and incentives. Drugs such as nicotine, cocaine, and opiates can artificially stimulate the mesolimbic reward pathway. In addition, the explosion of the limbic system during adolescence stimulates novelty seeking, thrill seeking, and risk taking. These personality changes are why the actuarial statisticians who work for car-insurance companies have long known that young people under the age of twenty-five are riskier to insure. Some rental-car companies won't even hand over the keys to this under-twenty-five crowd—even where the law allows them to drive.

While the limbic system fully flares up at puberty, the prefrontal cortex that constrains this teenage behavior does not finish maturing for another decade. Elizabeth McAnarney, a professor of pediatrics at the University of Rochester Medical Center, has described this "asynchrony" between the development of the limbic system and the prefrontal cortex as a "perfect storm" that generates "the seemingly reckless behavior of some adolescents."[9]

The novelty-seeking behavior that is greatly enhanced by the adolescent limbic system is also linked to the politically liberal personality trait of Openness. On the other hand, the cautionary function of the late-developing prefrontal cortex is related to the conservative personality trait of Conscientiousness. So as people around the world live through their twenties, Openness significantly drops and Conscientiousness significantly rises.[10]

Why does nature permit this pubertal disconnect? Novelty seeking, Openness, and risk taking may be involved in dispersal, and they may help in finding a mate. If the fitness advantages outweigh the risks, then evolution could have selected these mate-seeking personality traits to peak along with the sex hormones.

Another hypothesis for this risk-taking behavior involves our species's long history of war. Armies around the world prefer conscripting soldiers before the age of twenty-five, when they would develop more risk-averse long-term judgment. Drafts normally target men to first join a military around the ages of eighteen to twenty. In the United States, most males between the ages of eighteen and twenty-five are required to register for the Selective Service System (which would subject them to military conscription if a new draft were instated).

An army that tried to train all its combat soldiers beginning at age twenty-six would likely suffer a comparative disadvantage. Of course, older soldiers can fight well, too—but especially if they're trained before the prefrontal cortex matures, and before major synaptic pruning rigidifies the brain in the mid-twenties. Younger soldiers' greater sense of adventurousness and thrill seeking also compels them into military service, and it attracts them to faraway territories. As Aristotle said of young men: "Their hot tempers and hopeful dispositions make them more courageous than older men are; the hot temper prevents fear, and the hopeful disposition creates confidence."[11]

Perhaps there is some truth, then, to the perennially quoted saying about aging and political orientation: "If you're not a liberal when you're twenty, you have no heart. If you're not a conservative by the time you're thirty, you have no brain."

We could update this nineteenth-century aphorism as follows: "If you don't trustingly take risks and seek thrills when you're twenty, you have an underactive limbic system. If you don't cautiously control emotional impulses and warily consider the self-interest of others when you're thirty, you have an immature prefrontal cortex."

This isn't to say by any means that all teenagers are liberal and that all older people vote conservatively. At any given age, members of a population will fall into a fairly normal distribution on the political spectrum. This entire curve, however, appears to shift somewhat to the left—and then to the right—as people age.

22.

THE ALTRUISM THAT ISN'T

SELF-DECEPTION AMONG PEOPLE AND POLITICIANS

By contemplating different types of altruism, the last three chapters have sketched out a rudimentary portrait of human nature. As this image has developed, we've seen how people generally act altruistically toward kin to increase their inclusive fitness. In addition, people engage in reciprocal altruism with unrelated strangers. We're most attracted to helping those who help others, and we'll even incur personal risk to punish cheaters. Yet there is still a major element of complexity missing from the picture.

This missing element is especially apparent in the economic experiments we've reviewed. After each round of these games, the players learn whether others have cooperated or defected. Based on these revelations, the cheated players usually have a chance to retaliate (as in the iterated Prisoner's Dilemma). In reality, however, information about social interactions is often far from transparent. Not only do the real-life "players" occasionally defect, but they also *lie* about their cheating. Outside of the simplified microcosm of cooperation games, deceit can prevent people from developing the reputations that they actually deserve.

Although our model of human nature has thus far taken defection into account, it has all but ignored deception. Without the forces of deception at play, our social reality would look very much like a dark, satiric comedy. In fact, the world would resemble the film *The Invention of Lying* (2009), which was written, directed, and produced by the English comedian Ricky Gervais. Gervais opens his film with the following narration:

> The story you're about to see takes place in a world where the human race has never evolved the ability to tell a lie . . . where everyone tells the absolute truth. There's no such thing as deceit, or flattery, or fiction. People say exactly what they think, and sometimes that can come across as a bit harsh. But they've got no choice in the matter; it's their nature.

Since Gervais's characters speak aloud whatever's on their minds, their statements indeed strike the audience as blunt and cruel. For example, one woman exclaims, "Oh, your baby is so ugly! It's like a little rat." Even advertisements are artlessly honest. A bus rolls down the street with a billboard for a soft drink. The banner shows an unappetizing glass of brown cola and reads: "Pepsi: When they don't have Coke."

The protagonist of the film, Mark Bellison (Ricky Gervais), falls in love with the beautiful, charming, and wealthy young woman Anna McDoogles (Jennifer Garner). When Mark picks up Anna for a blind date, she tells him that she's not attracted to him because of his unsuccessful career, poor financial condition, and his physical appearance. Nonetheless, Anna agrees to go out with Mark once as a favor to one of Mark's friends. This favor shows that reciprocal altruism still exists in this guileless world.

As she gets ready to leave, Anna tells her embarrassed date: "I'm not really looking forward to tonight in general, but the thought of being alone the rest of my life scares both my mother and me equally." Throughout the film, Anna frequently submits to the desires of her mother, who constantly meddles in Anna's life to ensure the production of the fittest grandchildren possible. Thus, parental investment also exists in this dystopia.

As Mark continues his futile courtship of Anna, she tells him that, although she likes him as a friend, she can't consider him as a romantic partner because Mark would contribute half his genes to their children, and then she'd have "fat kids with snub noses." In a world where everyone tells the truth all the time, Mark has little opportunity to win Anna over with qualities such as morality or imaginativeness. Physical appearance and financial success are far more important under these circumstances.

People in Gervais's movie still cheat the rules once in a while, such as a character who drives under the influence of alcohol. But when a police officer risks his own safety to pull over the car (costly punishment), the driver simply tells the police, when asked, that he is drunk. The policeman then proceeds to punish the cheater, as in a transparent cooperation game where everyone finds out who has cheated after each round.

One day, however, just as Mark is losing his job, his apartment, and his dim prospects with Anna, the world suddenly changes. Some neurons connect in Mark's brain and he tells the world's first lie. Since no one else in the world has the ability to understand deceit, Mark's presumable genetic mutation gains him a remarkable fitness advantage. His first actions are selfish; he fools people into giving him money, and he tells a beautiful woman on the street that "the world's going to end unless we have sex right now!" She believes him.

But Mark's falsehoods quickly develop into white lies, which he benevolently whispers into people's ears. As those who listen become self-deceived, they are relieved of considerable anxiety about their futures. Self-deception prevents a neighbor from committing suicide; it encourages a demoralized worker to arrive at her office; and it prompts quarreling lovers to embrace. The world becomes a happier place with greater politeness and social harmony.

Meanwhile, Mark grows rich and famous as an artist and a moral guide. These leaps and bounds in fitness help Mark to find an extremely desirable mate and to successfully reproduce his genetic material. At the end of the film, his son inherits a couple of Mark's "undesirable" physical features, but also the highly adaptive ability to lie.

As outlandish as Gervais's film may be, it has some elements in common with the models that evolutionary theorists build: initially, all the members of a population have some particular trait (in this case, honesty); then, another type of social behavior enters the group through invasion or genetic mutation. If the novel behavior increases the fitness of those who carry the new genes, then the trait can spread. Extremely successful traits might even become fixed in the population (*fixation* means that everyone would eventually gain a trait, such as becoming adept at lying).

But how do biologists believe that deception has actually evolved? What about self-deception?

THE EVOLUTION OF SELF-DECEPTIVE ALTRUISM

In his landmark paper on reciprocal altruism, Robert Trivers characterized it as a "sensitive, unstable" system. In a world of indefinitely repeated Prisoner's Dilemmas, it can sometimes be adaptive to cheat, he argues, especially when:

- A partner will not find out about the cheating; or
- A partner won't interrupt the reciprocal relationship if he or she finds out about the cheating; or
- When one will likely die before having a chance to fully reciprocate.

There are essentially two types of cheating, then: (1) "gross" cheating occurs when an individual defects (as when a prisoner squeals on a criminal associate who keeps quiet—or if a freeriding Shuar doesn't help produce sugarcane but benefits from the sale of the produce nonetheless); (2) "subtle" cheating takes place when one returns a little bit less than the appropriate amount to a partner.

Subtle cheating can be difficult to detect, so it's quite an adaptive trick. Trivers argues that an arms race of sorts has occurred in which natural selection has favored subtler and subtler forms of cheating, on the one hand, and greater and greater abilities to perceive cheating, on the other. In the end, he believes that "individuals will differ not in being altruists or cheaters but in the degree of altruism they show and in the conditions under which they will cheat."[1] So although people still reciprocate with tit-for-tat behavior, it's not always clear if a partner has returned an appropriate amount of tat—or even none at all—in spite of communication to the contrary.

In a world of deceptive and counterdeceptive reciprocity, people keep track of altruistic exchanges not necessarily by measuring the precise amount of every tit and every tat, but also by perceiving emotional cues. According to philosopher of science Peter Railton:

> Devotion, loyalty, honesty, empathy, gratitude, and a sense of fairness are credible signs of value as a partner or friend precisely *because* they are messy and emotionally entangling, and so cannot simply be turned on and off by the individual to capture each marginal advantage.

Nonetheless, no one is perfectly safe in an arms race. There are always those who can emotionally out-manipulate the perception of others. Some people can, to borrow Railton's words, wear "a perfect camouflage of loyalty and reciprocity, but fine-tune [their emotions] underneath to turn self-sacrifice or cooperation on or off exactly as needed."[2]

Besides giving less than they owe, subtle cheaters can also use their emotions to extract more from partners than appropriate. Trivers has identified "sham moralistic aggression" as one coercive tactic for doing so.[3] For instance, a subtle cheater could scold a fair reciprocator into making disproportionately large contributions, thus manipulating the partner by creating an undeserved sense of guilt.

Humans don't have a monopoly on deception. Indeed, deceptive communication pervades the animal world. Female scorpionflies choose their mates based on the quality of prey that males present to them as gifts. Some males, however, pretend to be females in order to take food away from other males. Swallows make alarm calls to warn their fellow birds of an approaching predator. But when males see their mate near another male, they sometimes cry out as if a predator were about to strike; this false alarm call often causes the unsuspecting birds to dive for cover and can prevent extrapair copulations. Chimpanzees sometimes cover their faces, apparently to hide facial expressions, and they also hide objects behind their backs that they intend to throw at another.[4]

Among humans, however, deception has reached such an unparalleled level of complexity that it has become self-reflexive: to fool others, we fool even ourselves. Robert Trivers first proposed the evolution of self-deception in 1976. In the forward to Richard Dawkins's popular book *The Selfish Gene*, Trivers reasoned:

> If . . . deceit is fundamental to animal communication, then there must be strong selection to spot deception and this ought, in turn, to select for a degree of self-deception, rendering some facts and motives unconscious so as not to betray—by the subtle signs of self-knowledge—the deception being practiced. Thus, the conventional view that natural selection favors nervous systems which produce ever more accurate versions of the world must be a very naïve view of mental evolution.[5]

Trivers suggests here that distorted perceptions of reality can be better than accurate ones. And he greatly developed this idea in his 2011 book *The Folly of Fools: The Logic of Deceit and Self-Deception in Human Life*. But how is this possible? Although people's senses represent the physical world fairly faithfully, strong distortions can act on emotional and moral perceptions.

Randolph Nesse, a psychiatrist at the University of Michigan, is an expert on the evolutionary origin of emotions. According to Dr. Nesse, "People who can self-deceptively believe themselves to be altruistic while they are, in fact, pursuing selfish motives, will have higher Darwinian fitness, on average, than people who are consciously aware of their own motives."[6] Stated more elegantly: "People who incorrectly experience themselves as altruists will be better at exploiting others through deceptive means."[7]

Evolutionary psychologist Christopher Badcock of the London School of Economics has illustrated this concept with a metaphor: imagine that the whole of our consciousness is a company that has gotten into some trouble. But the operations department hides the truth from the company's public-relations department. This concealment allows the PR people to communicate more sincerely, and therefore more effectively, with the public.[8]

The same principle of information concealment applies to individuals. The fitness advantages of self-deception become especially apparent in the case of romantic love: "A man who believes that he will do anything for his new love and who can sincerely promise her his undying devotion," Nesse explains, "will be far more likely to garner her sexual favors than a man who says the same things without believing them."[9]

But why would there be any deception at all? Couldn't a man believe in his undying devotion, express it verbally, and truly intend it at all levels of consciousness? Of course he could, but the *self-deceived* altruist can convince himself and his lover of his devotion, while going on to impregnate other women. A morally blind natural selection would reward this second man by multiplying his genes more than those of the truly devoted lover. His brain, which has distorted the man's perceptions of his own altruism, has increased his fitness. So the self-deception trait can evolve in a population.

Evolutionary biologists, however, were not the first to explore these phenomena. In 1939, the Viennese psychiatrist Heinz Hartmann called his field a "science of self-deception."[10] Hartmann's teacher, the neurologist and psychiatrist Sigmund Freud, had also observed inner tensions in human psychology, which formed a cornerstone of his theories. When certain instincts cause moral anxiety, Freud believed, people react against them. Some types of reactions against these problematic instincts, Freud wrote in 1918, "take the deceptive form of a change in their content, as though egoism had changed into altruism."[11]

A textbook example of this phenomenon, which Freud called a "reaction formation," sometimes occurs when a person with homosexual instincts finds himself in a homophobic social environment. For example, in May 2010, an American Southern Baptist minister named George Rekers contracted the services of a male prostitute to accompany and sexually massage him on a daily basis during a vacation in Europe. When the details of Rekers's homoerotic holiday hit the news headlines, Rekers had to resign from the board of the National Association for Research & Therapy of Homosexuality (NARTH). Over a long career as an anti-gay-rights activist, Rekers had advocated for the formal exclusion of homosexuals from the Boy Scouts, and he had lobbied against gay adoption in Florida and Arkansas.[12]

Consciously, Rekers had dedicated himself to what, according to his political and religious moral beliefs, he considered altruism: helping young people avoid eternal damnation for sexual sins. His self-deceptive "altruism," however, originated as a reaction against deeper homosexual instincts. Before the rentboy.com escort outed Rekers to the media, the minister had succeeded in convincing his religious community that he was an altruist. Rekers had adapted to this social environment through an exceptionally powerful form of persuasion: self-deception.

As evolutionary theorists such as Trivers, Nesse, and Badcock have shown, far less distance exists than most people assume between biology and the fundamental assumptions of dynamic psychology. Few have demonstrated this close kinship between the two fields better than MacArthur Fellow Frank Sulloway, in his book *Freud, Biologist of the Mind: Beyond the Psychoanalytic Legend.* This monumental biography documents how Darwinian biology deeply influenced Freud's revolutionary theories. Although Freud built up a cult following and styled himself as a "psychoanalytic hero," Sulloway unmasks the legendary psychiatrist as a "crypto-biologist."[13]

Perhaps the main difference in perspective between the fields is that Freudian psychology has taken more time to catalogue the various distortions of reality, and it considers some of them to be more neurotic, immature, or potentially problematic than others. The larger field of evolutionary biology, on the other hand, has just sighted the tip of the Freudian iceberg. Because of their focus on fitness, evolutionary theorists also tend to concentrate on how healthy levels of self-deception help an individual to better adapt to the complex moral environment of human social life; they have spent far less time than psychiatrists trying to understand how and why this sensitive mechanism can fall into a maladaptive imbalance.

From either perspective, a human's perception of his or her social environment is *not* optimally fit when perfectly faithful to reality. As Gervais's *Invention of Lying* shows, a world without self-censorship and self-deception is clearly not the world we live in. Having a distorted perception of our own altruistic tendencies is actually quite adaptive. This condition also goes a long way toward explaining the substantial confusion, and the perennial philosophical disagreement, over the nature of human nature discussed in part V. After all, people's minds hide some of their most selfish motives from their owners.

The Return of the Repressed

The concealment of self-interest from consciousness poses the problem of how an ultimately rational actor can access this valuable information when needed. According to psychology, emotional and moral information that the mind hides from consciousness tends to reappear; but it resurfaces in a distorted, compromised form, in a process called the *return of the repressed*.[14] One of the most important avenues for accessing unconscious information is the ego-defense mechanisms.

The ego-defense mechanisms are some of the most widely accepted contributions of psychology. Specialists recognize more than two dozen defense mechanisms. A few of them have become so well known that they form part of common knowledge and language. We've already come across one defense mechanism: George Rekers's reaction formation against homosexuality concealed his own sexual orientation while improving his reputation as an altruist among his particular culture.[a]

In addition to deceiving others, defense mechanisms allow hidden information to return to consciousness in a compromised form that brings partial awareness of morally problematic self-interest. Self-deception, then, only hides one's drives incompletely. Eventually some defense mechanisms can even break down entirely—as in the case of George Rekers.

Rekers is by no means unique. Many others have gained reputations as altruists, only to be caught secretly laying their hands on the object of their public denunciations. As the attorney general of New York State, Eliot Spitzer gained an admirable moral status in the eyes of many by embracing what anti-prostitution groups called "the toughest and most comprehensive anti-sex-trade law in the nation."[15] In particular, this legislation targeted the *men* who patronize prostitutes. But in March 2008, Spitzer (who had by then become the governor of New York), was caught hiring the services of a prostitute named "Kristen" from the Emperors Club VIP escort agency.

The rights groups that Spitzer had worked with in prosecuting sex tourism felt betrayed. The director of Equality Now believed that Spitzer had been one of the *only* politicians or law-enforcement officials who really understood their cause. As Nesse points out, "it is surprising how reluctant people are to consider impure motives in loud moralists." In the morality arms race, however, there are some who have also developed sensitivity to self-deceived cheaters who "overdo" the moralizing. When people moralize in too loud a voice, some onlookers will think that "the lady doth protest too much" (as Queen Gertrude cynically remarked in Shakespeare's *Hamlet*).[16]

Another type of defense mechanism is called *projection*. This is similar to the previous phenomenon, in which people adopt exaggeratedly opposed attitudes toward socially problematic, "selfish" desires. Projection goes one step further, however, when someone ascribes such desires to someone else. So Rekers would be projecting if he were to attribute homosexuality to other people without rational justification. Similarly, Spitzer would be projecting if he were to casually accuse other men of being "Johns" without sufficient proof that they patronized sex workers.

In addition to fooling others, projection lifts hidden self-interests partially into consciousness. For example, a dishonest card player might "irrationally" accuse his fellow gamblers of cheating. This projection, however, has two advantages: (1) the other players will be more reluctant to accuse the real cheater of cheating, and (2) the self-deceived cheater has his cheating interests partially in mind, so he can anticipate cheating behavior more readily if others actually do attempt to break the rules.[17]

Numerous other defense mechanisms exist, which help individuals deal with guilt and anxiety without transparently communicating their inner self-interests to others. Many of these mechanisms conceal information by evoking the direct opposite idea.

Public opinion is a large aggregate of minds that often function on these same principles. Therefore, reaction formations and projections can deceive massive numbers of people. When these mechanisms occur in the political sphere, they're called *propaganda*.

The most effective propaganda does not tell people that an orange reality is actually yellow (an example of this would be if Spitzer had told his constituents that he was kind of sort of against prostitution). Rather, the deception works best with direct opposites: a government tells a mass audience that a white reality is actually black; or a company claims that a black reality is really white. Just as individuals rely on defense mechanisms when confronting problematic impulses, propaganda increases in society when social insecurity and anxiety rise.[18]

What does a specific black-and-white reality inversion look like? Suppose that a transnational mining company establishes a new mine in a poor, rural area of a foreign country. To increase profits, the company invests insufficiently in social and environmental safeguards. To make matters worse, a corrupt government absorbs the taxes and royalties, preventing any benefit to the impoverished communities in the mine's area of operation. The local communities begin to protest against water pollution, ill family members and livestock, rising prices in their town's economy, and prostitution. The demonstrations gather momentum, and the noise begins to echo in the national and international media. The communities call for the foreigners running the mine to "go home."

What does the mining company do? The foreign company *should have* invested more in preventing the mine's harmful social and environmental externalities. But those preventive measures would have cost a lot of money. Anyway, it's too late. So the company's public-relations team spins the story. The most effective propaganda for the mine in this situation is the direct opposite of the truth—not to claim that the protesters aren't entirely correct, but rather to completely invert reality.

The mine tells the media the following: a small group of self-interested foreigners have come from a third country to sow dissent among the local population, which overwhelmingly favors the presence of the mine. The outsider "green imperialists" benefit from the conflict, according to the propagandistic narrative, because it enables them to raise money for their environmentalist NGOs.

In reality, of course, the outside agitators are the foreign mining company, whose financial motive is *orders of magnitude* greater than that of even the largest NGO. Nonetheless, this propaganda is effective—and commonly used—because it conceals the situation with a perfectly inverted alternative reality.

SELF-DECEPTION IN POLITICIANS

In a well-developed democracy, people elect their leaders based on many criteria, especially ideology. But cynical swing voters often sway elections. This group—as well as others—votes for the individual whom they distrust the least. People focus so much attention on the moral fiber and biography of a candidate because they know that election to office will greatly empower the politician, and that this position of authority will constantly present temptations for the politician to place self-interests ahead of the public interest. Indeed, the potential for deception and self-deception increases as individuals gain political power.

One facet of a high-profile democratic election, then, is a morality contest. The competition to be perceived as the most altruistic candidate can be fierce. Exactly how much to exaggerate one's altruism, and how much to expose an opponent's current or past lies, becomes a kind of cooperation and defection game, not unlike the iterated Prisoner's Dilemma. Even if the opponent cooperates, the media are supposed to play the role of third-party costly punishers.

As every individual with a smartphone now has cheap access to recording tools and reliable fact-checking websites, it has become riskier than ever for politicians to stretch the truth. Still, subtle cheating offers clear advantages in the high-stakes game of elections. But here's what really tilts the odds in favor of scandal: for a candidate to truly believe that he is the best person to lead a nation of *millions* requires more than an average dose of narcissism; this belief usually involves a significant degree of self-deception. It's easy for such individuals to bend the truth just a bit too far.

In early 2011, Minnesota congresswoman Michele Bachmann was hoping to win the Republican nomination for the 2012 presidential race. Bachmann attended the "Rediscover God in America" conference in Iowa at the end of March. Winning over public opinion in this small, landlocked swing state is a coveted goal in American presidential elections, since Iowa holds the first caucus for choosing both major parties' presidential candidates.

Bachmann was so eager to gain the favor of her Iowan audience that she bent over backward to "Iowanize" her family history. The more Iowan the local people considered her, the more kin-selection altruism they might feel, which would translate into votes. The Minnesota congresswoman declared in her speech: "I'm actually even more than just an Iowan. I'm a seventh-generation Iowan. Our family goes back to the 1850s to the first pioneers that came to Iowa from Sognfjord, Norway." And at the end of the speech, Bachmann repeated the same family "fact" again: "I'm so thankful for the faith that they faithfully brought down through the family, and now to the seventh generation here in the United States."

A clever journalist skeptically did some basic math, and then checked the genealogical records. Bachmann's story didn't check out. She was, in fact, only a fourth-generation American (or fifth, if you count the immigrants born overseas, which most people don't). Rather than being among the first Norwegians to settle in Iowa, as she claimed, Bachmann's ancestors first settled in Wisconsin. And then they settled in the Dakota Territory. Although they did finally arrive in Iowa years later, Bachmann had definitely stretched the truth. In her great eagerness to establish an altruistic relationship with her local audience, Bachmann convinced the Iowans in attendance—and probably even herself—to believe in a distorted family history.[19] But by January 2012 a thoroughly discredited Bachmann had to drop out of the race after coming in sixth in the Iowa caucus.

Back in the 2008 US presidential race, Hillary Clinton was running for the nomination of the Democratic Party. During her unsuccessful bid to become the liberal candidate, Clinton told a tale about the trip she'd taken to Bosnia in 1996 as America's First Lady. She recalled that her airplane had landed under sniper fire in the midst of the Bosnian War, and that she had to be whisked away to safety. Hillary's heroic story painted her as a great altruist—as someone who had risked her own life to seek peace for strangers in a foreign land.

But then a videotape surfaced that showed Hillary's actual arrival. No bullets whizzed by. No urgent security measures had to be taken. And she could be seen calmly walking to her car. When asked how she got the story so backward, Clinton responded: "So I made a mistake. That happens."[20] Of course, Mrs. Clinton may have deliberately lied to manipulate the public's perception of her as an altruist. On the other hand, it's likely that a self-deceived Hillary harbored a grandiose self-image as an intrepid, selfless stateswoman.

Bachmann, the conservative, tried to impress her Iowa audience by self-deceptively strengthening a bond of kin-selection altruism. Clinton, the liberal candidate, self-deceptively emphasized her imaginary altruistic sacrifices during a xenophilic peace-making expedition abroad. Left or right, people are not as cooperative as they believe themselves to be. Political candidates have huge incentives to be seen as superhuman altruists; these incentives magnify the potential for self-deception.

When a politician finally reaches a position of great power, two changes occur: (1) the office-bearer's ability to cheat the system to increase his or her own fitness greatly rises; (2) at the same time, the public ratchets up its scrutiny on the politician's morality, since it has invested him or her with huge privileges and responsibilities. How does this pair of changes affect the morality of the powerful?

Dutch psychologist Joris Lammers decided to search for the answer to this question. He recruited colleagues to conduct a couple of experiments. In one of their games, they had students play a political simulation in which one group was randomly assigned to become prime ministers. The other group became groveling civil servants who had to submit to the commands of the high-status prime ministers.

After the game, all the participants faced a supposedly unrelated moral dilemma. Lammers asked them: "Is it acceptable to omit from one's tax declaration additional wages that one earned in one's spare time?" One group of students had to answer the question with regard to themselves, and another group answered with respect to others.

The students who had played the prime ministers answered these questions with significantly more moral hypocrisy. In other words, the powerful players thought it was all right to dodge taxes themselves, while judging other tax evaders harshly. The civil servants did the exact opposite: their attitudes regarding tax dodging were much more lenient toward others than toward themselves.

In another experiment, the researchers randomly primed one group to think of a time when they had experienced a high-power situation. The second group had to contemplate a low-power memory. Afterward, each participant went off to a private cubicle where they could roll a pair of dice to determine how many lottery tickets they could win. They were to self-report their scores, on an honor-system basis. Of course it was quite easy for the psychologists to calculate what the average score should have been for all the dice rollers. In comparison to the expected average, the high-power participants cheated in the lottery-ticket game significantly more than their low-power counterparts.[21]

Why do people in positions of power cheat more? Why are the powerful so lenient with themselves yet so strict with others? They cheat more because the incentives for personal gain are so much higher at the top. It's simply much easier to bend the rules at enormous personal profit. Greater gains lead to even greater power. And, as US Secretary of State Henry Kissinger once said, "Power is the ultimate aphrodisiac."[22] Power, it's true, bestows the potential and the temptation to multiply one's fitness. With these incentives, it becomes harder for group-serving ideologies to compete with the raw self-interests of extremely powerful people.

Because it's easier and more profitable to cheat, the powerful find their morality under intense public scrutiny. In these contradictory circumstances, powerful people's hyperstrict attitudes toward others are like the behavior of the dishonest gambler who unfairly accuses others of cheating (by projection). Loud moralization serves politicians as a defense mechanism against others who would question the moral integrity of those with the largest incentives to cheat. Thus, self-deception offers some degree of protection to the powerful.

"All right, I lied to you. All governments lie!"

There's one other reason that powerful people cheat—especially in relationships involving reciprocal altruism with much less powerful individuals. An enormous amount of help from a low-status person is equivalent in value to just a tiny amount of help from an extremely powerful one. For instance, a normal person could spend a year running a campaign in exchange for an easily made appointment to a powerful position. Or an intern could work for free for months on end in exchange for a letter of recommendation. As the imbalance grows between costs and benefits for each reciprocator, and one-way exchanges take place over longer time periods, the potential for exploitation (whether subtle cheating or defection) increases.

Power corrupts. It might not corrupt everyone all the time, but power dramatically changes the incentives to cheat. Despite all the measures and safeguards, all the trials and hoops that most individuals must pass through to legitimately reach the highest positions of authority, corruption endures like an obstinate force of nature. Even when a political candidate wins a national competition of trustworthiness, he or she may be under a state of self-deception with respect to how self-interested he or she really is. When self-deceptive altruism breaks down, the ensuing scandal often shocks the public and the perpetrator alike.

No matter how or who the public chooses to trust with high-powered positions of authority, corruption remains widespread around the world. In today's Latin America and the Caribbean, where heads of state are almost all democratically elected, corruption reduces GDP per capita by 58 percent. In South Asia and Sub-Saharan Africa, corruption siphons away 63 percent of the GDP per capita.[23]

Even in less corrupt developed democracies, self-deception blurs the vision of both politicians *and* public opinion. During every electoral cycle, candidates make great promises. Even though many of their intentions may be perfectly infeasible, a self-deceived candidate may really be convinced that he or she will be able to effect great changes. And much of the public shares this belief. Thus, presidents normally begin their terms with high approval ratings. When the public realizes that the superhuman candidate has "become" an all-too-human politician, the spell of self-deception is broken and approval ratings fall—often substantially.

SELF-DECEPTION AT THE EXTREMES OF THE POLITICAL SPECTRUM: HORSESHOE THEORY

"What's the difference between Communism and capitalism?" asks a Romanian political joke from the early 1970s. The answer: "In capitalism, man exploits man. In communism, it is vice versa."[24]

Irony aside, do the two ideological extremes share anything in common? In political science there's an idea called horseshoe theory, which is attributed to the French philosopher Jean-Pierre Faye. According to this concept, the political spectrum is not linear, but rather curved like a horseshoe. This arc shape brings the two extremes closer to one another, which symbolizes how both ends share certain traits in common.

In particular, governments on both extremes are considered authoritarian (while those closer to the center are more moderate and more frequently democratic). Authoritarianism means that power is highly concentrated in the leader or governing elite. Authoritarian governments on both extremes oppose the checks and balances on power that characterize centrist democracies. These democratic safeguards against the extreme consolidation of political power include free and fair elections, term limits, freedom of speech, freedom of the press, an independent judiciary and legislature, full participation of opposition parties, rule of law, freedom of religion, and the protection of minorities.

Throughout each section of the book, we've seen how the ideologies of the far left and the far right are polar opposites of one another. So how could it be that the spectral extremes share authoritarian traits in common?

These similarities begin with the fact that both extremes yearn for utopian worlds. The utopias sought after by the extreme right typically promise a better life for a particular ethnic group. In these worldviews, the ideal can only be achieved through the removal of out-groups. Thus, right-wing Hutu extremists envisioned a world without Tutsis, and the Nazis tried to create an Aryan world by destroying Jewish, Slavic, and Romani peoples.

The extreme left, on the other hand, promises drastic changes of another sort: "better-world ideologies" offer a utopia for all people under the premise that this world requires the eradication of socio-economic inequality, tribalism, and/or religion.[25] Some of those bearing the "undesirable" characteristics may be physically annihilated, although others may be subjected to "reeducation" and assimilation.

It's crucial to note that the utopias of the extreme left and the extreme right are *not* identical. The two ends of the horseshoe spectrum do not actually touch. The details of particular utopian *ideologies* differ according to the principles discussed throughout this book. The extreme right's ideology takes ethnocentrism and hierarchy to an extreme. In practice, the extreme right's policies are more likely to result in genocide. The ideology of the extreme left takes anti-ethnocentrism and egalitarianism to an extreme. In practice, these governments are more prone to ethnocidal assimilations and politicidal purges (such as in Stalinist Russia, or Cambodia under Pol Pot). Thus, when they come to power, extremist governments on both sides increase the probability of mass killings.

A similar common denominator of both ideological extremes is the belief in the perfectibility of humankind, or at least an ethnic group.[26] This faith in perfectibility is the reason why both sides have utopian convictions. Since the actual world is so different from the utopias they envision, extremists believe that the only way to achieve their drastic transformations is through *force*. For example, the late Venezuelan president Hugo Chavez once said: "Let's save the human race, let's finish off the US empire. This [task] must be assumed with strength by the majority of the peoples of the world."[27]

It is this belief in the use of force that truly bends the extreme left and the extreme right inward toward one another. One violent tactic used by extremists on both sides to advance their utopian objectives is terrorism. On the right, Al-Qaeda suicide bombers hope to achieve a worldwide caliphate and Islamic supremacy. On the extreme left, the Revolutionary Armed Forces of Colombia (FARC) has committed many acts of terrorism to forward its Marxist-Leninist agenda (Hugo Chavez covertly supported the FARC).

Farther to the south, the Communist Party of Peru (also known as the Shining Path) carried out numerous terrorist activities, such as car bombings, with the hopes of establishing a Maoist state. The Shining Path has lost its war, but during the height of the conflict, in the early 1990s, militants would cut off the fingers of voters to discourage participation in democratic elections (Peruvian voters dip their index fingers in indelible ink after voting to prevent electoral fraud).[28] The ultimate assumption behind the Shining Path's campaign of terror was that only brute force could bring about a Communist utopia in Peru—not democracy.

When an extremist movement does manage to install itself in the seat of political authority, extremist followers have greater tolerance for leaders who consolidate an enormous amount of power. The followers believe that concentrating all available power in the ruling party is necessary to enable the enormous changes required to "perfect" society, and thus to reach the promised utopia. If this process requires excluding opposition parties and scrapping democratic safeguards, extremists are willing to pay the price.

Usually, however, these utopian transformations do not take place entirely—or even at all. One reason why the changes do not unfold as planned in theory is because ideologies are a form of *group* morality. But the changes demanded by extremist ideologies require concentrating immense amounts of power in *individuals*. An individual's self-interest in strengthening and maintaining his or her own power can undermine the ideological ideals that originally empowered him or her. If left to their own devices, leaders who are superempowered by extremism inevitably become dictators because if "power tends to corrupt," then "absolute power corrupts absolutely."[b]

As a leader approaches absolute power, his or her ideology (that is, his or her commitment toward a particular group) can crumble away, revealing naked self-interest. As we'll see directly below, the self-interest of superempowered leaders on both ends of the political spectrum often conflicts with their own ideologies, in mind-boggling instances of hypocrisy.

The Self-Deception of Leaders on the Extreme Left

Extreme left-wing ideologies typically espouse radical egalitarianism. Yet in practice, Communist governments are famous for their authoritarian dictators. The hierarchy of Communist regimes presents an especially glaring moral hypocrisy. But this hierarchy has a function, which is to serve the self-interest of the dictator; without the protection of hierarchy, the leader wouldn't be able to remain on *top*.

In order to maintain popular support, however, left-wing dictators must continue towing a leftist ideology that exalts egalitarianism. Doing so successfully is easier with self-deception. After all, leftist revolutionary leaders would not command so much self-sacrifice from their followers if the followers suspected that their leaders had nefarious plans to sabotage the party's ideals once in power. Yet, as philosopher Peter Singer points out, "What egalitarian revolution has not been betrayed by its leaders? And why do we dream that the next revolution will be any different?"[29] The "dreaming" that Singer refers to suggests that utopian leftist followers are also self-deceived. And he is correct. According to political scientist Jim Sidanius, "There is not a single case in which an egalitarian transformation has actually succeeded."[30]

In practice, self-deceptive altruism allows leftist dictators to extol the egalitarian virtues of their ideology, while systematically violating every egalitarian promise. Below, we examine four of their broken promises.

The "First among Equals" Deception: Leader Worship

Dictators of the extreme left often try to present a "first among equals" image to obscure the fact that they sit on top of a steep hierarchy. To convince others (and perhaps themselves) that they are merely *primus inter pares*, these leaders are fond of wearing modest clothes. Some wear simple army fatigues. Others, such as Fidel Castro or Hugo Chavez, were also known for donning tracksuits.

Leftist dictators occasionally use titles to brandish their supposed anti-hierarchical credentials. Colonel Gaddafi's official title was "Brotherly Leader" of the Great Socialist People's Libyan Arab Jamahiriya.[c] Rather than taking titles such as president, prime minister, or king, some Communist leaders have assumed the role of "chairman" or "secretary," suggesting a first-among-equals status on a council.

The egalitarian ideal, however, belies a much more hierarchical reality. Indeed, leader worship pervades the cultures of extreme-leftist regimes. Vietnam's Marxist-Leninist revolutionary Ho Chi Minh is referred to simply as

"Uncle Hồ." Yet an almost god-like personality cult has sprung up around him in the Communist country. Hundreds of people constantly line up to pay homage to his embalmed body at a mausoleum in Hanoi.[31] And an image of Ho Chi Minh appears on *every denomination* of đồng banknotes circulating in Vietnam.

One of the world's most surreal leadership cults developed in Turkmenistan. Saparmurat Niyazov began his career in the Soviet political system in 1962. In 1985 he had risen to the top of the ranks to occupy the relatively humble-sounding position of First Secretary of the Turkmen Communist Party. But by the early 1990s, Niyazov had appointed himself "President for Life" and "Türkmenbaşy" (Leader of all Turkmen).

In the capital city of Ashgabat, Niyazov erected a two-hundred-fifty-foot-high monument called the "Neutrality Arch." The name supposedly celebrates Turkmenistan's political neutrality, however the giant tripod-shaped landmark was anything but politically neutral. At the very top stood a forty-foot-tall gold-plated statue of Niyazov, which rotated so as to always face the sun.[32] In addition to renaming towns, schools, and airports after himself, the Türkmenbaşy renamed months of the year after himself, his mother, and his book.

Niyazov intended his "Book of the Soul" (the *Ruhnama*) to serve the "spiritual guidance of the nation." Along with its moralistic content, the book combines autobiographical elements and revisionist history. Under the Niyazov regime, school syllabi reduced science and other material that fosters independent thinking, and instead imposed mandatory rote learning of the *Ruhnama*. Even acquiring a driver's license required a sixteen-hour course on Niyazov's "Book of the Soul." In some parts of the country, the Turkmen dictator banished all books except for the Qur'an and the *Ruhnama*. Mosques were ordered to display Niyazov's book as prominently as the Qur'an. Inscriptions praising the *Ruhnama* were carved into Central Asia's largest mosque in Niyazov's home village of Gypjak. In March 2006, Niyazov claimed that any student who read his book three times a day would go to heaven. The *Ruhnama* even has its own enormous shrine in Ashgabat. Every evening the cover opens up, and a recording reads a passage.[33]

In the New World, Hugo Chavez dominated politics in the Bolivarian Republic of Venezuela from 1999 until 2013. During this period, Chavez and his United Socialist Party lost no time in promoting a personality cult around their leader. Tens of thousands of billboards, buildings, and public spaces around the South American country have displayed Chavez's face along with slogans that sing the praises of his egalitarian reforms.

How did these cultish images appear? Some of the pictures were professionally drawn by officially sanctioned publicity organizations. Chavista Graffiti groups—some of which also received government sponsorship—produced somewhat less refined likenesses of their leader to express their support. But in November 2010, the Venezuelan government decreed that only artists who receive direct and prior permission from Chavez could lawfully use his image for propaganda purposes. The reason? Because painting Chavez required "controls which permit his identification as such, in the honorable role of first leader."[34]

Despite his unassuming, down-to-earth tracksuits, and in spite of trumpeting egalitarian reforms, a hierarchical leadership cult arose in the Bolivarian Republic. Why? Because Chavez's far-left supporters allowed him to concentrate an undemocratic amount of power in the presidency in exchange for short-term populist reforms. For a superempowered Chavez, the personal benefits of a cult following outweighed the ideals of a truly egalitarian democracy.

The Economic Egalitarianism Deception

But perhaps an exception must be made for a nation's leader. After all, groups cannot be absolutely egalitarian because they do not govern themselves. Do far-left leaders bring about greater socio-economic egalitarianism within their societies overall?

Not in Venezuela. According to Transparency International's 2010 Corruption Perceptions Index, Chavez's Bolivarian revolution created a country that ranked 164th out of the 178 nations surveyed—placing Venezuela below Haiti in perceptions of corruption. Unfairness in the Venezuelan public sector was at the level of the Congo, Guinea, and Kyrgyzstan.[35] Unfairness, of course, is the precise opposite of equality.

The most common measure of inequality that economists calculate is called the Gini coefficient. A score of zero means that the distribution of income or wealth in a country is perfectly equal (everyone would earn the same salary); a maximum score of one would indicate perfect inequality. Countries fall somewhere in between these hypothetical extremes. Democratic countries, which are politically moderate or centrist, tend to have the greatest income equality (and therefore the lowest Gini scores). According to the UNDP Human Development Report, Denmark, Japan, and Sweden scored around 0.25. In the United States, income inequality has risen from 0.34 in the mid-1980s to nearly 0.41 in 2010.

How do Communist countries compare? Venezuela's 2010 Gini score was 0.43—higher than that of the United States. Income inequality in the People's Republic of China has risen from 0.28 in the mid-1980s to 0.47 in 2010. Thus, China has beaten America in inequality even as socio-economic disparities soar in the United States.[36] Communist Laos and Vietnam scored 0.33 and 0.38. Scores were unavailable for Cuba and North Korea.

What we *do* know about North Korea and socio-economic inequality concerns the country's late "Dear Leader," Kim Jong-il. During various periods of his rule, this North Korean Communist leader also went by titles such as "Superior Person," "Dear Leader, who is a perfect incarnation of the appearance that a leader should have," "Shining Star of Paektu Mountain," "Sun of Socialism," "Highest Incarnation of the Revolutionary Comradely Love," and "Great Man, Who Descended From Heaven."[37] According to Hennessy, the company that produces Paradis cognac, Kim Jong-il was their largest buyer. The "Highest Incarnation of Revolutionary Comradely Love" purchased between $650,000 and $800,000 worth of cognac a year since 1992. This cognac budget alone amounted to 770 times the income of the average North Korean.[38]

The personality cult and profligate spending on North Korea's leader appears to have continued with the Dear Leader's son, Kim Jong-un. After he assumed power in December 2011, Jong-un was called "a great person born of heaven." And in November 2012, North Korea built a propaganda slogan on a hillside reading "Long Live General Kim Jong-un, the Shining Sun!" The sixty-five-foot-high characters stretch across the land for more than half a kilometer, and they can be easily read on satellite images.[39] Two months later, news stories of mass starvation and cannibalism escaped from the hermetic country.

The Anti-Ethnocentrism Deception

The extreme left promises egalitarian camaraderie for all peoples, regardless of ethnicity or religion. In practice, however, Communist governments have oppressed minority groups. University of British Columbia political psychologist Peter Suedfeld has pointed out numerous examples of minority groups treated as second-class citizens or worse under Communism. Some of them include the persecution of:

- Jews under the Stalinist Soviet Union;
- Other non-Russian minorities in the USSR's satellite republics;
- Non-Han minorities such as Tibetans in the People's Republic of China;
- Ethnic Chinese and hill tribes in Vietnam;
- Ethnic Hungarians in Ceaușescu's Romania; and
- Romani throughout Central Europe.[40]

Some part of the Communist persecution of minority groups might be internally coherent with the far-left's anti-ethnocentric ideology. The idea would be that to unite all people into a monolithic Communist society, all vestiges of ethnic and religious differences would need to be destroyed (this would explain the suppression of Yiddish literature in the Soviet Union, or the burning of the Uyghurs' Qur'ans during Mao's Cultural Revolution). But most persecution of minority groups under leftist dictatorships is not consistent with being a mere tactic to break down ethnocentrism and assimilate out-groups; rather, the politically dominant ethnicity of the dictator inevitably fairs better than the oppressed minority groups. Leftist dictators' kin-selective self-interest supersedes egalitarian ideology.

The Sexual-Equality Deception

What about equality between the genders? In theory, extreme leftist ideologies support a more egalitarian relationship between men and women, as well as between heterosexuals and homosexuals. The early Soviet Union did, in fact, liberalize Czarist prohibitions on divorce and homosexuality. Lenin's sexual convictions began as ideology—as an expression of an antiauthoritarian and antipatriarchal sentiment. But as Lenin's political power in Russia increased, his feminism decreased. By the time of the Stalinist dictatorship in the 1930s, male homosexuality and divorce were recriminalized.[41]

We find the same pattern in different cultures of the extreme left. In 1946, the North Korean revolutionary hero Kim Il-sung promulgated a "Law on Sex Equality." To commemorate the sixty-fifth anniversary of this law, the Korea News Service published a press release on July 30, 2001. According to Kim Jong-il's government, his father's sex equality law

> serves as guidelines for emancipation of women and a code based on respect for women. . . . In the period of building of a new society, [Kim Il-sung] took the women's social emancipation as a main task to be fulfilled during the anti-imperialist, antifeudal democratic revolution. In November 1945, soon after the country's liberation from the Japanese colonial rule, he organized the Democratic Women's Union of North Korea, a genuine political organization of women. Now the Korean women are living a worthwhile life and enjoying respect in society under deep care of General Secretary Kim Jong-il.[42]

But according to Jerrold Post, the founder of the Central Intelligence Agency's Center for the Analysis of Personality and Political Behavior, Kim Jong-il's "deep care" for North Korean women was not exactly in line with their supposed emancipation, as proclaimed by the "Law on Sex Equality." Intelligence reports indicate that Kim Jong-il recruited his country's prettiest junior-high-school girls into his "Joy Brigades" every July. To be selected, the girls had to be "virgins and have pale, unblemished skin." These adolescents received training on how to "entertain" Kim and his cronies, who also employed the services of strippers during their lavish parties.[43]

Why did Kim privately violate the same egalitarian ideals that he publicly broadcast? Because doing so was in his personal interest; he preferred to have the Joy Brigades. Although they contradicted his ideology, the Joy Brigades surely enhanced Kim's reproductive success. Self-deception likely helped him rationalize the fact that he was not as altruistic toward women as he proclaimed.

The Self-Deception of Leaders on the Extreme Right

Unlike the extreme left, the extreme right generally doesn't promise political, economic, ethnic, religious, and sexual equality. Moreover, right-wing followers feel less averse to hierarchy. These facts make life a bit more psychologically straightforward for right-wing authoritarian leaders. Nonetheless, ultra-conservative ideologies do not always align perfectly with the self-interests of leaders on the far right.

Rightist utopias promise to crack down on crime and corruption. They sometimes offer protection from moral hazards through strict religious values. Depending on the culture, such ideologies might fight against extramarital sexuality, prostitution, "undesirable" out-groups, or illegal drug use.

But not all of the leaders who espouse an ultra-conservative morality are faithful to their own ideals. Sometimes reaching power tears down their self-deceptive altruism; instead of upholding a group-based morality, they may succumb instead to self-interest. Extramarital sexuality could increase their personal fitness. Doing business with members of hated out-groups could be profitable. Consuming intoxicants could feel pleasurable. Leaders of the extreme right are not exempt from moral hypocrisy. Just like their counterparts on the left, they often break from the very values that have politically empowered them.

During the history of the papacy, not every pope was well behaved. Some of them—particularly during the tenth-century *Saeculum obscurum* ("dark age")—blatantly violated church doctrine and Christian morality. The *Patrologia Latina*, an enormous collection of the writings of the church fathers, accused Pope John XII (955–963) of adultery, fornication, incest with his niece, and turning the Lateran Palace into a brothel. It also alleges that he invoked the names of Roman gods, and murdered his confessor as well as a cardinal. Pope John XII was killed during the act of adultery by the hands of a jealous husband.[44]

Pope Victor III (1086–1087) accused Pope Benedict IX (1032, first term—1048, third term) of "rapes . . . and other unspeakable acts."[45] In 1051, Saint Peter

Damian published his *Book of Gomorrah: Homosexuality and Ecclesiastical Reform of the Church*. Among his attacks on the clergy's sex life, the treatise alleged that Benedict IX had regularly practiced sodomy and bestiality, and had sponsored orgies.[46] After his third term, Benedict IX sold the papacy to his godfather, John Gratian, who became Pope Gregory VI. Although no other pope has ever sold the chair of Saint Peter, a distinct minority of popes has, according to credible historical sources, committed grave transgressions. These cases highlight the morally corrosive strength of power. Not even popes are guaranteed immunity.

In the United States, loud moralists on the far right have often championed the war on drugs. The radio talk-show host and opinion leader Rush Limbaugh III, for instance, staunchly opposed illegal drug use. In 1995, he declared on his television show that

> drug use . . . is destroying this country. And we have laws against selling drugs, pushing drugs, using drugs, importing drugs. . . . And so if people are violating the law by doing drugs, they ought to be accused, and they ought to be convicted, and they ought to be sent up.[47]

But in October 2003, the *National Enquirer* reported allegations from Limbaugh's housekeeper that the radio host had been abusing opiate-based pharmaceuticals. Authorities soon issued a warrant against Limbaugh for prescription fraud and doctor shopping. Prosecutors eventually learned that he had received about two thousand painkillers from four doctors in only six months. Limbaugh was forced to admit to his listeners that he was addicted to narcotic painkillers and would seek in-patient treatment.

Despite his moral hypocrisy, Limbaugh did not lose his career as a moralist. A half-dozen years after the drug scandal, he aligned himself with the ultra-conservative Tea Party movement and enjoyed an audience of fourteen to twenty million listeners per week. Still, Limbaugh would be almost completely deaf had he not undergone cochlear-implant surgery. Limbaugh claimed that a rare autoimmune disease had selectively devastated his hearing; however, his cochlear implant doctor, Jennifer Derebery, said on *Good Morning America* that it was also possible that Limbaugh's pill abuse led to his sudden hearing loss. Indeed, severe hearing loss is a side effect of the long-term abuse of the opiate-based painkiller he was addicted to (hydrocodone).[48]

There is no shortage of ultra-conservative leaders who have broken the same principles that they preached. For years, Osama bin Laden decried the supposed sexual excesses and moral decadence of the West. But in May 2011, when US Navy SEALs raided bin Laden's secret compound in Abbottabad, Pakistan, they found a wooden box hidden in the Al-Qaeda leader's bedroom. Inside was a "huge" stash of pornography.[49] Needless to say, bin Laden's brand of Sunni Islam proscribes erotic films. The moral hypocrisy shocked many people who believed in bin Laden's absolute piety. The arch-terrorist's self-deceptive religiosity helped convince them.

Double Deception: Authoritarian Alliances across the Center

In December 2008, a group of over 350 Chinese intellectuals and activists published a document called "Charter 08." This manifesto called for human rights, and for multiparty democracy to replace sixty years of one-party Communist rule in China.

One of the prominent signatories of this document was the professor and human-rights advocate Liu Xiaobo (Liu had been a leader in the Tiananmen Square protests in 1989). Because of his active involvement with Charter 08, the Chinese government arrested Liu and found him guilty of "inciting subversion of state power." In December 2009, he received an eleven-year prison term for his political crime. It wasn't his first. Liu had already been incarcerated three times in the 1980s and 1990s for "counterrevolutionary behavior" and involvement in the prodemocracy movement. His previous jail term had consisted of three years' hard labor in a "re-education camp."

One year into his fourth prison sentence, Oslo awarded Liu Xiaobo the 2010 Nobel Peace Prize in absentia. A very displeased Beijing would clearly not attend the award ceremony. The Chinese Foreign Ministry spokesman derided the democracy movement, declaring that "interference by a few clowns" would not change her country. But the Norwegian Nobel Committee soon learned that eighteen other countries would also refuse to send representatives to the ceremony: Afghanistan, Colombia, Cuba, Egypt, Iran, Iraq, Kazakhstan, Morocco, Pakistan, the Philippines, Russia, Saudi Arabia, Serbia, Sudan, Tunisia, Ukraine, Venezuela, and Vietnam. The BBC reported that "countries such as India, South Africa, Brazil and Indonesia would attend."[50]

What do the boycotting countries have in common? The governments of Saudi Arabia and Cuba stand on opposite ends of the political spectrum (as do China, Sudan, and many of the other nonattending nations). So why did they show so much solidarity with one another? Fifteen of these nineteen countries score in the most authoritarian half of all countries. Excluding Colombia, the Philippines, Serbia, and Ukraine (which rank a bit above the midpoint), the boycotting countries' average ranking on the *Economist*'s 2010 Democracy Index was 131 out of 167 states. The sample of attending countries mentioned by the BBC, on the other hand (India, South Africa, Brazil, and Indonesia), all score in the democratic half of the Democracy Index. Their average score was 44. The controversy over Liu's Nobel Peace Prize clearly pitted democratic countries against authoritarian states of diverse ideologies.

Cooperation between extreme leftist and extreme rightist governments is quite common. They make "unlikely" bedfellows because, whichever end of the horseshoe spectrum they occupy, extremist governments share authoritarian characteristics. Thus, they also share similar hierarchical moralities and similar governance challenges. Authoritarian countries actually have even more interests in common with one another than they do with centrist, democratic states. Therefore, it's not hard to find "paradoxical" authoritarian alliances across the two ends of the horseshoe.

Hitler and Stalin represented political polar opposites. Communism was Nazism's sworn ideological enemy, and vice versa. Nonetheless, the two superempowered dictators' self-interests led them to form an alliance in one of history's highest-stakes cooperation games.

Germany and Russia signed the Molotov–Ribbentrop Pact in Moscow in August 1939 (see figure 44). Officially, this nonaggression agreement established that the two nations would not turn against each other if either one were attacked by third parties. But a secret protocol also carved up Northern and Eastern Europe into German and Russian spheres of influence.

A cooperative tit-for-tat ensued for two years. Each side reduced propaganda against the other's ideology. Foreign trade between the two countries strengthened them both militarily. Russian imports especially helped Hitler to overcome a British naval blockade.

But in June 1941, Hitler finally defected by invading the Soviet Union. The ensuing war on the Eastern Front was the largest military conflict in human history; it cost over thirty million lives. Despite the nightmarish aftermath, the fact that Communist and Nazi dictators cooperated extensively for a couple years speaks volumes about authoritarianism, self-interest, and self-deception.

Seventy years later, authoritarian governments on opposite ends of the horseshoe were still forming alliances. Mahmoud Ahmadinejad, the president of Iran, led the "Alliance of Builders," a coalition of far-right political parties. Ideologically, Ahmadinejad and his coalition can be defined as Islamist, populist, and anti-Communist. His anti-Communist credentials, however, didn't stop Ahmadinejad from allying with precisely the most extreme leftist governments in the Western hemisphere.

In October 2010, Ahmadinejad and Venezuelan leader Hugo Chavez promised to strengthen their "strategic alliance" against US "imperialism." Chavez regularly visited Tehran, and the two men referred to each other as "brothers."

The two oil-exporting countries had cooperated closely in OPEC, and Iran had invested billions of dollars in developing industries in Venezuela.

In July 2006, Iran awarded Chavez its highest honor—the Islamic Republic Medal—for supporting Tehran in its conflict with the international community concerning its nuclear-energy program. Five months earlier, Venezuela had opposed the International Atomic Energy Agency's decision to report Iran's suspected nuclear-weapons program to the UN Security Council.[51]

Figure 44. Map Showing the Partition of Poland, according to the Soviet-Nazi Agreement Made in 1939 (from the Collection of the Archives of Modern Records in Warsaw; oval added for emphasis).

Iran's Ahmadinejad also developed an especially cozy relationship with the Bolivian head of state Evo Morales, the leader of his country's Movement for Socialism Party. Under Ahmadinejad's government, Iran pledged $1.1 billion toward modernizing Bolivia. The two countries signed "memos of understanding" pertaining to their agriculture and energy sectors. Morales expressed interest in purchasing planes and helicopters from Iran, while the Islamic Republic has sought materials for its nuclear-weapons program in Bolivia's uranium and lithium reserves. Morales joked on various occasions that he forms part of the "axis of evil" (the term used by George W. Bush to refer to Iraq, Iran, and North Korea).[52]

Thus, the most extreme leaders on the left and right ends of the horseshoe have been willing to ignore vast ideological differences in pursuit of economic self-interest and military advantage. To the extent that leaders such as Ahmadinejad and Chavez found common political ground in their authoritarianism, it was to blame powerful democratic countries for their problems. When Chavez visited Tehran in 2010, Ahmadinejad said: "The enemies of our nations [Venezuela and Iran] will go one day. This is the promise of God, and the promise of God will definitely be fulfilled."[53] Whether theocratic or socialist, authoritarians of all stripes find it useful to divert attention to external enemies, often without sufficient justification; doing so distracts their followers' attention from excessively self-interested governance at home.

Another pair of close allies is the People's Republic of China and Omar al-Bashir's Sudan. Bashir, who has ruled Sudan since 1989, presides over an Islamist, Arab-nationalist, right-wing political party. Despite their apparent ideological differences, Beijing has helped Khartoum develop Sudan's oil industry in exchange for energy. Al-Bashir's government also imports arms from China.

In October 2010, China tried to prevent the release of a UN report that identified a dozen types of Chinese bullet casings in Darfur. In addition to being fired at peacekeepers, the bullets would have been used in the genocide that has killed three hundred thousand black Sudanese. Even though Sudan's dictator is wanted by the International Criminal Court for war crimes in Darfur, Beijing and Bashir's shared hierarchical values place economic and military self-interests above all else.[54]

This chapter has shown why it's evolutionarily adaptive for humans to have a distorted perception of reality, such that people believe themselves to be more altruistic than they actually are. Our "self-deceptive altruism" helps explain the substantial confusion over the nature of human nature. It also sheds light on why political scandals perpetually shock both the public and the perpetrators themselves. Finally, self-deception over self-interest explains the similarities in authoritarian governments on both extremes of the spectrum. Although on paper and in public opinion the ideology of one extreme differs greatly from that of the other, both extremes use utopian justifications to concentrate an enormous amount of power in a small number of people. The more power a leader consolidates, the more his or her self-deceptive group-based ideology melts away and reveals self-interest. Therefore, extremism on both ends produces similar outcomes, including the potential for authoritarian alliances.

23.

THE ENIGMATIC ALTRUISM OF HEROIC RESCUERS

On January 2, 2007, a construction worker and Navy veteran named Wesley Autrey was waiting with his six- and four-year-old daughters to catch a subway train in New York City. Suddenly, Mr. Autrey noticed a young man having a seizure. He and two nearby women rushed over to help the stranger, a twenty-year-old film student named Cameron Hollopeter. When Hollopeter tried to get up, however, he stumbled and fell down onto the subway tracks.

As Autrey saw the headlights of the No. 1 train approach the 137th Street–City College station, he jumped down to save the seizure victim. But when Autrey wasn't able to hoist Hollopeter up to safety in time, he pushed Hollopeter into a drainage trench between the rails and then threw himself over the student's body. Five cars went by before the conductor managed to bring the massive vehicle to a halt.

Autrey called out to the astonished bystanders to let them know the two men had safely survived. He had pressed their bodies into a two-foot vertical space as the train passed so close over Autrey's head that it left grease stains on his cap.[1]

Autrey later told reporters: "I don't feel like I did something spectacular; I just saw someone who needed help."[2] The media called the humble rescuer "The Subway Samaritan," "The Subway Superman," and "The Hero of Harlem."

What kind of altruism could explain Mr. Autrey's selfless act? Not kin selection, because the victim and the rescuer weren't related. In fact, Mr. Autrey was black, and Mr. Hollopeter was white. Reciprocal altruism doesn't seem to fit either, because the potential fitness cost of the rescue was enormous. Moreover, the beneficiary was unlikely to reciprocate—and far less likely to ever fully return the favor.

When the public learned of Mr. Autrey's well-deserved reputation as a hero, however, torrential amounts of *indirect* reciprocity began to flow toward him. That is, uninvolved third parties acted altruistically toward the exceptional altruist. The Subway Samaritan received $5,000 in cash and $5,000 in scholarships for his daughters from the president of the New York Film Academy (where Hollopeter studied). The hero also got a check for $10,000 from the real-estate tycoon Donald Trump (in honor of construction workers). Because Autrey was wearing a cap with a Playboy Bunny logo on it during the rescue (the one stained by oil), Playboy gave Autrey a new cap and a lifetime subscription to its magazine. Additionally, the Subway Superman received a trip to Walt Disney World Resort, a $5,000 gift certificate to the Gap, backstage passes to a Beyoncé concert, season tickets to the New Jersey Nets, two brand-new Jeep Patriots from two different donors, two years of car insurance from Progressive, a year of free parking in New York City, a year of free subway rides, and new computers for his daughters, to be updated periodically until their graduation from high school.

The indirect kindness of strangers didn't end there. David Letterman, Charlie Rose, and Ellen DeGeneres invited Mr. Autrey to be a guest on their television shows. CNN presented him with an "Everyday Hero" award. New York City mayor Michael Bloomberg awarded him the Bronze Medallion. And George W. Bush invited Mr. Autrey to his 2007 State of the Union address, in which the president lavished praise on the national hero.

But surely none of these benefits would have crossed Autrey's mind as he pressed Hollopeter between the tracks underneath an oncoming train. Risking one's life to save an unrelated stranger very often does not turn out so favorably for the rescuer as it did for Wesley Autrey.

Only three months after the daring subway rescue, a mentally ill college student named Seung-Hui Cho killed thirty-two students and faculty at Virginia Tech before taking his own life. It was the deadliest one-man shooting spree in US history. During Cho's rampage, a seventy-six-year-old professor and Holocaust survivor named Liviu Librescu was teaching a class on solid mechanics. When the gunman attempted to enter Professor Librescu's classroom, the elderly man blocked the door and urged his students to escape through the windows. By doing so, Librescu took five bullets, including a fatal shot to the head. But twenty-two of his twenty-three students survived.[3] Although the media celebrated Librescu's heroic act of altruism, neither he nor his family received any flood of gifts and privileges. Aside from a posthumous award and a building named after him, there was no one left to receive indirect reciprocity.

Georgetown University philosopher Judith Lichtenberg has pointed out that evolutionary theory does not seem to account for the actions of altruists like Autrey and Librescu.[4] Indeed, there are even more extreme cases in which heroic rescuers certainly hoped *not* to attain any reputation for their deeds, and therefore could expect *nothing* from others in return. One such secret hero was Irena Sendler (1910–2008), a Polish Catholic social worker who served during World War II in Żegota, the underground Polish Council to Save Jews.

With help from her undercover colleagues, Sendler saved approximately two thousand five hundred Jewish children from disease and deportation to the Treblinka death camp. To accomplish this extraordinary act of altruism, Sendler smuggled the children out of the Warsaw Ghetto in ambulances, trams, and even in packages. As a provider of medical care, the Nazis allowed Sendler access into and out of the tightly guarded Ghetto because they feared that typhus (a disease fostered by the population density, sanitary conditions, and starvation of the Ghetto) would contaminate Germans as well. Once they escaped, Żegota provided the children with false identification documents and placements in Polish homes and institutions.

Rescuing or harboring Jews in German-occupied Poland was punishable by death. Obviously Sendler hoped that her altruism would remain secret. But in 1943, the Gestapo discovered Sendler's activities, arrested her, and tortured her. On the way to her execution, Żegota managed to rescue her by bribing a German guard. For the next two years, she had to live in hiding.

Despite her near demise, Sendler continued her altruistic work on behalf of Jewish children. The task remained to reunite them with their surviving family members. In anticipation of their reunification with relatives, Żegota members had buried the true identities of the rescued children in jars. By the end of the war, however, the Nazis had orphaned almost all of them.

Peacetime brought little positive recognition to Sendler—let alone reward. The new Communist authorities in Poland jailed and tortured her for having served in the Żegota underground rather than in the Communist one. During this period, she miscarried her second child. When finally released, Sender lived for decades as a second-class citizen in the eyes of her government.

At the age of fifty-five, the Israeli holocaust museum awarded her a "Righteous among the Nations" honor. But it wasn't until her ninety-fourth year that Sendler received any recognition as a hero from her own country or continent.[5] In Sendler's ninety-eighth year—the year before she passed away—a campaign emerged to nominate her for the Nobel Peace Prize. But Oslo awarded the Prize instead to former US vice president Al Gore for his work on climate change.

How can we explain why Irena Sendler sacrificed so much for so little self-benefit? She wasn't a particularly religious woman; rather, she said she acted "from the need of my heart."

Perhaps Irena had inherited a constitution with an especially low tolerance for inequality—or rebellious genes. Czarist Russia deported her great-grandfather to Siberia for participating in the 1863–64 January Uprising.[6] And Sendler's father had been a physician who cared for Jewish typhoid victims whom other doctors refused to treat. When he himself died from the disease in 1917, the Jewish community donated money to his widow to pay for Irena's education.

Even though the Sendler family had a reciprocal relationship with the Jewish community, the great sacrifice made by the Sendler family doesn't fit the mold of reciprocal altruism; in the textbook cases of this phenomenon that we reviewed in chapter 20, the potential cost to reciprocal altruists is far less than the benefit to the recipient. Moreover, reciprocal altruism occurs in the context of an ongoing relationship. But Sendler rescued many people with whom she had no previous relationship. Likewise, Autrey jumped in front of a train for a complete stranger.

Psychologist Ervin Staub, an expert on genocide and altruism, has discovered an interesting characteristic of heroic rescuers: many were not well integrated into their in-group, or they had mixed families. Some were new to their community. Others had a parent of foreign birth, or came from a different religious background. One example is Oskar Schindler, the German businessman who bribed officials in Berlin to save twelve hundred Jewish workers from deportation to death camps. Schindler was born in Czechoslovakia.[7] Paul Rusesabagina, the Hutu hotel manager who saved the lives of nearly thirteen hundred Tutsis and moderate Hutus during the Rwandan genocide, was married to a Tutsi woman.

Still, xenophilia or being a semi-outsider to one's community could hardly explain these rare instances of remarkably bravery. The compassion of heroic rescuers seems to transcend the categories of altruism currently recognized by evolutionary theory.

Kristen Monroe, a political scientist at the University of California, Irvine, certainly thinks so. Monroe has systematically studied heroic rescuers who have risked their lives to save strangers from the Nazis. To do so, she has sampled names provided by the Carnegie Hero Fund, as well as the Israeli holocaust museum (Yad Vashem). After extensive questionnaires and interviews, Monroe concluded that these acts of extreme altruism could not be explained by age, gender, educational level, socioeconomic background, birth order, or religion.

Neither did Monroe's rescuers have an unusual perception of human nature. They believed people's nature to be mixed—sometimes good, and sometimes bad, but mostly self-interested.

All of them had known at the time of their rescues that they were risking their own lives. They were also aware that their bravery would receive no recognition (at least not for years); so reputation could not have been an incentive. In fact, it did not seem important to the heroes that they in particular should have been the rescuers. A Berliner rescuer named Beth explained: "It was not important that we [helped them]. But because they came to us and had nobody else, we helped them."

Rather, the trait that *was* shared by heroic rescuers concerned how they viewed themselves in relation to other people. According to Monroe, these selfless individuals perceived themselves as "strongly linked to others through a shared humanity." A Dutch rescuer best explained this extraordinary sense of interconnectedness:

> I was to learn to understand that you're part of a whole, and that just like cells in your own body altogether makes your body, that in our society and in our community that we all are like cells of a community that is very important . . . I mean the human race, and that you should always be aware that every other person is basically you.[8]

CONCLUDING THOUGHTS

In exploring the evolutionary origins of political orientation, we've covered enormous ground. This journey has taken us beyond the social sciences and into the worlds of our hunter-gatherer ancestors, nonhuman animals, and microscopic pathogens. We've seen how evolutionary conflicts have selected for a spectrum of key personality traits. In some cases, we've even traced individual variations in political behavior to differences in brain physiology and genes.

Let's review how all of this fits together. Everything that we've learned reflects two principles:

(1) The Personality Principle: Human political orientation has an underlying logic defined by three clusters of measurable personality traits. These three clusters consist of varying attitudes toward: (I) tribalism, (II) inequality, and (III) different perceptions of human nature.

(2) The Evolutionary-Origins Principle: Each of the three personality clusters has roots that reach even deeper into evolutionary origins. The first cluster (tribalism) results primarily from centripetal and centrifugal pressures that act on mate choice; the second cluster stems from conflicts within the nuclear family; and the evolution of various types of altruism explains cluster three. Thus, *political orientations throughout time and space systematically and predictably reflect much deeper biological conflicts*.

Figure 12 shows how these evolutionary roots fit together with their personality clusters.

Another way to think about these three clusters is in terms of resources. **Tribalism** involves conflict over which genetic, economic, and political resources are available to which populations. **Egalitarian and hierarchical moralities** justify the equal or the unequal distribution of resources, both within and between groups. And different forms of **altruism** describe people's tendencies to share or to horde resources under various circumstances.

This resource perspective explains why economics has had such a difficult time accounting for the "irrationality" of political behavior: because economics defines resources as wealth *outside* of the body, owned primarily by individuals. An evolutionary viewpoint, in contrast, expands the definition of resources to include material *inside* the body, which belongs to groups as well as to individuals.

Thus, politics isn't "just about money," but also about who gains authority over a population's minds and bodies. Controlling education, healthcare, economic policies, and morally controversial laws can influence how people recombine and transmit DNA in various ways, at different rates, and to what consequence. The rise to power of political extremists greatly magnifies this control over both genetic and economic resources (see chapter 13).

People fight fiercely over which individuals attain positions of political authority because the outcomes influence the public's perception of entire groups. Political power is therefore owned by groups as well as by individuals. By broadening the definitions of "resource" and "ownership," an evolutionary perspective shows that political conflict follows somewhat rational principles (although not necessarily moral ones).

This explains why there are many rich liberals in New York who vote to pay more taxes, and there are many poor conservatives in Oklahoma who vote to receive fewer benefits from social spending—because, just as often as not, people's moral values are dearer to them than dollars. And they run deeper. The strong identification that many people feel with their political orientation is understandable, considering that the heritability of our political attitudes is between 40 and 60 percent.

Still, the environment also makes a substantial contribution to these attitudes, so devoted partisans on the right and the left fight with each other in public spaces. Even if they don't persuade one another, there are many political souls up for grabs in the center. Since the distribution of left-right political attitudes forms a bell curve (see figure 2), most people are actually in the middle, at the higher parts of this curve. This very important, moderate segment of the electorate tends to have a lower interest in politics and lower ideological coherence (see appendix B). Some of them are swing voters who are more easily swayed one way or another, such that they often play a decisive role in elections.

But evolution obviously didn't select for particular voting habits. We're here with the political orientations we have because our ancestors' personalities helped them survive and reproduce successfully over thousands of generations. Their political personalities were instrumental in the regulation of inbreeding and outbreeding. These dispositions helped them mediate biological conflicts between parents, offspring, and siblings. And their moral emotions also balanced various types of altruism against self-interest in countless social interactions. In some types of social or ecological environments, more extreme personality traits were adaptive. In most cases, moderate personality solutions proved fit. That's one reason why there are many moderates among us. Another reason for moderates and flexibility is that environments change, so it wouldn't make sense for our genes to rigidly determine our personalities. They just influence them based on the "memory" of our ancestors' success.

Now, recall that the strongest correlations of any physical or personality traits measured between spouses concern their ideological attitudes. It makes sense that people mate assortatively by political orientation; after all, the three personality clusters affect who people are attracted to and who they dislike, how people think children should be raised, and how people manage their relationships with others in general. As we learned in chapter 10, Americans have been segregating into more and more politically homogeneous counties because more mating options for blue people lie in blue territories, and red people have more people to mate with in red ones.

In addition to moving geographically, partisans have the ability to move others ideologically—particularly the moderates in the middle of the curve. We could think of political behavior, then, as the way in which individuals try to increase their fitness by widening the genetic, economic, and political resources available to themselves and their group. Newspaper and media elites of different colors, religious leaders, politicians, and even many interpersonal interactions attempt to persuade others to adopt similar attitudes. That is, partisans try to tilt people's political orientations closer to their own.

Conservatives want more conservatives in the world because greater numbers would boost their fitness: there would be more potential mates for themselves and their family members, more coreligionists to provide altruism, fewer immigrants of other ethnic groups to compete with for resources, greater support for pro-natalist, pro-family lifestyles, and a common code of morality for raising children and doing business. Liberals would also benefit from more liberals in their population; greater secularism, sexual freedom, and more immigrants would break down reproductive barriers and expand mate choices for xenophiles. And more people would share an egalitarian approach to childrearing, have compatible views on gender equality between spouses, and practice a liberal code of altruism in society and between countries.

WHAT CHANGES PUBLIC OPINION?

Political rhetoric aimed at persuading others is easy to notice. But there are also many types of larger changes taking place around us that can alter the distribution of political orientations and attitudes in our country and world. Some happen on a much slower timescale than our perception is normally attuned to. Others occur much more quickly. Here we'll briefly consider these crucial moving parts that cause fluctuations in collective attitudes. Then in the following section we'll quantify exactly how stable public opinion actually remains over time.

Demographics constantly change, and often quite unevenly. A population transformation can push public opinion beyond a tipping point. Migrations, for instance, can alter the ethnic face of a country. According to the 2010 US Census, more than half of California's children are now Hispanic. At the same time, the Asian population grew by 31 percent during the century's first decade. By comparison, the proportion of whites and blacks dropped by 5 percent and 1 percent.[1] At the national level, non-Hispanic white births became a minority in the United States in 2011.[2] Changes in the ethnic fabric of society affect voting patterns. In particular, America's cultural conservatives are being forced to reevaluate their stance on immigration, for the very practical reason that they need more votes to capture a conservative majority of today's increasingly diverse electorate.

Migrations also take place *within* countries. According to the UN Population Fund, approximately half the world's population lived in cities in the year 2007, for the first time in history. But by 2030 almost 80 percent of the world's people will be urbanites.[3] Urbanization can profoundly change cultures. Cities bring diverse groups of people into contact with one another. They also provide greater anonymity and foster individualism, which may be more welcoming to alternative lifestyles and sexual minorities.[4] So urbanization can liberalize societies. This trend has likely contributed to the major shift in public opinion that's led to the legalization of gay marriage in many states and countries.

Greater contact between diverse peoples can change marriage patterns over time, especially after younger generations are exposed to and imprinted by new out-groups (see chapter 11). Interethnic unions have more than doubled over the past thirty years in the United States—from 7 percent of new marriages in 1980, to 15 percent in 2010. Almost three-quarters of people with English, Irish, or Polish backgrounds marry into other communities, as do about half of all American Jews. Marriage rates between blacks and whites have quadrupled.[5] Trends such as these liberalize public opinion as a whole (and reflect an already liberalized society).

As time passes, age cohorts come and go. Each generation is exposed to different periods of history, different cultures, and different values. They receive names such as the Greatest Generation, Baby Boomers, Generation X, the MTV Generation, or Digital Natives. Younger generations may have been parented with less hierarchical discipline, and they may have acquired more years of formal education.[6] These particular age-cohort effects could swing public opinion in a liberal direction.

Now, consider the age structure within a population. There could be a disproportionate number of young people or elderly people. As we explored in chapters 18 and 21, age can influence how cynical people are regarding the self-interest of others, and also how averse people feel toward risk. Numerous political analysts have pointed out that most of the countries heavily affected by the Arab Spring have some of the youngest populations in the world outside of sub-Saharan Africa. The median ages in Egypt, Libya, Syria, and Yemen in 2011 were 24.0, 24.2, 21.5, and 17.9 years.[7]

In contrast, the population of developed countries is quickly aging. By 2050, these nations will have two times as many people over the age of sixty as those under fifteen, which could push public opinion to the right. At midcentury, China's over-sixty-five population will triple from 2010 levels, reaching 300 million people. The Chinese leadership worries that the country will age before it finishes growing wealthy. The old require enormous wealth transfers from the young to cover the costs of their healthcare and pensions. When a population ages too quickly, the money runs out. Either the old aren't taken care of or the system won't be able to afford to care for the young when they grow old. Thus, rapidly aging countries can decline economically and lose political power on the world stage.[8]

One reason why developing countries age is because women attain more education and extract more resources from the economy (whether to invest in increasingly expensive offspring, or to gain greater career options and independence). So gender equality increases, and fertility drops (see chapter 12). In almost all developed countries, women's total fertility rates are below the replacement level of 2.1 children. Higher gender equality and lower fertility shift public opinion in a liberal direction.

When women have fewer babies, this also means that their children have fewer younger siblings. So the proportion of first-born children in society grows. From 1960 to 1984, for instance, the proportion of first-born children in Canada rose from 25 to 44 percent.[9] In areas where China enforced its one-child policy, first-borns have become virtually 100 percent of their generation. These only children have been called "Little Emperors" because they have supposedly been overindulged with parental investment. These shifts toward first-borns could exert a slightly conservative force on a population's average personality, as explained in chapter 15.

Although the average woman's total fertility rate drops in a developed country, not all women have fewer children. Due to pro-birth attitudes among the most religious communities, these populations multiply comparatively faster than secular ones do. In fact, the correlation between religiosity and fertility is much stronger in developed countries than it is in poor ones.[10] When religious communities continue to reproduce at high rates in rich countries (where the *average* fertility level is below replacement level), this proportion of the population can change quickly. The populations of Israel and parts of Europe, for example, are becoming proportionately more religious—and potentially more conservative.

New technologies can affect political trends in quite unexpected ways. In 1960, the birth-control pill went on the US market. "The pill" increased gender equality and liberalized society. Other medical technologies increase life expectancy, which alters the age structure of the population. Although people live longer lives, they also develop more complicated chronic health conditions, which contribute to the rising cost of healthcare. Since liberals and conservatives have different ideas about how to insure the population, political polarization can rise.

New communications technologies increase interconnectivity and economic efficiency, especially when developing countries first acquire mobile devices and the Internet. But these disruptive innovations also create new and bigger markets and fiercer competition. The ways in which these economic forces interact with demographic trends can contribute to the swelling or shrinking of the middle class. Because the globalization of information and higher education also raises expectations, even a quickly growing middle class can cause political instability, as reflected by protesting students in Chile. Moisés Naím explains these and other seismic shifts in his insightful book *The End of Power* (2013).[11] Declining middles classes, on the other hand, are particularly dangerous because they can broaden the political spectrum and lead to political extremism (see chapter 3).

Social-networking applications like Facebook and Twitter greatly multiply the number of news sources in the world, since mainstream media outlets can pick up stories from micro-bloggers. An empowered freedom of the press erodes government power in general and threatens the very existence of authoritarian regimes. These social-networking applications, running on ubiquitous smartphones, have played a significant role in the 2009–2010 Iranian election protests, as well as during the 2011 Arab Spring.

The introduction of these communications technologies into the Arab world has had a particularly large impact. In 2002, the number of books translated in Spain in one year equaled the number of books translated into Arabic in the past thousand years.[12] But by the end of the decade, the region received an onslaught of new information via Internet-equipped cell phones and satellite television channels. These technologies have likely transformed the consciousness and expectations of the Arab world. And they've played a role in the way "Facebook revolutions" transpire.

The new hyperconnectivity afforded by smartphones has also benefited Islamic militants, who have used mobiles to disseminate multi-media propaganda. Bluetooth short-range radio waves allow users to share this information anonymously, away from the eyes of spying security services. According to Nico Prucha, an expert on online jihadists, cell-phone technology has been "the only way for them to remain part of the debate. Without it they would be isolated like they were in the 1980s." Before the wide spread of the Internet, people had to take greater risk to acquire information about Islamist organizations. And the quality of VHS tapes, which jihadist groups previously used to circulate propaganda, would quickly degenerate each time a cassette was copied. Now, the quality of digital jihadi videos, songs, and speeches remains undamaged. Islamist organizations have developed "brigades" specialized in producing high-production-value material.[13] The Boston Marathon terrorists apparently copied their bomb designs from Al-Qaeda's online English-language magazine (although Tamerlan Tsarnaev may have also received training from militants in the North Caucuses in 2012).[14] Technology does not guarantee liberalization.

Finally, new technology revolutionizes political scandals. The textbook example was once Richard Nixon's Watergate affair in the early 1970s. Nixon's men broke into the Watergate complex in Washington, DC, to spy on the political opposition. But the public eventually learned of Nixon's direct involvement in the crime—from the very tape-recording system that the president had used to document conversations in his offices. As news emerged in the media about the Watergate affair (including the tape in which the president gave the order to "Do it"), the nation's desire to impeach Nixon rose by a whopping 34 percentage points in eleven months.[15]

Today, the magnitude, geographical scope, and speed of scandals have increased tremendously due to technology like WikiLeaks. Between 2010 and 2011, the WikiLeaks whistleblowing website released into the public domain over a quarter million classified diplomatic cables from hundreds of US embassies, consulates, and missions around the world. Almost all countries were affected in some way or another. Accusations of high-level corruption and narcotics smuggling in Afghanistan particularly shocked public opinion in the United States because of the blood and treasure that Americans had spent there.[16]

Public opinion also holds political leaders accountable for other quickly developing phenomena, such as natural disasters,[a] wars,[b] and especially the complex fluctuations in today's fast-moving economic environment. In our globalized economy, local resources linked to our survival can rise and fall depending on competition on distant continents. A meta-economy of financial trading and gambling shifts around enormous resources through complex financial vehicles. Markets move at the speed of the news cycle. And computers execute high-frequency trades within microseconds.

An important element of public opinion holds leaders accountable—whether justly or otherwise—for the health of the economy. People who have a low interest in politics, or low ideological coherence, are more likely to associate good and bad economic conditions with the incumbent. This key fraction of the population does not vote according to party loyalty; rather, it swings back and forth to reward or punish incumbents based on the current health of the economy. As mentioned above, these "nature of the times" voters often have a decisive impact on the small margin that determines elections (see appendix B).

WHAT REMAINS CONSTANT?

The moving parts mentioned above account for a great deal of the changes that occur within public opinion. Yet some things simply don't change (at least not on a timescale that affects our lives). Specifically, the three personality clusters discussed throughout this book remain the universal constants of what divides us. And so do their evolutionary roots. As New York University political psychologist John Jost has pointed out: "ideology does not 'advance' at the rate of other technologies" because "it is constrained in fundamental ways by human nature."[17]

The constraints on political ideology are indeed considerable. In spite of all the changing variables pointed out above, public opinion actually shifts less than many people assume. Political scientists Benjamin Page and Robert Shapiro have determined the precise flexibility of public opinion in their monumental book *The Rational Public: Fifty Years of Trends in Americans' Policy Preferences* (1992).

To measure the stability of public opinion, Page and Shapiro collected what they believed to be "the largest and most comprehensive set of information about Americans' policy preferences ever assembled." This sample comprised all the surveys they could find from 1935 through 1990. The data came from polling companies, prestigious universities, television networks, and newspapers.

Then, the researchers selected only the questions asked more than once over the fifty-five-year period *with identical wording*. They found 1,128 such questions, typically repeated every one or two years. The specific questions spanned the entire range of political issues, both foreign and domestic.

Of these 1,128 identically worded questions repeated more than once over half a century, 58 percent of the responses did not change significantly. Domestic-policy opinions proved more stable, since they showed no significant change 63 percent of the time.

Even a large proportion of the statistically significant changes were not very dramatic. If we were to consider shifts in opinion to be "considerable" only if attitudes changed by 10 percentage points or more, then 76 percent of the repeated questions would fall below this mark. From another perspective, however, 24 percent of the long-term political attitudes measured *did* change in important and meaningful ways that have affected our lives.

Still, why does it often seem as though public opinion is even more dynamic than these figures suggest? According to Page and Shapiro, most of the graphs that the public is accustomed to seeing in the news media chop off the tops and bottoms of the scales (that is, they truncate the Y-axis measuring percentage points). This cutting of the graphs prevents the waste of empty space. Yet stretching out the graphs creates a misleading impression by magnifying small changes. In addition, these polls often cover short time periods. *The Rational Public*, in contrast, is full of relatively flat-lined graphs covering long periods of time, spanning important historical and political changes.

For example, surveys frequently asked whether "*the U.S. should take an active part in world affairs*." Between the years 1942 and 1956, the Japanese attacked Pearl Harbor, America became involved in World War II, and the Cold War began. Nevertheless, nearly every one of the eighteen times this question was asked over the tumultuous fourteen-year period, support for taking "*an active part in world affairs*" remained between 70 and 80 percent.[18]

Other political scientists have also been impressed by the stability of public opinion—even during dramatic, widely publicized, and high-stakes ideological clashes. David Sears, a UCLA political scientist, has remarked:

> While Republicans strove mightily to frame [the offenses that led to Bill Clinton's impeachment] as a flagrant violation of the law against perjury, and Democrats as a case of a minor sexual indiscretion, public opinion did not budge an iota over the course of a year of some of the most intensive political propaganda in the nation's history. It remained sharply polarized along party

> lines, strongly tilted toward support for Clinton, despite massively negative elite communication about him.[19]

Most people's political attitudes conform to their partisanship and ultimately to their orientations. When general elections are free and fair, they never end in complete landslides. From 1935 (when Page and Shapiro began their sample) through 2013, no US presidential candidate has won by more than 24.3 points.[c] The average margin over this period is just 9 points. Only in the rigged elections of authoritarian regimes are the margins of election "victories" substantially larger.

THE FUTURE OF GOVERNMENTS AND KNOWLEDGE

Great thinkers have been studying politics for thousands of years. Over the centuries some things have changed, but many facets of politics have remained the same. Reading through the content of a treatise on statecraft from the fourth century BCE really brings these similarities and contrasts to life. In his great work the *Arthaśāstra* (or *The Science of Politics*), the political philosopher Kauṭilīya recorded for posterity nearly everything that a political leader of his day had to know. With Kauṭilīya as his advisor, the ruler Chandragupta founded the Maurya Empire, thus becoming the first unifier of the Indian subcontinent.

Kauṭilīya's 2,300-year-old *Science of Politics* covers matters of personal importance to a king, such as his personal safety and his "duty towards the harem." The *Arthaśāstra* instructs on how to manage government servants, including a superintendent of weaving, a superintendent of prostitutes, and a superintendent of elephants. We learn about fiscal policy, such as "the business of collection of revenue by the collector-general," the "detection of what is embezzled by government servants out of state-revenue," and the "examination of gems that are to be entered into the treasury."

Kauṭilīya writes about judicial concerns, including marriage law, inheritance, and rules regarding slaves and laborers. His discussions of criminal law cover "trial and torture to elicit confession," "death with or without torture," "fines in lieu of mutilation of limbs," and "sexual intercourse with immature girls" (punishments depended on caste and the time elapsed since the girl's first period).

In the field of diplomacy, the Mauryan minister advised how to make alliances with kings of equal, inferior, or superior "character"; how to send envoys; and about the making and breaking of peace. Military strategy was also covered in some depth.

The greatest difference between political science during the Mauryan

Empire and in today's democracies is that public opinion was a rather peripheral topic for Kauṭilīya. The attitudes of the masses concerned him only in relation to the art of espionage. To learn the political thoughts of others, Kauṭilīya advised the creation of "spies in the guise of householders, merchants, and ascetics," as well as "spies with weapons, fire, and poison." These undercover agents could detect the "criminal tendency" of youths. Counterintelligence spies could try to entice ministers into supporting a supposed coup.

Kauṭilīya sought to learn public opinion only to confer "honors and rewards . . . upon those who are [politically] contented," and to dish out patronage or punishment to people who were not. He counseled:

> Those who are inebriated with feelings of enmity may be put down by punishment in secret or by making them incur the displeasure of the whole country. Or having taken the sons and wives of such treacherous persons under State protection, they may be made to live in mines, lest they may afford shelter to enemies.[20]

Thus, Kauṭilīya's politician of the fourth century BCE would have monitored public opinion only to better govern as a supreme disciplinarian. Indeed, the very first book of his *Science of Politics* is titled "Concerning Discipline."

Today, political scientists like Stanford's Larry Diamond have observed that democracy is on the rise around the world, in spite of a recent "democratic recession."[21] In more and more countries, the centrality of authoritarian tactics to governing is subsiding and being replaced by the importance of public opinion. The acquisition of political power is increasingly difficult to attain by brute force or through appeals to divine right or a genealogical link to a deity. Rather, more and more heads of state have risen to their positions by shaping public opinion. And to legitimately remain in office for more than a few years requires a constant monitoring of mass political attitudes, both before and during elections.

Although democracy may not technically be a brand-new institution, this form of government has existed in its current form for only a couple hundred years. Universal suffrage arose only in the late nineteenth century. Full enfranchisement still continues to spread to democratizing countries. So even though the ancient arts of diplomacy, spy craft, and warfare remain relevant, public opinion has become paramount.

The scientific study of public opinion in democratic societies is also relatively new. Yet individual differences in political attitudes have prehistoric, evolutionary origins. As democracy is promoted, exported, demanded, and fought for across a globalizing world, it's more important than ever to continue studying the forces that increasingly determine the selection of our leaders.

This research effort requires neuroscientists, geneticists, and biologists to collaborate with social scientists. Pollsters should consider gathering new kinds of data, and not only measuring what is conventional or convenient. Of course, the need will always remain for area specialists to explain how the particular geography, languages, cultures, religions, history, and idiosyncrasies of a region influence its population's political behavior. But regionalists would also do well to expand their awareness of the evolutionary origins and the universal characteristics of political orientation.

Not everything we learn in the future about our nature as political animals may be to our liking, or even something that we can change. But gaining greater knowledge of political psychology is still beneficial. As political scientist C. Fred Alford has pointed out, "A sailor can do nothing about the weather. If, however, the sailor learns about weather and seamanship, he may keep his ship afloat, working with the sea, not against it."[22]

A deeper knowledge of our political nature may help us to foresee the future more accurately. If this knowledge also provides us with a more enlightened understanding of what divides us, then maybe that future will be a little bit brighter.

FOUR STEPS TOWARD POLITICAL ENLIGHTENMENT

- No matter what your own political orientation happens to be, step back and contemplate the ultimate causes of the diversity of personalities we live among. Try to understand those across the aisle not as naïve or stupid or ignorant or evil, but rather on their own moral terms and also on nature's terms. Try to appreciate your own values on nature's terms.
- When you come across inequalities in life, observe your emotional reaction to them. Which hierarchies are fair and which ones are unfair? If you mostly believe in a just world, try to see some injustice; and if you typically believe in an unjust world, try to acknowledge some justice. A truly objective person should be careful before passing judgment. Often we cannot know.
- Cut down on your consumption of political junk food, or force yourself to balance biased media with other sources inclined in the opposite direction. Better still, read the in-depth explanations of professional geopolitical analysts (like those produced by Stratfor). These are meant not to gratify the partisan emotions, but rather to provide a solid understanding of current events. Governments, large companies, and others with a lot at

stake in the future read these to forecast the likelihood of events. You're just as important as they are.
- Share this book with a friend in another political party, as well as with someone who votes with your own. Discuss it with anyone you hold dear. Together we can work toward solutions to our common challenges that are compatible with our values and in harmony with the constraints of public opinion.

For in the end, the political dispositions described in this book need not be walls of stone or causes for fear or further polarization. Seen in their fullest light, these are just leanings, hurdles that each of us can learn to cross. The beauty of our democracy is that, at its heart, we are a nation of people, of individuals, each of us with our own character, our own will, and our own capacities to learn and grow and follow our dreams. And the more we learn and the deeper the understandings we gain of ourselves and of others, the more easily we will be able to transcend any attitudes that still divide us.

Speaking personally, I always gain hope and strength from the enduring words and feelings of the Bengali poet Rabindranath Tagore, as he set them forth in his timeless 1912 poem, "Where the Mind Is without Fear."[23] As the great poet and sage reminds us all:

Where the mind is without fear and the head is held high;
Where knowledge is free;
Where the world has not been broken up into fragments by narrow domestic walls;
Where words come out from the depth of truth;
Where tireless striving stretches its arms towards perfection;
Where the clear stream of reason has not lost its way into the dreary desert sand of dead habit;
Where the mind is led forward by thee into ever-widening thought and action—
Into that heaven of freedom, my Father, let my country awake.

ACKNOWLEDGMENTS

Anything in these pages that's politically incorrect is my fault alone. And for any content that might be inadvertently misinterpreted, I'm likely to blame. But for everything that came out right, I have many people to thank.

I'm forever thankful to my wonderful loving family, Mark, Jana, and Eva, for their encouragement and for giving me opportunities to learn, explore, and write. I give my deepest gratitude to journalist and author Paul Chutkow for his expert and selfless mentorship throughout the journey of this book. I'm also profoundly grateful to Francis Fukuyama for his invaluable support, and for being an inspiration as a scholar, a thinker, and a professor.

I'm especially thankful to Robert Mindelzun for sharing his lucid writing advice and his vast knowledge of medicine and history, to Glenn Lyons for reading through the entire manuscript and offering his insightful feedback, and to Cecilia Sedano for her tutoring in genetics and microbiology. I'm grateful to my agent Don Fehr for taking on a rookie with a challenging project, and to Steven L. Mitchell, Brian McMahon, Jill Maxick, Meghan Quinn, Melissa Shofner, Catherine Roberts-Abel, Jade Zora Scibilia, and the entire team at Prometheus Books.

For their reviews of technical sections and their help in accurately communicating their research, I owe a great debt to Jüri Allik, Jeremy Bailenson, Sir Patrick Bateson, Alan Bittles, Thomas Bouchard, Carl-Johan Dalgaard, Richard Ebstein, Alejandro Feged-Rivadeneira, James Fowler, R. Chris Fraley, George Garratty, Chris Gonzalez, Barbara Harff, Weijing He, Agnar Helgason, Robert Hoshowsky, Hillard Kaplan, Kathleen Knight, Rodrigo Labouriau, Joris Lammers, Neil Van Leeuwen, Jason E. Lewis, Marco Lopez, Elainie Madsen, Nicholas G. Martin, Stewart McCann, Ola Olsson, Peter Parham, Geraint Rees, Peter Rentfrow, Camilo Andrés Rivera, Tomás Rodríguez Barraquer, Gidi Rubinstein, David Sears, Robert Shapiro, William Shields, Michael Shifter, Kevin B. Smith, Randy Thornhill, Jacob Miguel Vigil, Alexander Weiss, Robin Weiss, Arthur Wolf, and Richie Zweigenhaft.

For sharing their research or their knowledge, I'm grateful to Daphna Canetti, Shirley Chen, David Cohen, Tülin Daloglu, Chengdiao Fan, Christopher

Federico, Marcus Feldman, Israel Gershoni, Sandeep K. Giri, Arash Hazeghi, Miguel Hilario, James Jankowski, Matthew Jobin, Tim King, Charles Laidley, Noam Lupu, Rahul Madhavan, Jim Moore, Lee Nelson, Sven Oskarsson, Raj Patel, Mark Potok, Elly Power, Andrew T. Shepard, Paul M. Sniderman, Frank Sulloway, Sarah Tishkoff, Derya Tokdemir, Polly Weissner, and Vanessa Yorke.

I'd like to thank Lynore Banchoff, Tom Banchoff, Bruce Beron, Daniel Brehon, Yosem Companys, David Cotacachi, Larry Diamond, Linda Donley-Reid, Ben Dooley, Lusio Filiba, Diane Jordan Wexler, Kevin Mack, Moshe Malkin, Lynn Meisch, Moisés Naím, David Oestreicher, Vincent Philip, Zach Pogue, Jerrold M. Post, David Rubenstein, Lorena Sander, Atul Singh, and Robert Trivers for their encouragement on this project.

My appreciation goes to Paul Chutkow, Mary Bisbee-Beek, Jeannette Boudreau, Tyler Bridges, Blake Charlton, Yael Goldstein Love, Sara Houghteling, Don Jacobs, Daniel Mason, Bill Murphy Jr., and indexer Ellen Sherron for advising me on the publishing process. I thank coach Brian Callaghan, coach Allison Shapira, Roy Blitzer, Paul Rovner, Ceevah Sobel, Irwin Sobel, Sarah Sobel, and Joe Tuschman for their help with public speaking and presentations. For their advice on graphics and design, I'm grateful to Joan Hausman, Robert Kato, Larry Stueck, and Richard Tuschman; although I haven't been able to apply their advice in full, they're to thank for the absence of numerous indecipherable diagrams that I wrongly thought made sense.

APPENDIX A

A BRIEF WORD ON CORRELATIONS AND LEVEL OF SIGNIFICANCE FOR LAYPEOPLE

CORRELATIONS

So what exactly is a correlation? Correlations are the most common type of comparison that scholars of public opinion make. In fact, correlations help all kinds of scientists to identify relationships between the myriad interacting parts of our surroundings and ourselves. By taking measurements of different phenomena, we can begin to tease apart which elements change in relation to one another; doing so oftentimes enables us to see the world in a much more objective way than we might if left to our own biases.

It's easiest to illustrate the concept with a specific relationship. In the 1880s, Sir Francis Galton, who was the half-cousin of Charles Darwin, decided to study what determines adult height. He hypothesized that we inherit it from our parents. Galton found a correlation of 0.57 between the average height of parents and that of their offspring. What does 0.57 mean?

A perfect correlation is that of 1.0, which would have occurred if every child had grown to a height that was exactly proportional to the average height of his or her parents. In this case, every data point would fall on a straight line. A correlation of 0.0, on the other hand, would have occurred if one measurement had no linear predictive power whatsoever over the other; the graph would be a dispersed scattering of dots, with no linear trend. A perfect negative correlation of –1.0 would occur if people's stature were to *decrease* in a proportional manner to increases in their parents' average height, and vice-versa, creating a straight line with a downhill slope.

As it turns out, Galton's correlation did not reach 1.0 because a number of other independent variables weakened the relationship; these may have included different levels of malnutrition and stress between the generations compared, misattributed paternity, acquired or genetic illnesses, maternal health and age during pregnancy, or the combination of the many height-influencing genes that happened to be inherited.

Thus, a correlation between two sets of measurements indicates the strength of the linear relationship between them when plotted on a graph. It's easiest to understand correlations visually (as in figure 45): *the fatter the scatter, the weaker the correlation*. Positive correlations mean that both measurements increase together, while a negative sign means that one measurement increases while the other one decreases, creating a downhill slope. The second row shows that even the strongest possible correlation of ±1.0 can have different slopes; this means that a perfectly predictable relationship can exist where the dependent variable changes at a slower, equal, or faster rate than the independent variable—as long as the proportional relationship remains the same.[a]

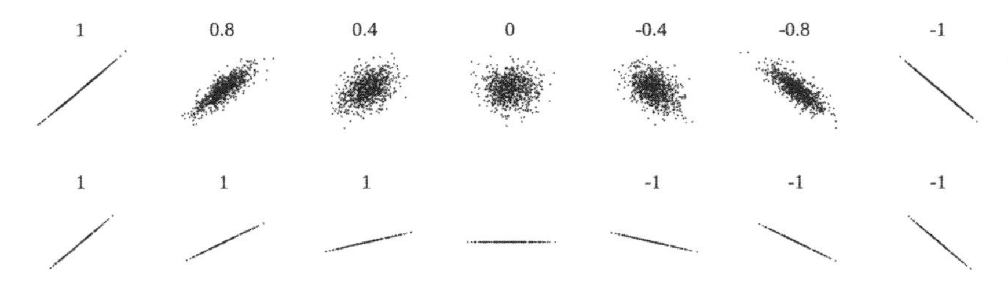

Figure 45. Correlation Coefficients and Graphs of the Two Variables Compared.

No explanation of correlations could be complete without the following warning: correlation does not prove causation! Consider this relationship: if an American individual disapproves of black-white dating, there is a good chance that this person approves of schools firing homosexual teachers. The correlation between these two attitudes, according to the Pew Research Center's 1987–2003 Values Survey, is –0.77.[1]

But what does one view have to do with the other? It could be that homophobia directly causes disapproval of black-white dating (ethnocentrism); alternatively, maybe ethnocentrism is the direct cause of homophobia. Perhaps ethnocentrism *indirectly* brings about homophobia through some third factor. Or it could be that "conservatism" is the independent cause of all three factors. More often than not, especially in the social sciences, multiple factors correlate with the phenomenon of interest.

To avoid cluttering the pages with numbers, this book uses words to denote the strength of a correlation:

A "very weak" correlation is between ±0.00 and ±0.19
A "weak" correlation is between ±0.20 and ±0.39
A "moderate" correlation is between ±0.40 and ±0.59
A "strong" correlation is between ±0.60 and ±0.79
A "very strong" correlation is between ±0.80 and ±1.0

More specific information can be found in the original studies cited in the bibliographic notes.

LEVEL OF SIGNIFICANCE

Suppose now that a scientist wants to compare two attitudes to determine if and how well changes in one attitude may covary with the second attitude. Let's assume that our scientist isn't biased, and uses a proper random sample that represents the important segments of the population in question. If the scientist chooses a small handful of subjects, it may occur by chance that the strength or weakness of the relationship between the two attitudes is greater than it would be if the entire population's attitudes were measured.

To keep one another honest about the likelihood that a measurement (such as a correlation) does not happen by chance, and that enough evidence in the sample exists to confidently generalize the results to the entire population, scientists depend on the concept of "statistical significance." The most common measurement of significance is called the *p*-value, which is used to determine if a measurement or correlation did not simply occur by chance.

If there is less than a 5 percent chance that the results occur by chance, then we can say that the result is "significant" (the notation in a study would look like this: $p < 0.05$). The general convention is that a 5 percent probability of chance results is the maximum amount tolerable to call the measurement significant. In studies with very large samples, however, scientists are able to know that results are significant with a much greater degree of confidence. So the value of *p* can drop to 1.0 percent, 0.1 percent, 0.01 percent, or even lower. In one incredible study discussed in chapter 11, an enormous sample means that the probability that the study's conclusions occurred by chance is mind-bogglingly minuscule; however, the correlation they found is fairly weak. Nevertheless, the phenomenon they researched is so critical that even a well-proven weak correlation changes the way the world works.

To illustrate the idea of the *p*-value and significance with a much more intuitive example, imagine that you have a coin in your hand. You and your greatest rival have decided to settle a dispute of great consequence by flipping this coin. However, your rival provided an antique piece of silver from another country, and you don't readily trust that the coin is fair. If you are allowed to flip the coin twenty times before the decisive flip, how many times would the coin have to land on the same side for you to know that there is more than a 95 percent chance that your rival is trying to cheat you? If the coin lands on "heads" fourteen times out of twenty, then there is a 5.8 percent chance that the coin is fair ($p = 0.058$). Though this result may be convincing, to publish a study about the unfairness of the coin in the peer-reviewed *Journal of Coin-Flipping Studies*, the bogus piece of money must land on the same side fifteen or more times out of twenty to reject the assumption that the coin is fair. Of course, the study would be far more convincing, and achieve a far lower p-value (in the case of an unfair coin), if you had the time to flip the coin one thousand times.

The purpose of this book is to share with you amazing scientific findings about the hidden world of political conflict—*not* to bore you with mathematical notations constantly qualifying the statistics that you *will* be interested to learn about. Therefore, unless specified otherwise, all findings mentioned in the book are significant at the level where p is less than 0.05 (5 percent). In many studies, p is less than 1 percent ($p < 0.01$), or even less than a tenth or a hundredth of a percent ($p < 0.001$ or $p < 0.0001$). In the last case, there would be less than one chance in ten thousand that such extreme results would occur due to chance.

Congratulations! That's as bad as it gets, and now we can continue with the interesting findings.

APPENDIX B

JOE THE PLUMBER, POLITICAL ELITES, AND THE CONTROVERSIAL EXISTENCE OF PUBLIC OPINION

Does a reliable public opinion endure over time? Does the public have meaningful political views, or is the majority easily manipulated by a small number of self-interested elites? Statistics show that, at least in the United States, a great many people know very little about basic political facts.

Surveys taken from the early 1950s through the 1990s reveal that a quarter of Americans are consistently unable to name the vice president of their country.[1] In 1986, for example, 24 percent of Americans could not even recognize the name of George H. W. Bush or his office, even though Bush Sr. had already been vice president for six years.[2] In 1998, only 41 percent of American teenagers could name the three branches of government, yet 59 percent of them could name the Three Stooges (all of whom had already died twenty-three years earlier—before the participants had been born).[3]

Pollsters have found the same political unawareness even amid intense media coverage of scandals. At the height of the Iran-Contra affair in the summer of 1987, after extensive televised hearings, only 54 percent of the American public knew that the United States was supporting the rebels in Nicaragua—as opposed to the Nicaraguan government. In fact, only about a third of the country even knew that Nicaragua was located in Central or Latin America.[4]

A SPECTRUM OF COHERENCE

In light of findings such as these, a group of political scientists at the University of Michigan began in the 1960s and 1970s to question perceptions about public opinion. One of these professors, Philip Converse, argued that most people respond to questions about politically emotive issues in a surprisingly unstable, almost random manner over time. Moreover, the Michigan team discovered that many people had inconsistent attitudes toward political issues that were supposedly related by ideology. That is, most people did not have as nearly a coherent belief system as did political elites.

On the other hand, the researchers found a substantial minority of sophisticates. These people, who tended to have a high level of education, interest in, or knowledge of politics, showed much more stable attitudes. And their reliable attitudes also grouped together with greater ideological coherence. If they took a conservative or liberal position on one controversial issue, they consistently did likewise for other issues.[5]

Philip Converse, Angus Campbell, and colleagues published a book in 1960 called *The American Voter*, which would create much debate in the field of public opinion.[6] The book categorized Americans into several groups, defined by the ideological coherence of their political attitudes. The political scientists ranked each category in hierarchical levels of coherence, and calculated the percentage of their sample that best fit within each category.

In the most sophisticated and coherent level were the "ideologues." The Michigan professors found that only 2.5 percent of their respondents belonged to this group. One rung lower were the "near-ideologues," who were less confident or articulate about liberal-versus-conservative principles yet still knew the difference. These respondents made up 9.5 percent of the sample.

Further down was the "group benefits" level, in which people weighed political issues in terms of how their group might benefit or not with respect to others. A substantial 42 percent of people fell into this category. Campbell derisively claimed that members of this group lacked "basic philosophies rooted in postures toward change" and a "comprehension of long-range plans."[7]

The 24 percent of people *The American Voter* placed in the fourth level, called "nature of the times," didn't understand ideology or even recognize group interests; they merely associated good and bad economic times with whichever party happened to be in power, then punished or rewarded its politicians accordingly. Robert Reich, the former US Secretary of Labor, has argued (in his own words) that "nature of the times" votes determine presidential elections. Reich believes that Jimmy Carter lost in 1980 because of double-digit inflation, while Ronald Reagan won reelection in 1984 with the help of a surging economy. According to this logic, George Bush senior lost in 1992 when Alan Greenspan tried to fend off inflation by raising interest rates—which also raised unemployment. By 1996, when jobs were on the increase again, Bill Clinton won a reelection. But John McCain's chances were spoiled in 2008 by partisan association with George W. Bush's Great Recession, which allowed his rival, Barack

Obama, to win.[8] When put to statistical scrutiny, in fact, a combined measure of real GDP growth and inflation can predict the electoral fate of incumbent parties in 76 percent of the last twenty-five US presidential elections, according to statistician wunderkind Nate Silver.[9]

Finally, *The American Voter*'s lowest category, which encompassed 22.5 percent of Americans, was called "absence of issue content." In other words, this group knew absolutely nothing about politics, with the possible exception of affiliation with a party or an attraction to a candidate's looks or personality. Most in this group didn't bother to vote.[10]

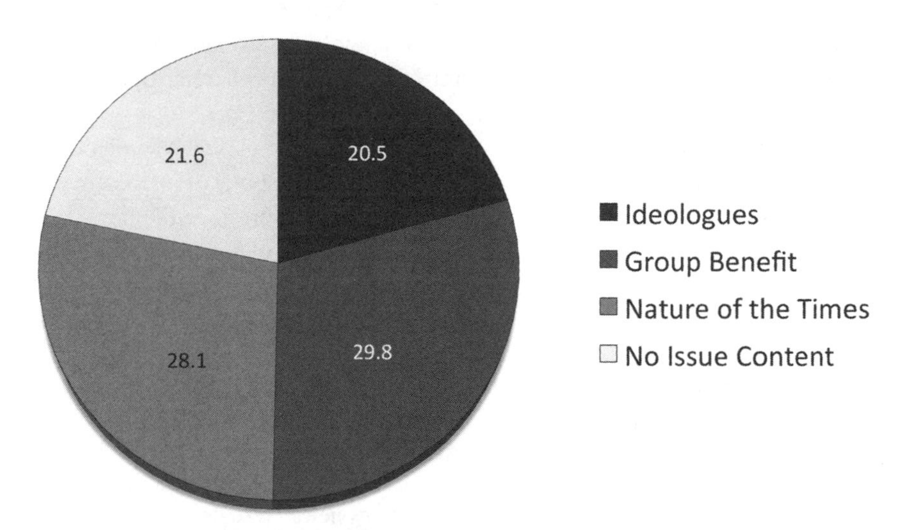

Figure 46. Percentages of the American Electorate in Each Level of Ideological Coherence (1956–2000 averages).

In retrospect, *The American Voter* made a critical point: differences between individuals in political awareness and attitude coherence were as large and important as the left-right ideological differences that divide us.[11] At the same time, we now have better data for categorizing the American electorate into these groups, thanks to political scientists like Kathleen Knight. And the cumulative results paint a less pessimistic picture for democracies. By averaging ten samples from the elections between 1956 and 2000, the pie chart in figure 46 shows what proportion of Americans tested into each category.[12]

These categories help break down what is really a spectrum of ideological coherence. At one end, about 22 percent of the population doesn't know or care very much about politics. This apathy seems to correspond with the percentage of Americans who perennially can't name their vice president. The next level up cares a bit more, and holds the president responsible for the economy to some degree. In countries without obligatory voting, such as the United States, most people who don't show up at the ballot box belong to one of these two groups.

At the other end of the spectrum, ideologues and political elites have the greatest coherence of attitudes, and consequently they are the most politically polarized. Politicians represent the very extreme of this half of the spectrum. It may be that politicians have such notable attitude coherence because of the great amount of time they spend surrounded by elites, advisors, and other highly aware and educated people. Of course, strong incentives also motivate politicians to adopt attitudes that are in line with as large a segment of the electorate as possible. Over half the population *is* moderately or strongly polarized, so successful politicians in democratic countries must have coherent logic in their political positions.

The fact that political elites run the news media also explains some degree of the media's polarization. As the group with the highest level of interest in politics and the highest ideological consistency, those who produce the news very often choose a particular focus and magnify the elements of conflict according to their ideological colors. According to John Zaller, a political scientist at UCLA and author of *The Nature and Origins of Mass Opinion*:

> The information that reaches the public is never a full record of important events and developments in the world. It is, rather, a highly selective and stereotyped view of what has taken place. It can hardly be otherwise. But even if it could, the public would have little desire to be closely informed about the vast world beyond its personal experience. It requires news presentations that are short, simple, and highly thematic—in a word, stereotyped.[13]

This is not to say that all media are equally biased; of course there are professional journalists who strive for objectivity. But even in the case of more objective media, the topics covered in political headlines are chosen precisely because the issues are politically controversial, and often sensational. Moreover, commercially funded news media ultimately face the same challenge as political leaders in democratic countries: how to appeal to the largest market segment possible. Again, the easiest way to do so is through predictable political polarization and sensationalism.

The force of public opinion is quite likely even stronger than the political-coherence pie chart suggests, because the Michigan study could have easily underestimated the ideological coherence of people in the bottom two categories. Even though someone may be ignorant of complex policy issues, foreign geography, or the structure of government, the same person could have strong social opinions and a high level of knowledge about his or her own town or different church congregations.

Consider "Joe the Plumber," who achieved brief fame during the 2008 US election. Three days before the last presidential debate, Barack Obama made a campaign stop to speak with the residents of Joe Wurzelbacher's Ohio neighborhood. It all began when Joe skeptically asked Obama how his tax plan would affect the small business that he hoped to start one day. Obama's opponent, the McCain-Palin campaign, jumped on the opportunity to make "Joe the Plumber" into an icon—the symbol of ordinary Americans just looking out for their economic interest against a tax-raising Democrat. In other words, the Republican campaign sought to portray Joe as a normal, unbiased guy because he didn't belong to the "ideologues" group, but rather to the "group benefits" category, which could be perceived as more rational.

"Joe the Plumber" turned out to be quite a bit more ideological than McCain and Palin had bargained for. Joe called Obama's plan to raise taxes three points on incomes higher than $250,000 "one step closer to socialism." He also "greatly questioned" Obama's "loyalty to our country." Finally, "Joe the Plumber" criticized McCain's support of the financial-industry bailout and backed the far-right Tea Party movement. Neither Joe nor the company he worked for would have even been subject to the three-point tax increase, further detracting credibility from the implication that he was a "group benefits" voter—at least not in the sense of economic self-interest. In short, not only did Joe demonstrate high ideological coherency, he also showed some signs of political extremism.

So who exactly are the people who have high ideological coherence? What does it mean to be a political elite? The answer is people who (1) attain a high level of education, (2) score highly on a test of political facts (if you're American, then you have a good chance of qualifying if you know the name of the vice president and the number of senators from each state), and/or (3) express a high interest in politics. In figure 47, we can see quite clearly how education impacts a specific political attitude. People with different levels of education were asked:

> *Some people believe that the government should make an active effort to see that blacks can live anywhere they choose, including white neighborhoods. Others believe that this is not the government's business and it should stay out of this. How do you feel? Is this an area the government should stay out of or should government make an active effort to see that blacks can live anywhere they can afford to—including white neighborhoods?*[14]

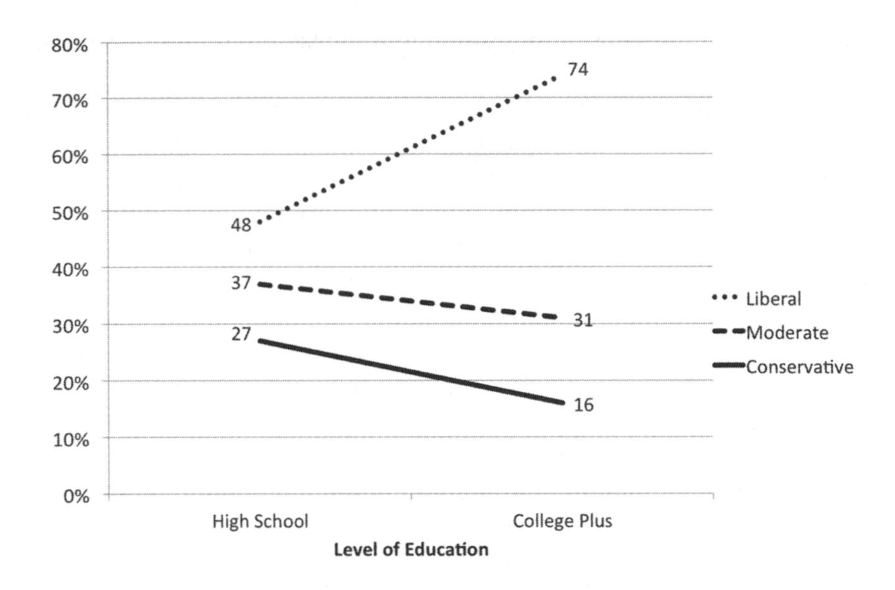

Figure 47. Percentage Who Agree with Government Intervention for Racial Desegregation of Neighborhoods, by Ideology and Education (1991).

The chart suggests that exposure to higher levels of education may make people more liberal if they are liberal to begin with; conservatives, overall, seem to change less with education, although attaining more degrees can increase conservatism as well. Political scientists have found that higher education levels in Europe have the same polarizing effect on political ideologies, as do higher levels of political awareness in America.[15]

NOTES

PREFACE

1. Nicholas Wade, "Depth of the Kindness Hormone Appears to Know Some Bounds," *New York Times*, January 10, 2011; Patrick Goldstein and James Rainey, "Colin Firth on Conservatives: Is There Something Biologically Wrong with Them?" *Los Angeles Times*, January 4, 2011; Tom Feilden, "Are Political Beliefs Hard-Wired?" *BBC Radio 4: Tom Feilden's Blog*, December 28, 2010; Nicholas D. Kristof, "Our Politics May Be All in Our Head." *New York Times*, February 13, 2010; Nicholas D. Kristof, "Would You Slap Your Father? If So, You're a Liberal," *New York Times*, May 27, 2009; Allison Aubrey, "How Much Does Birth Order Shape Our Lives?" National Public Radio, November 23, 2010; Adam Gorlick, "Researchers Say Voters Swayed by Candidates Who Share Their Looks," *Stanford Report*, October 22, 2008; Frans B. M. de Waal, "Bonobos, Left & Right: Primate Politics Heats up again as Liberals & Conservatives Spindoctor Science," *Skeptic: Promoting Science and Critical Thinking*, August 8, 2007; Robert Sapolsky, "Are the Desert People Winning?" *Discover*, August 2005, p. 38.

INTRODUCTION: SEARCHING FOR THE ORIGIN OF POLITICAL IDENTITY

Chapter 1. The Burning Man of Tunisia

a. More precisely, "basboosa" is Arabic for a dessert made of semolina grains soaked in floral-scented nectar.

b. The Egyptian parliamentary monarchy (1922–1952) did have several vigorously contested multi-party elections, but women were barred from voting, and the government often tampered with the ballot boxes.

1. Sadiki Larbi, "Bin Ali Baba Tunisia's Last Bey?" *Al Jazeera*, September 27, 2010.

2. Tom Jensen, "Voters Moving against Occupy Movement," Public Policy Polling, November 16, 2011.

3. Lydia Saad, "Tea Partiers Are Fairly Mainstream in Their Demographics,"

Gallup Politics, April 5 2010, http://www.gallup.com/poll/127181/tea-partiers-fairly-mainstream-demographics.aspx (accessed April 4, 2013).

Chapter 2. The Universal Political Animal

a. In the years since the Blocks conducted their seminal research, at least one other longitudinal study replicated their findings (in 2012). The results of this much larger experiment, whose sample included over seven hundred individuals, were consistent with the Blocks's original work: children who were rated as having fearful behavior at the age of four-and-a-half were more likely to become politically conservative at the age of eighteen; on the other hand, children with "high levels of activity and restlessness" were more likely to become liberals when they reached voting age. See R. Chris Fraley, Brian N. Griffin, Jay Belsky, and Glenn I. Roisman, "Developmental Antecedents of Political Ideology: A Longitudinal Investigation from Birth to Age 18 Years," *Psychological Science* 20, no. 10 (2012): 1–7.

b. The first scientists to compare the political attitudes of twins, and to discover that a substantial portion of these attitudes had a genetic origin, were psychologist Hans Eysenck and geneticist Lindon Eaves. Although their findings were published in *Nature* in 1974, and they published similar findings again in 1986 in the *Proceedings of the National Academy of Sciences*, virtually no one in the political sciences took notice; the assumption in political science at the time was that ideological orientation was a purely cultural phenomenon, transmitted through the social environment. But the stubborn findings on the hereditary component kept returning through increasingly sophisticated studies, such as those described here.

c. The graph approximates a normal distribution if one disregards the raised tails at the extremes and the skewing of the two center bars to the left. Where do the little peaks at "left" and "right" come from? On the far right, the country that had the greatest percentage of people who chose the most conservative option possible was Communist Vietnam (fully 64.2 percent of Vietnamese chose the extreme right). The greatest outlier on the extreme left was Zimbabwe, under the one-man dictatorial rule of Robert Mugabe, where 27.3 percent of Zimbabweans chose the most extreme left-wing stance possible. Countries like Vietnam and Zimbabwe show how the curve's elevated tails represent a popular rejection of the ruling regime. That is, public opinion can polarize away from the ideology of extremist governments.

The uneven double peaks in the middle (at "5" and "6") are actually an artifact caused by a technical mistake. If the World Values Survey had consulted a (better) political psychologist to design this question, the scale would have consisted of an odd number of points—but not ten. Why? Because five, seven, or even nine points would create something called a standard Likert scale.

Likert scales are well-studied standards in psychology, proven to avoid various kinds of skewing and errors. They typically have answers such as "very liberal," "liberal," "moderate," "conservative," and "very conservative." The key elements of Likert scales are their symmetry and their neutral middle choice. But the World Values Survey had no neutral middle choice for this question. Many people who selected "5" surely intended to identify as "moderate" or "neutral" (rather than as "slightly to the left"). Therefore, the center peak is artificially skewed to the left. If the survey used a Likert scale for this question, we would see an even more symmetrical curve.

1. J. Block and J. H. Block, "Nursery School Personality and Political Orientation Two Decades Later," *Journal of Research in Personality* 40, no. 5 (2006): 734–49.

2. J. C. Flugel, *Man, Morals and Society: A Psycho-Analytical Study* (New York: International Universities Press, 1945).

3. Bob Altemeyer, *The Authoritarian Specter* (Cambridge, MA: Harvard University Press, 1996).

4. J. R. Alford, C. L. Funk, and J. R. Hibbing, "Are Political Orientations Genetically Transmitted?" *American Political Science Review* 99, no. 2 (2005): 153–67.

5. J. R. Alford and J. R. Hibbing, "The Origin of Politics: An Evolutionary Theory of Political Behavior," *Perspectives on Politics* 2, no. 4 (2004): 707–23; P. K. Hatemi, Sarah E. Medland, Robert Klemmensen, Sven Oskarsson, Levente Littvay, Chris Dawes, Brad Verhulst, et al. "Genetic Influences on Political Ideologies: Genome-Wide Findings on Three Populations, and a Mega-Twin Analysis of 19 Measures of Political Ideologies from Five Western Democracies," *Behavior Genetics* (January 10, 2013); L. J. Eaves and H. J. Eysenck, "Genetics and the Development of Social Attitudes," *Nature* 249 (1974): 288–89; N. G. Martin, L. J. Eaves, A. C. Heath, R. Jardine, L. M. Feingold, and H. J. Eysenck, "Transmission of Social Attitudes," *Proceedings of the National Academy of Sciences of the United States of America* 83, no. 12 (1986): 4364.

6. P. K. Hatemi and R. McDermott, "The Genetics of Politics: Discovery, Challenges, and Progress," *Trends in Genetics* (2012); P. K. Hatemi, N. A. Gillespie, L. J. Eaves, B. S. Maher, B. T. Webb, A. C. Heath, S. E. Medland, et al., "A Genome-Wide Analysis of Liberal and Conservative Political Attitudes," *Journal of Politics* 73, no. 1 (2011): 271–85.

7. J. C. Gewirtz and M. Davis, "Second-Order Fear Conditioning Prevented by Blocking NMDA Receptors in Amygdala," *Nature* 388, no. 6641 (1997): 471–74.

8. David O. Sears, "Long-Term Psychological Consequences of Political Events," in *Political Psychology*, ed. Kristen R. Monroe (Mahwah, NJ: L. Erlbaum, 2002), pp. 249–69; D. O. Sears and C. L. Funk, "Evidence of the Long-Term Persistence of Adults' Political Predispositions," *Journal of Politics* 61, no. 1 (1999): 1–28.

9. Peter Mair, "Left-Right Orientations," in *The Oxford Handbook of Political Behavior*, ed. Russell J. Dalton and Hans-Dieter Klingemann (Oxford: Oxford University Press, 2007), pp. 206–22.

10. J. T. Jost, "The End of the End of Ideology," *American Psychologist* 61, no. 7 (2006): 651.

11. Mair, "Left-Right Orientations."

12. Ibid.; Kenneth Benoit, *Party Policy in Modern Democracies*, ed. Michael Laver (London: Routledge, 2006).

13. J. Allik and R. R. McCrae, "Toward a Geography of Personality Traits: Patterns

of Profiles across 36 Cultures," *Journal of Cross Cultural Psychology* 35, no. 1 (2004): 13–28.

14. Patrick Goldstein and James Rainey, "Colin Firth on Conservatives: Is There Something Biologically Wrong with Them?" *Los Angeles Times*, January 4, 2011.

15. R. Kanai, T. Feilden, C. Firth, and G. Rees, "Political Orientations Are Correlated with Brain Structure in Young Adults," *Current Biology* (accepted manuscript, 2011).

16. Frans B. M. de Waal, *Chimpanzee Politics: Power and Sex among Apes* (Baltimore: Johns Hopkins University Press, 2007).

17. A. Weiss, J. E. King, and W. D. Hopkins, "A Cross Setting Study of Chimpanzee (Pan Troglodytes) Personality Structure and Development: Zoological Parks and Yerkes National Primate Research Center," *American Journal of Primatology* 69, no. 11 (2007): 1264–77; Alexander Weiss, Miho Inoue-Murayama, James E. King, Mark James Adams, and Tetsuro Matsuzawa, "All Too Human? Chimpanzee and Orang-Utan Personalities Are Not Anthropomorphic Projections," *Animal Behaviour* 83 no. 6 (2012); J. E. King and A. Weiss, "Personality from the Perspective of a Primatologist," in *Personality and Temperament in Nonhuman Primates*, ed. A. Weiss, J. E. King and L. Murray (New York: Springer, 2011), pp. 77–99.

Chapter 3. What Economics Can and Cannot Predict

a. Appendix A, which is completely optional, includes an explanation of statistical significance for those who may be interested.

1. "Why Do People Often Vote against Their Own Interests?" BBC News, January 30, 2010.

2. J. T. Jost, "The End of the End of Ideology," *American Psychologist* 61, no. 7 (2006): 651.

3. William G. Domhoff, "Wealth, Income, and Power," *Who Rules America?* September 2005 (updated September 2010), http://iactnow.org/2010/11/11/who-rules-america-wealth-income-and-power-september-2005-updated-september-2010/ (accessed January 2011).

4. Thomas Frank, *What's the Matter with Kansas? How Conservatives Won the Heart of America*, 1st ed. (New York: Metropolitan Books, 2004).

5. Edward L. Glaeser and Bryce A. Ward, "Myths and Realities of American Political Geography," National Bureau of Economic Research, 2006.

6. Ibid.

7. D. J. Levinson, "Politico-Economic Ideology and Group Memberships in Relation to Ethnocentrism," in *The Authoritarian Personality*, ed. T. W. Adorno, Else Frenkel-Brunswik, Daniel J. Levinson, and R. Nevitt Sanford (New York: Harper & Row, 1950), pp. 151–221.

8. David O. Sears and Carolyn L. Funk, "The Role of Self-Interest in Social and Political Attitudes," *Advances in Experimental Social Psychology* 24, no. 1 (1991): 1–91.

9. Benjamin I. Page, *The Rational Public: Fifty Years of Trends in Americans' Policy Preferences*, ed. Robert Y. Shapiro (Chicago: University of Chicago Press, 1992); David O. Sears, John J. Hetts, Jim Sidanius, and Lawrence Bobo, "Race in American Politics: Framing the Debates," in *Racialized Politics: The Debate about Racism in America*, ed. David O. Sears, Jim Sidanius, and Lawrence Bobo (Chicago: University of Chicago Press, 2000), pp. 1–43.

10. Glaeser and Ward, "Myths and Realities of American Political Geography."

11. "Keiko Quita Respaldo En Lima a Castañeda," Peru21, http://peru21.pe/noticia/611098/fujimori-le-quita-electores-lima-luis-castaneda (accessed July 20 2010).

12. Michael Shifter, Javier Corrales, and Eduardo Posada-Carbó, "Latin America Goes Centrist," *Journal of Democracy* 22 (2011): 107–21.

13. Michael Shifter, "Latin America's Shift to the Center," *Foreign Policy*, August 6, 2010.

14. Noam Lupu, "Electoral Bases of Leftist Presidents in Latin America," Department of Politics, Princeton University, 2009.

15. Carl-Johan Dalgaard and Ola Olsson, "Why Are Rich Countries More Politically Cohesive?" in *Department of Economics, University of Copenhagen: Discussion Papers, No. 09-23*, 2009.

16. Jos D. Meloen, "The Political Culture of State Authoritarianism," in *Political Psychology: Cultural and Cross-Cultural Foundations*, ed. Stanley Allen Renshon and J. H. Duckitt (New York: New York University Press, 2000), pp. 108–27.

17. Albert Somit, *Darwinism, Dominance, and Democracy: The Biological Bases of Authoritarianism*, ed. Steven A. Peterson (Westport, CN: Praeger, 1997).

18. J. T. Jost, J. Glaser, A. W. Kruglanski, and F. J. Sulloway, "Political Conservatism as Motivated Social Cognition," *Psychological Bulletin* 129, no. 3 (2003): 339.

19. Meloen, "The Political Culture of State Authoritarianism"; Neil Jeffrey Kressel, *Mass Hate: The Global Rise of Genocide and Terror*, updated ed. (Boulder, CO: Westview Press, 2002); Daniel Chirot and Clark McCauley, *Why Not Kill Them All? The Logic and Prevention of Mass Political Murder* (Princeton, NJ: Princeton University Press, 2006).

20. J. T. Jost, B. W. Pelham, O. Sheldon, and B. Ni Sullivan, "Social Inequality and the Reduction of Ideological Dissonance on Behalf of the System: Evidence of Enhanced System Justification among the Disadvantaged," *European Journal of Social Psychology* 33, no. 1 (2003): 13–36.

21. Peter J. Rentfrow, John T. Jost, Samuel D. Gosling, and Jeffrey Potter, "Statewide Differences in Personality Predict Voting Patterns in 1996–2004 U.S. Presidential Elections," in *Social and Psychological Bases of Ideology and System Justification*, ed. John T. Jost, Aaron C. Kay, and Hulda Thorisdottir (Oxford: Oxford University Press, 2009), pp. 314–47.

22. T. J. Bouchard Jr. and M. McGue, "Genetic and Environmental Influences on Human Psychological Differences," *Journal of Neurobiology* 54, no. 1 (2003): 4–45.

23. Rentfrow et al., "Statewide Differences in Personality Predict Voting Patterns in 1996–2004 U.S. Presidential Elections"; Martha L. Cottam, Beth Dietz-Uhler, Elena

Mastors, and Thomas Preston, *Introduction to Political Psychology* (Mahwah, NJ: Lawrence Erlbaum Associates, 2004).

24. Jost et al., "Political Conservatism as Motivated Social Cognition."

25. G. V. Caprara, S. Schwartz, C. Capanna, M. Vecchione, and C. Barbaranelli, "Personality and Politics: Values, Traits, and Political Choice," *Political Psychology* 27, no. 1 (2006): 1–28.

26. S. E. Ha, S. Kim, and S. H. Jo, "Personality Traits and Political Participation: Evidence from South Korea," *Political Psychology* 34, no. 2 (2013); J. J. Mondak, *Personality and the Foundations of Political Behavior* (Cambridge: Cambridge University Press, 2010).

Chapter 4. The Invention of the Political Litmus Test

a. Appendix B ("Joe the Plumber, Political Elites, and the Controversial Existence of Public Opinion") discusses their important findings.

1. Roger Brown, "The Authoritarian Personality and the Organization of Attitudes," in *Political Psychology: Key Readings*, ed. John T. Jost and Jim Sidanius (New York: Psychology Press, 2004), pp. 39–68.

2. R. D. Smither, "Authoritarianism, Dominance, and Social Behavior: A Perspective from Evolutionary Personality Psychology," *Human Relations* 46, no. 1 (1993): 23; Martha L. Cottam, Beth Dietz-Uhler, Elena Mastors, and Thomas Preston, *Introduction to Political Psychology* (Mahwah, NJ: Lawrence Erlbaum Associates, 2004).

3. Brown, "The Authoritarian Personality and the Organization of Attitudes."

4. T. W. Adorno, E. Frenkel-Brunswik, D. J. Levinson, and R. Nevitt Sanford, *The Authoritarian Personality* (New York: Harper & Row, 1950).

5. J. M. Steiner and J. Fahrenberg, "Die Ausprägung Autoritärer Einstellung Bei Ehemaligen Angehörigen Der SS Und Der Wehrmacht," *Kölnischer Zeitschrift für Soziologie und Sozial Psychologie* 22 (1970): 551–66.

6. Neil Jeffrey Kressel, *Mass Hate: The Global Rise of Genocide and Terror*, updated ed. (Boulder, CO: Westview Press, 2002); J. M. Steiner, "The S. S. Yesterday and Today: A Sociopsychological View," in *Survivors, Victims, and Perpetrators: Essays on the Nazi Holocaust*, ed. Joel E. Dimsdale (New York: Hemisphere, 1980), pp. 405–56.

7. Kressel, *Mass Hate: The Global Rise of Genocide and Terror.*

8. Brown, "The Authoritarian Personality and the Organization of Attitudes."

9. R. Christie and P. Cook, "A Guide to Published Literature Relating to the Authoritarian Personality through 1956," *Journal of Psychology* 45, no. 2 (1958): 171–99.

10. Richard Christie, *Studies in the Scope and Method of "the Authoritarian Personality*, ed. Marie Jahoda (Glencoe, IL: Free Press, 1954).

11. Smither, "Authoritarianism, Dominance, and Social Behavior: A Perspective from Evolutionary Personality Psychology."

12. Brown, "The Authoritarian Personality and the Organization of Attitudes"; Cottam et al., *Introduction to Political Psychology*.

13. Brown, "The Authoritarian Personality and the Organization of Attitudes."

14. Smither, "Authoritarianism, Dominance, and Social Behavior: A Perspective from Evolutionary Personality Psychology."

15. Else Frenkel-Brunswik, "Parents and Childhood as Seen through the Interviews," in *The Authoritarian Personality*, ed. T. W. Adorno, Else Frenkel-Brunswik, Daniel J. Levinson, and R. Nevitt Sanford (New York: Harper & Row, 1950), pp. 337–89.

16. J. T. Jost, J. Glaser, A. W. Kruglanski, and F. J. Sulloway, "Political Conservatism as Motivated Social Cognition," *Psychological Bulletin* 129, no. 3 (2003): 339.

17. Smither, "Authoritarianism, Dominance, and Social Behavior: A Perspective from Evolutionary Personality Psychology."

18. William J. McGuire, "The Poly-Psy Relationship: Three Phases of a Long Affair," in *Explorations in Political Psychology*, ed. Shanto Iyengar and William J. McGuire (Durham, NC: Duke University Press, 1993), pp. 9–35; J. T. Jost, "The End of the End of Ideology," *American Psychologist* 61, no. 7 (2006): 651; Hulda Thorisdottir, John T. Jost, and Aaron C. Kay, "On the Social and Psychological Bases of Ideology and System Justification," in *Social and Psychological Bases of Ideology and System Justification*, ed. John T. Jost, Aaron C. Kay, and Hulda Thorisdottir (Oxford: Oxford University Press, 2009), pp. 3–23 .

19. McGuire, "The Poly-Psy Relationship: Three Phases of a Long Affair"; Milton Lodge and Patrick Stroh, "Inside the Mental Voting Booth: An Impression-Driven Process Model of Candidate Evaluation," in *Explorations in Political Psychology*, ed. Shanto Iyengar and William J. McGuire (Durham, NC: Duke University Press, 1993), pp. 225–63; Roger D. Masters and Denis G. Sullivan, "Nonverbal Behavior and Leadership: Emotion and Cognition in Political Information Processing," in *Explorations in Political Psychology*, ed. Shanto Iyengar and William J. McGuire (Durham, NC: Duke University Press, 1993), pp. 150–82.

20. Thorisdottir, Jost, and Kay, "On the Social and Psychological Bases of Ideology and System Justification"; John L. Sullivan, Wendy M. Rahn, and Thomas J. Rudolph, "The Contours of Political Psychology: Situating Research on Political Information Processing," in *Thinking About Political Psychology*, ed. James H. Kuklinski (Cambridge: Cambridge University Press, 2002), pp. 23–47; Jon A. Krosnick, "The Challenges of Political Psychology: Lessons to Be Learned from Research on Attitude Perception," in *Thinking about Political Psychology*, ed. James H. Kuklinski (Cambridge: Cambridge University Press, 2002), pp. 115–52.

21. David Van Nuys, "The Authoritarian Personality with Dr. Robert Altemeyer," http://shrinkrapradio.com/127-the-authoritarian-personality/ (accessed April 8, 2013).

22. Robert Altemeyer, *Enemies of Freedom: Understanding Right-Wing Authoritarianism*, 1st ed. (San Francisco: Jossey-Bass Publishers, 1988).

23. S. Feldman, "Enforcing Social Conformity: A Theory of Authoritarianism," *Political Psychology* 24, no. 1 (2003): 41–74.

24. Altemeyer, *Enemies of Freedom: Understanding Right-Wing Authoritarianism*.

25. Robert Altemeyer, *The Authoritarian Specter* (Cambridge, MA: Harvard University Press, 1996).

26. Altemeyer, *Enemies of Freedom: Understanding Right-Wing Authoritarianism.*

27. Altemeyer, *The Authoritarian Specter.*

28. Altemeyer, *Enemies of Freedom: Understanding Right-Wing Authoritarianism.*

29. Altemeyer, *The Authoritarian Specter.*

30. Gidi Rubinstein, "Two Peoples in One Land: A Validation Study of Altemeyer's Right-Wing Authoritarianism Scale in the Palestinian and Jewish Societies in Israel," *Journal of Cross-Cultural Psychology* 27, no. 2 (1996): 216.

31. Ibid.

Chapter 5. Unearthing the Three Roots of Political Orientation

a. Since most of the thirty statements refer to more than one category, the percentages add up to more than one hundred. Ideally, each item on the RWA test would pertain to only one of the six categories. Then the percentages would add up to one hundred.

1. J. T. Jost, J. Glaser, A. W. Kruglanski, and F. J. Sulloway, "Political Conservatism as Motivated Social Cognition," *Psychological Bulletin* 129, no. 3 (2003): 339; J. T. Jost, "The End of the End of Ideology," *American Psychologist* 61, no. 7 (2006): 651.

PART I. TRIBALISM ON THE POLITICAL SPECTRUM

Chapter 6. Ethnocentrism vs. Xenophilia

a. "Xenophilia" means an attraction to members of out-groups.

b. Rural peoples from the Oklahoma region, often of mixed Native American heritage, who migrated to the West Coast in the 1930s.

c. In chapter 11, we'll learn about "sexual imprinting," by which humans (and other animals) pattern mate choices after early memories of their familiar environment. Tribal clothing and markings may play some role in amplifying perceptions of ethnic differences, thereby increasing attraction to similarly clad in-groupers.

1. Glenn Beck, "What's Driving President Obama's Agenda?" http://www.glennbeck.com/content/articles/article/198/28330/ (accessed April 19, 2013).

2. "Fox Host Glenn Beck: Obama Is a 'Racist,'" *Huffington Post*, May 25, 2011, http://www.huffingtonpost.com/2009/07/28/fox-host-glenn-beck-obama_n_246310.html (accessed April 19, 2013).

3. Mark Leibovich, "Being Glenn Beck," *New York Times Magazine*, September 29, 2010.

4. Robert Altemeyer, *Enemies of Freedom: Understanding Right-Wing Authoritarianism*, 1st ed. (San Francisco: Jossey-Bass Publishers, 1988).

5. Robert Altemeyer, *The Authoritarian Specter* (Cambridge, MA: Harvard University Press, 1996).

6. Ibid.

7. J. T. Jost, "The End of the End of Ideology," *American Psychologist* 61, no. 7 (2006): 651.

8. J. T. Jost, J. Glaser, A. W. Kruglanski, and F. J. Sulloway, "Political Conservatism as Motivated Social Cognition," *Psychological Bulletin* 129, no. 3 (2003): 339.

9. Barry Goldwater, *The Conscience of a Conservative* (New York: Hillman Books, 1960).

10. John Stossel, "Racism and Rand Paul," *Fox Business*, May 21, 2010.

11. Ibid.

12. Paul M. Sniderman and Thomas Leonard Piazza, *The Scar of Race* (Cambridge, MA: Belknap Press of Harvard University Press, 1993).

13. Ibid.

14. "Black Americans in Congress: Member Profiles," Office of the Clerk, United States House of Representatives, http://baic.house.gov/member-profiles/ (accessed April 8, 2013).

15. Thomas F. Pettigrew, "Systematizing the Predictors of Prejudice," in *Racialized Politics: The Debate about Racism in America*, ed. David O. Sears, Jim Sidanius, and Lawrence Bobo (Chicago: University of Chicago Press, 2000), pp. 280–301.

16. Guillermo C. Jiménez, *Red Genes, Blue Genes: Exposing Political Irrationality* (Brooklyn, NY: Autonomedia, 2009).

17. James H. Kuklinski, "Introduction," in *Citizens and Politics: Perspectives from Political Psychology*, ed. James H. Kuklinski (Cambridge, UK: Cambridge University Press, 2001), pp. 243–53.

18. Jennifer L. Hochschild, "Where You Stand Depends on What You See: Connections among Values, Perceptions of Fact, and Political Prescriptions," in *Citizens and Politics: Perspectives from Political Psychology*, ed. James H. Kuklinski (Cambridge, UK: Cambridge University Press, 2001), pp. 313–40.

19. Ibid.

20. Daniel Chirot and Clark McCauley, *Why Not Kill Them All? The Logic and Prevention of Mass Political Murder* (Princeton, NJ: Princeton University Press, 2006).

21. Imogen Foulkes, "Swiss Voters Back Expulsion of Foreign Criminals," BBC News Europe, November 28, 2010.

22. Frank Rich, "The Rage Won't End on Election Day," *New York Times*, October 16, 2010.

23. Sniderman and Piazza, *The Scar of Race*.

24. Roger Brown, "The Authoritarian Personality and the Organization of Attitudes," in *Political Psychology: Key Readings*, ed. John T. Jost and Jim Sidanius (New York: Psychology Press, 2004), pp. 39–68.

25. Paul M. Sniderman, Peri Pierangelo, Rui J. P. de Figueiredo Jr., and Thomas Piazza, *The Outsider: Prejudice and Politics in Italy* (Princeton, NJ: Princeton University Press, 2000).

26. D. Canetti-Nisim and A. Pedahzur, "Contributory Factors to Political Xenophobia in a Multi-Cultural Society: The Case of Israel," *International Journal of Intercultural Relations* 27, no. 3 (2003): 307–33.

27. Sniderman and Piazza, *The Scar of Race*.

28. Charlie Savage, "On Nixon Tapes, Ambivalence over Abortion, Not Watergate," *New York Times*, June 23, 2009.

29. Pew Social Trends, "The Decline of Marriage and Rise of New Families," *Social and Demographic Trends* (2010).

30. R. Boyd and P. J. Richerson, "The Evolution of Ethnic Markers," *Cultural Anthropology* 2, no. 1 (1987): 65–79.

31. Jos D. Meloen, "The Political Culture of State Authoritarianism," in *Political Psychology: Cultural and Cross-Cultural Foundations*, ed. Stanley Allen Renshon and J. H. Duckitt (New York: New York University Press, 2000), pp. 108–27.

32. Raphael S. Ezekiel, *The Racist Mind: Portraits of Neo-Nazis and Klansmen* (New York: Viking, 1995).

33. Adolf Hitler, *Mein Kampf*, ed. and transl. Ralph Manheim (Boston: Houghton Mifflin, 1971).

34. "Hitler Book Bestseller in Turkey," BBC News, March 18, 2005.

35. "U.S. Elections 2010: Craziest Campaign Moments," BBC News, October 27, 2010.

36. Daniel. Hernandez, "Journalists Protest Proposed Law in Bolivia That Would Outlaw Reporting on Acts, Opinions Deemed Racist," *Los Angeles Times*, October 5, 2010.

37. Shulah Mendelson and Unknown, "What Does It Mean to Be a Leftist? (translated from the Hebrew, "Mah Zeh Lihiot Smolani?" a political humor presentation circulated on the Internet, 2010).

38. Simon Romero, "In Venezuela, a New Wave of Foreigners," *New York Times*, November 6, 2010.

39. David M. Levy, "Anti-Nazis: Criteria of Differentiation," in *Personality and Political Crisis: New Perspectives from Social Science and Psychiatry for the Study of War and Politics*, ed. Alfred H. Stanton (Glencoe, IL: Free Press, 1951), pp. 151–227.

40. Jost, "The End of the End of Ideology."

41. Jost et al., "Political Conservatism as Motivated Social Cognition."

42. Edward L. Glaeser and Bryce A. Ward, "Myths and Realities of American Political Geography," National Bureau of Economic Research, 2006.

43. Jerrold M. Post, *Leaders and Their Followers in a Dangerous World: The Psychology of Political Behavior*, 1st ed., ed. Alexander George (Ithaca: Cornell University Press, 2004).

44. Ibid.

45. Ibid.

Chapter 7. Religiosity vs. Secularism

a. When personal income, the Consumer Price Index, work stoppages, and expectations and attitudes were all significantly better.

b. Also, vegetarianism reflects the ethical principle of nonviolence against animals.

c. About half of the "Other Religions" category represents indigenous religions, and the other half, traditional Chinese beliefs. "Non-Religious" includes agnostics, secular individuals, and those who are "theistic" or "spiritual" but do not identify with a religion; the "spiritual" proportion represents a substantial part of this category.

1. Mohandas K. Gandhi, *An Autobiography, or, the Story of My Experiments with Truth*, ed. Mahadev H. Desai (Ahmedabad: Navajivan Publishing House, 1996).

2. Kenneth D. Wald, *Religion and Politics in the United States* (New York: St. Martin's Press, 1987).

3. Orla Guerin, "Pakistani Christian Asia Bibi 'Has Price on Her Head,'" BBC News South Asia, December 6, 2010.

4. "Mumtaz Qadri Charged with Salman Taseer Murder," BBC News South Asia, February 14, 2011; Aleem Maqbool, "Salman Taseer: Thousands Mourn Pakistan Governor," BBC News South Asia, January 5, 2011.

5. "Pakistan Minorities Minister Shahbaz Bhatti Shot Dead," BBC News South Asia, March 2, 2011.

6. "Pakistan Judge Pervez Ali Shah 'Flees Death Threats,'" BBC News South Asia, October 25, 2011.

7. Yilmaz Esmer and Thorleif Pettersson, "The Effects of Religion and Religiosity on Voting Behavior," in *The Oxford Handbook of Political Behavior*, ed. Russell J. Dalton and Hans-Dieter Klingemann (Oxford: Oxford University Press, 2007), pp. 481–503.

8. Bernard Lewis, "The Roots of Muslim Rage: Why So Many Muslims Deeply Resent the West, and Why Their Bitterness Will Not Easily Be Mollified," *Atlantic*, September 1990.

9. Robert Altemeyer, *Enemies of Freedom: Understanding Right-Wing Authoritarianism*, 1st ed. (San Francisco: Jossey-Bass Publishers, 1988).

10. Robert Altemeyer, *The Authoritarian Specter* (Cambridge, MA: Harvard University Press, 1996).

11. Gidi Rubinstein, "Two Peoples in One Land: A Validation Study of Altemeyer's Right-Wing Authoritarianism Scale in the Palestinian and Jewish Societies in Israel," *Journal of Cross-Cultural Psychology* 27, no. 2 (1996): 216.

12. Mark Leibovich, "Being Glenn Beck," *New York Times Magazine*, September 29, 2010.

13. Anne Barnard and Alan Feuer, "Outraged, and Outrageous," *New York Times*, October 8, 2010.

14. Edward L. Glaeser and Bryce A. Ward, "Myths and Realities of American Political Geography," National Bureau of Economic Research, 2006.

15. Melissa J. Ferguson, Travis J. Carter, and Ran R. Hassin, "On the Automaticity of Nationalist Ideology: The Case of the USA," in *Social and Psychological Bases of*

Ideology and System Justification, ed. John T. Jost, Aaron C. Kay, and Hulda Thorisdottir (Oxford: Oxford University Press, 2009), pp. 53–82.

16. Esmer and Pettersson, "The Effects of Religion and Religiosity on Voting Behavior."

17. David M. Levy, "Anti-Nazis: Criteria of Differentiation," in *Personality and Political Crisis: New Perspectives from Social Science and Psychiatry for the Study of War and Politics*, ed. Alfred H. Stanton (Glencoe, IL: Free Press, 1951), pp. 151–227.

18. Esmer and Pettersson, "The Effects of Religion and Religiosity on Voting Behavior."

19. L. B. Koenig, M. McGue, R. F. Krueger, and T. J. Bouchard Jr., "Genetic and Environmental Influences on Religiousness: Findings for Retrospective and Current Religiousness Ratings," *Journal of Personality* 73, no. 2 (2005): 471–88; G. Saucier, "Isms and the Structure of Social Attitudes," *Journal of Personality and Social Psychology* 78, no. 2 (2000): 366–85.

20. Oliver Craig, "Extreme World: God," BBC World News, November 25, 2010; Steve Crabtree, "Religiosity Highest in World's Poorest Nations," Gallup, August 31, 2010.

21. Paul H. Rubin, *Darwinian Politics: The Evolutionary Origin of Freedom* (New Brunswick, NJ: Rutgers University Press, 2002); R. Inglehart, M. Basanez, and A. M. Moreno, *Human Values and Beliefs: A Cross-Cultural Sourcebook: Political, Religious, Sexual, and Economic Norms in 43 Societies; Findings from the 1990–1993 World Value Survey* (Ann Arbor: University of Michigan Press, 1998).

22. Nicholas Wade, "The Evolution of the God Gene," *New York Times*, November 14, 2009.

23. Koenig et al., "Genetic and Environmental Influences on Religiousness: Findings for Retrospective and Current Religiousness Ratings."

24. Crabtree, "Religiosity Highest in World's Poorest Nations."

25. Esmer and Pettersson, "The Effects of Religion and Religiosity on Voting Behavior."

26. Richard M. Doty, Bill E. Peterson, and David G. Winter, "Threat and Authoritarianism in the United States, 1978–1987," in *Political Psychology: Key Readings*, ed. John T. Jost and Jim Sidanius (New York: Psychology Press, 2004), pp. 69–84.

27. Titus Lucretius Carus, *De Rerum Natura Libri Sex*, ed. Cyril Bailey (Oxford: Clarendon Press, 1963).

28. Bronislaw Malinowski, *Coral Gardens and Their Magic* (Bloomington: Indiana University Press, 1965).

29. V. Florian and M. Mikulincer, "The Impact of Death-Risk Experiences and Religiosity on the Fear of Personal Death: The Case of Israeli Soldiers in Lebanon," *OMEGA—Journal of Death and Dying* 26, no. 2 (1992): 101–11.

30. Bernard Spilka, *The Psychology of Religion: An Empirical Approach*, ed. Ralph W. Hood and Richard L. Gorsuch (Englewood Cliffs, NJ: Prentice-Hall, 1985).

31. Robb Willer, "No Atheists in Foxholes: Motivated Reasoning and Religious

Belief," in *Social and Psychological Bases of Ideology and System Justification*, ed. John T. Jost, Aaron C. Kay, and Hulda Thorisdottir (Oxford: Oxford University Press, 2009), pp. 241–64; K. Franks, D. I. Templer, G. G. Cappelletty, and I. Kauffman, "Exploration of Death Anxiety as a Function of Religious Variables in Gay Men with and without Aids." *OMEGA—Journal of Death and Dying* 22, no. 1 (1990): 43–50.

32. Jonathan Haidt and Jesse Graham, "Planet of the Durkheimians, Where Community, Authority, and Sacredness Are Foundations of Morality," in *Social and Psychological Bases of Ideology and System Justification*, ed. John T. Jost, Aaron C. Kay, and Hulda Thorisdottir (Oxford: Oxford University Press, 2009), pp. 371–401.

33. Emile Durkheim, *The Elementary Forms of the Religious Life* (New York: Free Press, 1965).

34. "Amnesty International Urges Iran to Stop Kurd Execution," BBC News Middle East, December 25, 2010.

35. Valerio Valeri, *Kingship and Sacrifice: Ritual and Society in Ancient Hawaii* (Chicago: University of Chicago Press, 1985).

36. David Malo, *Hawaiian Antiquities (Moolelo Hawaii)*, 2d ed. (Honolulu: Museum, 1951).

37. Nicholas Thomas, *Marquesan Societies: Inequality and Political Transformations in Eastern Polynesia* (Oxford: Clarendon Press, 1990); Elizabeth Wood-Ellem, *Queen Sālote of Tonga: The Story of an Era, 1900–1965* (Auckland: Auckland University Press, 1999).

38. Aletta Biersack, "Blood and Garland: Duality in Tongan History," in *Tongan Culture and History: Papers from the 1st Tongan History Conference, Held in Canberra, 14–17 January 1987*, ed. Phyllis Herda, Jennifer Terrell, and Niel Gunson (Canberra: Australian National University, Department of Pacific and Southeast Asian History, 1990).

39. Douglas L. Oliver, *The Pacific Islands*, 3rd ed. (Honolulu: University of Hawaii Press, 1989).

40. Geeta Pandey, "Indian Community Torn Apart by 'Honour Killings,'" BBC News, June 16, 2010.

41. M. Bamshad, T. Kivisild, W. S. Watkins, M. E. Dixon, C. E. Ricker, B. B. Rao, J. M. Naidu, et al. "Genetic Evidence on the Origins of Indian Caste Populations," *Genome research* 11, no. 6 (2001): 994; I. Thanseem, K. Thangaraj, G. Chaubey, V. K. Singh, L. V. K. S. Bhaskar, B. M. Reddy, A. G. Reddy, and L. Singh, "Genetic Affinities among the Lower Castes and Tribal Groups of India: Inference from Y Chromosome and Mitochondrial DNA," *BMC Genetics* 7, no. 1 (2006): 42.

42. S. Sabir, "Chimerical Categories: Caste, Race, and Genetics," *Developing World Bioethics* 3, no. 2 (2003): 170–77.

43. Daniel Chirot and Clark McCauley, *Why Not Kill Them All? The Logic and Prevention of Mass Political Murder* (Princeton, NJ: Princeton University Press, 2006); Israel Finkelstein and Neil Asher Silberman, *The Bible Unearthed: Archaeology's New Vision of Ancient Israel and the Origin of Its Sacred Texts*, ed. Neil Asher Silberman, 1st Touchstone ed. (New York: Simon and Schuster, 2002).

44. "Israel to Allow in 8,000 Falash Mura from Ethiopia," BBC News Middle East, November 14, 2010.

45. "Ultra-Orthodox Jews in Mass Protest over School Ruling," BBC News, June 17, 2010.

46. "Religions," CIA World Factbook, https://www.cia.gov/library/publications/the-world-factbook/geos/xx.html (accessed April 8, 2013).

47. C. R. Badcock, *Psychodarwinism: The New Synthesis of Darwin & Freud* (London: Harper Collins, 1994).

48. Leibovich, "Being Glenn Beck."

49. Raphael S. Ezekiel, *The Racist Mind: Portraits of Neo-Nazis and Klansmen* (New York: Viking, 1995).

50. Finlo Rohrer, "What Are You Not Allowed to Say?" BBC News Magazine, November 29, 2010.

51. "Harry Potter Actress's Brother Jailed for Attacking Her," BBC News, January 21, 2011.

52. Thilo Sarrazin, "Thilo Sarrazin Schreibt Über Den Islam," *Bild*, August 26, 2010; Jana Randow and Christian Vits, "Weber to Debate Next Sarrazin Steps as Merkel Condemns Comments," Bloomberg, August 31, 2010.

53. "18 Prozent Der Deutschen Würden Sarrazin Wählen," *Berliner Morgenpost*, September 6, 2010.

54. George Lakoff, *Moral Politics: How Liberals and Conservatives Think*, 2nd ed. (Chicago, IL: University of Chicago Press, 2002).

55. Esmer and Pettersson, "The Effects of Religion and Religiosity on Voting Behavior."

56. Soeren Kern, "Germany Debates Muslim Immigration," Gatestone Institute, International Policy Council, September 15, 2010.

57. Shulah Mendelson and Unknown, "What Does It Mean to Be a Leftist? (translated from the Hebrew, "Mah Zeh Lihiot Smolani?" a political humor presentation circulated on the Internet," 2010).

58. F. J. Sulloway, Philip T. Starks, and Michael B. Shermer, "The Adaptive Significance of Religion: A Mutualistic Relationship between Genes and Memes" (Berkeley: University of California, Berkeley; Tufts University; Claremont Graduate University, 2011).

59. Ibid.

60. Rachel Harvey, "Thailand Foetus Find Breaks Abortion Taboo," BBC News Asia-Pacific, November 27, 2010.

61. C. S. Prebish and D. Keown, *Buddhism the Ebook* (Journal of Buddhist Ethics: Online Books, Ltd., 2005).

62. Benjamin I. Page, *The Rational Public: Fifty Years of Trends in Americans' Policy Preferences*, ed. Robert Y. Shapiro (Chicago: University of Chicago Press, 1992).

63. Esmer and Pettersson, "The Effects of Religion and Religiosity on Voting Behavior"; Pippa Norris, *Sacred and Secular: Religion and Politics Worldwide*, ed. Ronald Inglehart (Cambridge, UK: Cambridge University Press, 2004).

64. T. J. Bouchard, "Authoritarianism, Religiousness, and Conservatism: Is 'Obedience to Authority' the Explanation for Their Clustering, Universality and Evolution?" in *The Biological Evolution of Religious Mind and Behavior*, ed. E. Voland and W. Schiefenhövel (Berlin: Springer-Verlag, 2009), pp. 165–80; M. Hout, A. Greeley, and M. J. Wilde, "The Demographic Imperative in Religious Change in the United States," *American Journal of Sociology* 107, no. 2 (2001): 468–500.

Chapter 8. Attitudes toward Sexuality, Homosexuality, and Gender Roles

a. Surely it's sexist to say that women's sexuality is more malleable to cultural influence than men's. But consider a case where a culture tries to control both male and female sexuality equally: the clergy of the Catholic Church. Sex scandals involving priests are far, far more common than those involving nuns.

b. Bride stealing is prevalent in the Caucuses and Central Asia but has also occurred in many other regions around the world, including in East Africa, among the Roma, the Hmong, and in the Amazon. Sometimes bride kidnapping is arranged and consensual, and other times the abductions are involuntary and violent.

1. Karim Sadjadpour, *Reading Khamenei: The World View of Iran's Most Powerful Leader* (Washington, DC: Carnegie Endowment for International Peace, 2009).

2. R. Flacks, "The Liberated Generation: An Exploration of the Roots of Student Protest," *Journal of Social Issues* 23, no. 3 (1967): 52–75; J. C. Flugel, *Man, Morals and Society: A Psycho-Analytical Study* (New York: International Universities Press, 1945).

3. Robert Altemeyer, *Enemies of Freedom: Understanding Right-Wing Authoritarianism*, 1st ed. (San Francisco: Jossey-Bass Publishers, 1988).

4. "Iranian Cleric Blames Quakes on Promiscuous Women," BBC News, April 20, 2010.

5. "Dubai Jails Indian Pair for 'Sexy Texts,'" BBC News, March 17, 2010; "Jailed Dubai Kissing Pair Lose Appeal over Conviction," BBC News, April 4, 2010.

6. "Venezuela Politician Offers Breast Implant in $6 Raffle," BBC News Latin America and Caribbean, August 27, 2010.

7. Seema Mehta, "GOP Ticket Split over Condom Use," *Los Angeles Times*, September 6, 2008.

8. Ibid.

9. Moisés Naím, "Los Bebés Del Tea Party" *El País* [Spain], November 7, 2010.

10. Aleem Maqbool, "Fury over Doctor's Book on Sex Education for Muslims," BBC News South Asia, January 9, 2011.

11. Ewen MacAskill, "Christine O'Donnell Interview: 'The Republicans Have Lost Their Way,'" *Guardian* [UK], September 15, 2010.

12. *Catechism of the Catholic Church* (Citta del Vaticano: Libreria Editrice Vaticana, 1993).

13. Jason Tedjasukmana, "Indonesia's Skin Wars," *Time*, April 3, 2006.

14. "Indonesia Sex Tape Star Is Jailed," BBC News Asia-Pacific, January 31, 2011.

15. "Tokyo Introduces Manga Restrictions," BBC News Magazine, December 15, 2010.

16. Pew Social Trends Staff, "The Decline of Marriage and Rise of New Families," *Social and Demographic Trends* (2010).

17. Najlaa Abou Mehri and Linda Sills, "The Virginity Industry," BBC News, April 24, 2010.

18. Jeffrey Fleishman and Amro Hassan, "Gadget to Help Women Feign Virginity Angers Many in Egypt," *Los Angeles Times*, October 7, 2009.

19. "Taliban 'Kill Adulterous Afghan Couple' in Marketplace," BBC News South Asia, August 16, 2010.

20. "Malaysia Canes Three Women over Extramarital Sex," BBC News, February 17, 2010.

21. "Dubai Jails Indian Pair for "Sexy Texts.""

22. Finlo Rohrer, "Why Does the West Love the Dalai Lama?" BBC News Magazine, February 18, 2010.

23. "Vatican: Pope Did Not Back Condom Contraception Use," BBC News Europe, December 21, 2010.

24. Augusto Álvarez Rodrich, "Condón: La Intolerancia Inaceptable Del Cardenal Cipriani," *La República* [Peru], November 19, 2010.

25. "Spain 'Orgasm' Video Criticised by Politicians," BBC News Europe, November 19, 2010.

26. Jeffrey Gettleman, "Ugandan Who Spoke up for Gays Is Beaten to Death," *New York Times*, January 27, 2011.

27. "Uganda Gay Rights Activist David Kato Killed," BBC News Africa, January 27, 2011.

28. "Uganda Anti-Homosexuality Bill: MPs Drop Death Penalty," BBC News Africa, November 23, 2012.

29. Altemeyer, *Enemies of Freedom: Understanding Right-Wing Authoritarianism*.

30. Neil Jeffrey Kressel, *Mass Hate: The Global Rise of Genocide and Terror*, updated ed. (Boulder, CO: Westview Press, 2002).

31. Martha L. Cottam, Beth Dietz-Uhler, Elena Mastors, and Thomas Preston, *Introduction to Political Psychology* (Mahwah, NJ: Lawrence Erlbaum Associates, 2004).

32. Frank Rich, "The Rage Won't End on Election Day," *New York Times*, October 16, 2010.

33. Mark Leibovich, "Being Glenn Beck," *New York Times Magazine*, September 29, 2010.

34. Pew Social Trends Staff, "The Decline of Marriage and Rise of New Families."

35. Nicholas Confessore, "Agendas of Paladino and Rabbi Meld," *New York Times*, October 11, 2010.

36. Elizabeth A. Harris, "Paladino Laces Speech with Antigay Remarks." *New York Times*, October 10, 2010.

37. "U.S. Elections 2010: Craziest Campaign Moments," BBC News, October 27, 2010; Confessore, "Agendas of Paladino and Rabbi Meld.

38. Michael D. Shear, "Both Parties Seek Slightest Edge in Colorado," *New York Times*, October 26, 2010.

39. Raphael S. Ezekiel, *The Racist Mind: Portraits of Neo-Nazis and Klansmen* (New York: Viking, 1995).

40. Lisa Keen, "Gay Support for Obama Similar to Dems in Past Elections," Williams Institute, UCLA School of Law, November 26, 2008.

41. Ibid.

42. "UN Restores Gay Clause to Killings Resolution," BBC News US and Canada, December 21, 2010.

43. Mark Lowen, "Scores Arrested in Belgrade after Anti-Gay Riot," BBC News Europe, October 10, 2010; "Anti-Gay Parade Rioters Get Jail Time," *B 92, Crime & War Crimes*, April 20, 2011.

44. Rohrer, "Why Does the West Love the Dalai Lama?"

45. "Toledo Proposes Legalization of Same Sex Marriage," LivinginPeru.com, http://archive.peruthisweek.com/news-13922-2011-elections-toledo-campaigns-peru-elections-miami (accessed January 20, 2011).

46. Altemeyer, *Enemies of Freedom: Understanding Right-Wing Authoritarianism.*

47. Ibid.

48. Ibid.; Robert Altemeyer, *The Authoritarian Specter* (Cambridge, MA: Harvard University Press, 1996).

49. Michael Paul Rogin, *"Ronald Reagan," the Movie: And Other Episodes in Political Demonology* (Berkeley: University of California Press, 1987).

50. Dave Barry, "It's Time for the Red and Blue States to Start a Cultural Exchange Program and Begin the Healing," *Baltimore Sun*, December 12, 2004.

51. J. T. Jost, "The End of the End of Ideology," *American Psychologist* 61, no. 7 (2006): 651.

52. Paul M. Sniderman and Thomas Leonard Piazza, *The Scar of Race* (Cambridge, MA: Belknap Press of Harvard University Press, 1993).

53. Alexander Zaitchik, "Glenn Beck Becomes Damaged Goods," Salon.com, http://www.salon.com/2009/09/22/glenn_beck_two/ (accessed September 22, 2009).

54. Ezekiel, *The Racist Mind: Portraits of Neo-Nazis and Klansmen.*

55. Jos D. Meloen, "The Political Culture of State Authoritarianism," in *Political Psychology: Cultural and Cross-Cultural Foundations*, ed. Stanley Allen Renshon and J. H. Duckitt (New York: New York University Press, 2000), pp. 108–27; M. Kidron, R. Segal, and Swanston Graphics Limited, *The New State of the World Atlas* (New York: Simon & Schuster, 1991).

56. Salim Mia, "Bangladesh 'Eve Teasing' Craze Takes a Terrible Toll," BBC News, June 11, 2010.

57. "Sexual Harassment in Egypt," *PRI's The World: Global Perspectives for an American Audience* (February 17, 2011).

58. Lucy Ash, "'Exorcisms' Performed on Chechen Stolen Brides," BBC News Europe, August 9, 2010.

59. Ibid.; Lucy Ash, "Can Chechen President Kadyrov Stamp out Bride-Stealing?" BBC News Europe, October 8, 2010.

60. "Uproar over Saudi Women's 'SMS Tracking,'" BBC News Middle East, November 23, 2012.

61. "Yemen Child Bride 'Bleeds to Death,'" BBC News, April 8, 2010.

62. "Mutilated Afghan Girl Aisha in US for New Nose," BBC News South Asia, August 6, 2010.

63. Kathryn Westcott, "Can You Get Married over the Phone?" BBC News, April 8, 2010.

PART II. THE BIOLOGY OF TRIBALISM

a. For example, you can see the rise and collapse of Middle Eastern empires over five thousand years in only ninety seconds at Maps of War, http://www.mapsofwar.com/images/EMPIRE17.swf (accessed April 8, 2013).

Chapter 9. When Outbreeding Is Fit and Inbreeding Isn't

a. For comparison's sake, we share about 94 percent of our genes with chimpanzees. Whether paleoanthropologists classify Neanderthals as a sub-species of modern humans or as a separate species is a matter of debate. Proponents of the "sub-species" argument would denote them as *Homo sapiens neanderthalensis*.

b. In 2013, however, paleoanthropologists discovered the remains of an individual who died in Italy approximately forty thousand years ago, who they believe to be the first known human-Neanderthal hybrid. The mitochondrial DNA extracted from the lower jaw suggests that some Neanderthal females did mate with anatomically modern human males. But for whatever reason, their children did not succeed in passing on this maternally inherited DNA to living populations of humans. See Silvana Condemi, Aurélien Mounier, Paolo Giunti, Martina Lari, David Caramelli, and Laura Longo, "Possible Interbreeding in Late Italian Neanderthals? New Data from the Mezzena Jaw (Monti Lessini, Verona, Italy)," *PloS One* 8, no. 3 (2013): e59781.

c. The shared proportion of genetic material could be higher if their lineage has a history of inbreeding.

1. Edward M. East, *Inbreeding and Outbreeding: Their Genetic and Sociological Significance*, ed. Donald Forsha Jones (Philadelphia: J. B. Lippincott Company, 1919).

2. A. Gibbons, "Close Encounters of the Prehistoric Kind," *Science* 328, no. 5979 (2010): 680; Sean B. Carroll, "Hybrids May Thrive Where Parents Fear to Tread," *New York Times*, September 13, 2010.

3. Pallab Ghosh, "Ancient Humans, Dubbed 'Denisovans,' Interbred with Us," BBC News Science and Environment, December 22, 2010.

4. Nickolas M. Waser, "Sex, Mating Systems, Inbreeding, and Outbreeding," in *The Natural History of Inbreeding and Outbreeding: Theoretical and Empirical Perspectives*, ed. Nancy Wilmsen Thornhill (Chicago: University of Chicago Press, 1993), pp. 1–13.

5. M. Szulkin and B. C. Sheldon, "Dispersal as a Means of Inbreeding Avoidance in a Wild Bird Population," *Proceedings of the Royal Society B: Biological Sciences* 275, no. 1635 (2008): 703.

6. "City Life 'Boosts Bug Resistance,'" BBC News Health, September 25, 2010.

7. Jeffry B. Mitton, "Theory and Data Pertinent to the Relationship between Heterozygosity and Fitness," in *The Natural History of Inbreeding and Outbreeding: Theoretical and Empirical Perspectives*, ed. Nancy Wilmsen Thornhill (Chicago: University of Chicago Press, 1993), pp. 17–41; William M. Shields, "The Natural and Unnatural History of Inbreeding and Outbreeding," in *The Natural History of Inbreeding and Outbreeding: Theoretical and Empirical Perspectives*, ed. Nancy Wilmsen Thornhill (Chicago: University of Chicago Press, 1993), pp. 143–69.

8. Susan E. Riechert and Rose Marie Roeloffs, "Evidence for and Consequences of Inbreeding in the Cooperative Spiders," in *The Natural History of Inbreeding and Outbreeding: Theoretical and Empirical Perspectives*, ed. Nancy Wilmsen Thornhill (Chicago: University of Chicago Press, 1993), pp. 283–303.

9. P. R. Whitehorn, M. C. Tinsley, M. J. F. Brown, B. Darvill, and D. Goulson, "Genetic Diversity, Parasite Prevalence, and Immunity in Wild Bumblebees," *Proceedings of the Royal Society B: Biological Sciences* 278, no. 1709 (2011): 1195.

10. Mark Kinver, "Inbred Bumblebees 'Face Extinction Threat,'" BBC News Science and Environment, September 6, 2010.

11. P. P. G. Bateson, "Optimal Outbreeding," in *Mate Choice*, ed. P. P. G. Bateson (Cambridge: Cambridge University Press, 1983), pp. 257–77.

12. Robert C. Lacy, Ann Petric, and Mark Warneke, "Inbreeding and Outbreeding in Captive Populations of Wild Animal Species," in *The Natural History of Inbreeding and Outbreeding: Theoretical and Empirical Perspectives*, ed. Nancy Wilmsen Thornhill (Chicago: University of Chicago Press, 1993), pp. 352–74.

13. Nickolas M. Waser, "Sex, Mating Systems, Inbreeding, and Outbreeding," in *The Natural History of Inbreeding and Outbreeding: Theoretical and Empirical Perspectives*, ed. Nancy Wilmsen Thornhill (Chicago: University of Chicago Press, 1993), pp. 1–13.

14. Jim Moore, "Primates: What's Wrong with 'the Dispersing Sex'?" in *The Natural History of Inbreeding and Outbreeding: Theoretical and Empirical Perspectives*, ed. Nancy Wilmsen Thornhill (Chicago: University of Chicago Press, 1993), pp. 392–426.

15. Y. Song, S. Endepols, N. Klemann, D. Richter, F. R. Matuschka, C. H. Shih, M. W. Nachman, and M. H. Kohn, "Adaptive Introgression of Anticoagulant Rodent Poison Resistance by Hybridization between Old World Mice," *Current Biology* 21, no. 15 (2011): 1296–1301; Matt McGrath, "'Super' Mouse Evolves Resistance to Most Poisons," BBC News, Science and Environment, August 9, 2011.

16. J. Mallet, "Hybrid Speciation," *Nature* 446, no. 7133 (2007): 279–83.

17. Frank Hailer, Verena E. Kutschera, Björn M. Hallström, Denise Klassert, Steven R. Fain, Jennifer A. Leonard, Ulfur Arnason, and Axel Janke, "Nuclear Genomic Sequences Reveal That Polar Bears Are an Old and Distinct Bear Lineage." *Science* 336, no. 6079 (April 20, 2012): 344–47.

18. Carroll, "Hybrids May Thrive Where Parents Fear to Tread"; "Bear Shot in N. W. T. Was Grizzly-Polar Hybrid," CBC News, April 30, 2010.

19. William H. Durham, *Coevolution: Genes, Culture, and Human Diversity*, (Stanford, CA: Stanford University Press, 1991).

20. L. A. García-Escudero, E. A. Arruza, N. J. Padilla, and G. R. Puig, "Charles II: From Spell to Genitourinary Pathology," *Archivos españoles de urología* 62, no. 3 (2009): 181.

21. Ibid.

22. G. Alvarez, F. C. Ceballos, and C. Quinteiro, "The Role of Inbreeding in the Extinction of a European Royal Dynasty," *PloS One* 4, no. 4 (2009): e5174.

23. Alan H. Bittles, "Genetic Aspects of Inbreeding and Incest," in *Inbreeding, Incest, and the Incest Taboo: The State of Knowledge at the Turn of the Century*, ed. Arthur P. Wolf and William H. Durham (Stanford, CA: Stanford University Press, 2004), pp. 38–60.

24. Emma Wilkinson, "Cousin Marriage: Is It a Health Risk?" BBC News, May 16, 2008.

25. Ben Marshall, "Asian Infant Death Rate 'Higher,'" BBC News, November 16, 2009.

26. Cindi John, "Campaign Tackles Asian Child Deaths," BBC News, February 21, 2001.

27. Wilkinson, "Cousin Marriage: Is It a Health Risk?"; Richard Conniff, "Go Ahead, Kiss Your Cousin. Heck, Marry Her If You Want To," *Discover* 24, no. 8 (August 2003).

28. A. H. Bittles, personal communication on global statistics (updated to 2011) for first-cousin-marriage excess mortality and morbidity rates, 2013.

29. Carole Ober, Terry Hyslop, and Walter W. Hauck, "Inbreeding Effects on Fertility in Humans: Evidence for Reproductive Compensation," *American Journal of Human Genetics* 64, no. 1 (1999): 225–31.

30. A. Gibbons, "A Denisovan Legacy in the Immune System?" *Science* 333, no. 6046 (2011): 1086; Matt McGrath, "Neanderthal Sex Boosted Immunity in Modern Humans," BBC News, Science and Environment, August 26, 2011.

31. Peter Parham and Ashley Moffett, "Variable NK Cell Receptors and Their MHC Class I Ligands in Immunity, Reproduction and Human Evolution," *Nature Reviews Immunology* 13, no. 2 (2013): 133–44.

32. Mitton, "Theory and Data Pertinent to the Relationship between Heterozygosity and Fitness."

Chapter 10. When Inbreeding Is Fit and Outbreeding Isn't

a. Just as this book was going to press, another study was published that showed that pathogen prevalence is also strongly correlated (at 0.65) with cross-national scores on the F-scale (see Damian R. Murray, Mark Schaller, and Peter Suedfeld, "Pathogens and Politics: Further Evidence That Parasite Prevalence Predicts Authoritarianism," *PLOS ONE* 8, no. 5 (2013): e62275).

b. A "self-antigen," as in this case, is a molecule found on the surfaces of blood cells. These molecules help the body distinguish between the self and foreign entities. "Antigen" can also refer to a (fragment of a) molecule from a pathogenic substance (such as a virus, bacterium, or fungus), which the body recognizes as a potential foreign threat. The name comes from the term "**anti**body **gen**erator," because nonself antigens that enter the body stimulate the creation of antibodies. These antibodies, in turn, allow the immune system to neutralize or kill the intruder. Autoimmune diseases can occur if the body confuses self-antigens for foreign intruders.

c. Genetic research suggests that SIVs, which today infect dozens of species of African monkeys and apes, arose between five and twelve million years ago. See Alex A. Compton and Michael Emerman, "Convergence and Divergence in the Evolution of the APOBEC3G-Vif Interaction Reveal Ancient Origins of Simian Immunodeficiency Viruses," *PLOS Pathogens* 9, no. 1 (January 2013): e1003135.

d. In fact, the only single trait that showed a higher concordance between spouses was church attendance (with a correlation of 0.714). Church attendance, of course, is one of the greatest predictors of Republicanism in the United States. And religiosity in general is universally linked to political conservatism.

1. William M. Shields, "Optimal Inbreeding in the Evolution of Philopatry," in *The Ecology of Animal Movement*, ed. Ian R. Swingland and Paul J. Greenwood (Oxford: Clarendon Press, 1983), pp. 132–59. Emphasis added.

2. Nickolas M. Waser, "Sex, Mating Systems, Inbreeding, and Outbreeding," in *The Natural History of Inbreeding and Outbreeding: Theoretical and Empirical Perspectives*, ed. Nancy Wilmsen Thornhill (Chicago: University of Chicago Press, 1993), pp. 1–13.

3. Shields, "Optimal Inbreeding in the Evolution of Philopatry."

4. Andrew T. Smith, "The Natural History of Inbreeding and Outbreeding in Small Mammals," in *The Natural History of Inbreeding and Outbreeding: Theoretical and Empirical Perspectives*, ed. Nancy Wilmsen Thornhill (Chicago: University of Chicago Press, 1993), pp. 329–51; Robert C. Lacy, Ann Petric, and Mark Warneke, "Inbreeding and Outbreeding in Captive Populations of Wild Animal Species," in *The Natural History of Inbreeding and Outbreeding: Theoretical and Empirical Perspectives*, ed. Nancy Wilmsen Thornhill (Chicago: University of Chicago Press, 1993), pp. 352–74.

5. Linda Partridge, "Non-Random Mating and Offspring Fitness," in *Mate Choice*, ed. P. P. G. Bateson (Cambridge: Cambridge University Press, 1983), pp. 227–55.

6. Ibid.

7. Waser, "Sex, Mating Systems, Inbreeding, and Outbreeding"; Richard E.

Michod, "Inbreeding and the Evolution of Social Behavior," in *Mate Choice*, ed. P. P. G. Bateson (Cambridge: Cambridge University Press, 1983), pp. 74-96.

8. Susan E. Riechert and Rose Marie Roeloffs, "Evidence for and Consequences of Inbreeding in the Cooperative Spiders," in *Mate Choice*, ed. P. P. G. Bateson (Cambridge: Cambridge University Press, 1983), pp. 283–303.

9. Alan H. Bittles, "Genetic Aspects of Inbreeding and Incest," in *Inbreeding, Incest, and the Incest Taboo: The State of Knowledge at the Turn of the Century*, ed. Arthur P. Wolf and William H. Durham (Stanford, CA: Stanford University Press, 2004), pp. 38–60; D. Thiessen and B. Gregg, "Human Assortative Mating and Genetic Equilibrium: An Evolutionary Perspective," *Ethology and Sociobiology* 1, no. 2 (1980): 111–40.

10. W. D. Hamilton, "Innate Social Aptitudes of Man: An Approach from Evolutionary Genetics," *Biosocial Anthropology* (1975): 133–55.

11. Partridge, "Non-Random Mating and Offspring Fitness."

12. P. P. G. Bateson, "Optimal Outbreeding," in *Mate Choice*, ed. P. P. G. Bateson (Cambridge: Cambridge University Press, 1983), pp. 257–77.

13. Waser, "Sex, Mating Systems, Inbreeding, and Outbreeding"; Bateson, "Optimal Outbreeding."

14. Shields, "Optimal Inbreeding in the Evolution of Philopatry."

15. William M. Shields, *Philopatry, Inbreeding, and the Evolution of Sex* (Albany: State University of New York Press, 1982).

16. Ibid.

17. William M. Shields, "The Natural and Unnatural History of Inbreeding and Outbreeding," in *The Natural History of Inbreeding and Outbreeding: Theoretical and Empirical Perspectives*, ed. Nancy Wilmsen Thornhill (Chicago: University of Chicago Press, 1993), pp. 143–69.

18. Lacy, Petric, and Warneke, "Inbreeding and Outbreeding in Captive Populations of Wild Animal Species."

19. Jim Moore, "Primates: What's Wrong with 'the Dispersing Sex'?" in *The Natural History of Inbreeding and Outbreeding: Theoretical and Empirical Perspectives*, ed. Nancy Wilmsen Thornhill (Chicago: University of Chicago Press, 1993), pp. 392–426.

20. Neil Jeffrey Kressel, *Mass Hate: The Global Rise of Genocide and Terror*, updated ed. (Boulder, CO: Westview Press, 2002).

21. Raphael S. Ezekiel, *The Racist Mind: Portraits of Neo-Nazis and Klansmen* (New York: Viking, 1995).

22. Ibid.

23. "HIV among African Americans," US Centers for Disease Control and Prevention, http://www.cdc.gov/hiv/topics/aa/PDF/HIV_among_African_Americans_final.pdf (accessed April 8, 2013).

24. Marc Sheppard, "Come on Cosby: It's Time to Come Clean about AIDS," *American Thinker*, November 1, 2007.

25. Ibid.

26. Kate Zernike, "The Persistence of Conspiracy Theories," *New York Times*, April 30, 2011.

27. Martha L. Cottam, Beth Dietz-Uhler, Elena Mastors, and Thomas Preston, *Introduction to Political Psychology* (Mahwah, NJ: Lawrence Erlbaum Associates, 2004).

28. Palestinian Media Watch, "Libel: Israel Spreads AIDS and Drugs," December 2, 2012, http://palwatch.org/main.aspx?fi=766&all=1 (accessed April 29, 2013).

29. Kressel, *Mass Hate: The Global Rise of Genocide and Terror.*

30. Nicholas D. Kristof, "Would You Slap Your Father? If So, You're a Liberal," *New York Times*, May 27, 2009.

31. K. Smith, D. Oxley, M. Hibbing, J. Alford, and J. Hibbing, "The Ick Factor: Physiological Sensitivity to Disgust as a Predictor of Political Attitudes," paper presented at the Annual Meeting of the Midwest Political Science Association, Chicago, Illinois, 2008.

32. C. D. Navarrete and D. M. T. Fessler, "Disease Avoidance and Ethnocentrism: The Effects of Disease Vulnerability and Disgust Sensitivity on Intergroup Attitudes," *Evolution and Human Behavior* 27, no. 4 (2006): 270–82.

33. C. L. Fincher, R. Thornhill, D. R. Murray, and M. Schaller, "Pathogen Prevalence Predicts Human Cross-Cultural Variability in Individualism/Collectivism," *Proceedings of the Royal Society B: Biological Sciences* 275, no. 1640 (2008): 1279.

34. M. Schaller and D. R. Murray, "Pathogens, Personality, and Culture: Disease Prevalence Predicts Worldwide Variability in Sociosexuality, Extraversion, and Openness to Experience," *Journal of Personality and Social Psychology* 95, no. 1 (2008): 212–21.

35. Monte Reel, "The Most Isolated Man on the Planet," *Slate*, August 20, 2010.

36. "Brazil: Land for Last Survivor of Unknown Amazon Tribe," *Survival International*, November 9, 2006.

37. Arthur C. Aufderheide, *The Cambridge Encyclopedia of Human Paleopathology*, ed. Conrado Rodríguez-Martín and Odin Langsjoen (Cambridge: Cambridge University Press, 1998).

38. G. Garratty, "Blood Groups and Disease: A Historical Perspective," *Transfusion Medicine Reviews* 14, no. 4 (2000): 291.

39. A. E. Mourant, "Associations between Hereditary Blood Factors and Diseases," *Bulletin of the World Health Organization* 49, no. 1 (1973): 93.

40. A. E. Mourant, Ada C. Kopeć, and Kazimiera Domaniewska-Sobczak, *The Distribution of the Human Blood Groups, and Other Polymorphisms*, 2nd ed. (London: Oxford University Press, 1976).

41. Ibid.

42. Garratty, "Blood Groups and Disease: A Historical Perspective."

43. Mourant, Kopeć, and Domaniewska-Sobczak, *The Distribution of the Human Blood Groups, and Other Polymorphisms*.

44. Garratty, "Blood Groups and Disease: A Historical Perspective"; Mourant, "Associations between Hereditary Blood Factors and Diseases"; J. M. Moulds and J. J. Moulds, "Blood Group Associations with Parasites, Bacteria, and Viruses," *Transfusion Medicine Reviews* 14, no. 4 (2000): 302–11; A. M. Hutson, R. L. Atmar, D. Y. Graham, and M. K. Estes, "Norwalk Virus Infection and Disease Is Associated with Abo Histo-Blood Group Type," *Journal of Infectious Diseases* 185, no. 9 (2002): 1335–37; S. A.

Berger, N. A. Young, and S. C. Edberg, "Relationship between Infectious Diseases and Human Blood Type," *European Journal of Clinical Microbiology and Infectious Diseases* 8, no. 8 (1989): 681–89.

45. Garratty, "Blood Groups and Disease: A Historical Perspective."

46. Mourant, "Associations between Hereditary Blood Factors and Diseases."

47. John E. Bennett, Raphael Dolin, R. Gordon Douglas, and Gerald L. Mandell, ed., *Mandell, Douglas, and Bennett's Principles and Practice of Infectious Diseases*, 7th ed. (Philadelphia: Churchill Livingstone/Elsevier, 2010).

48. Mourant, "Associations between Hereditary Blood Factors and Diseases"; Paul H. Rubin, *Darwinian Politics: The Evolutionary Origin of Freedom* (New Brunswick, NJ: Rutgers University Press, 2002).

49. Moulds and Moulds, "Blood Group Associations with Parasites, Bacteria, and Viruses."

50. W. He, S. Neil, H. Kulkarni, E. Wright, B. K. Agan, V. C. Marconi, and M. J. Dolan, "Duffy Antigen Receptor for Chemokines Mediates Trans-Infection of HIV-1 from Red Blood Cells to Target Cells and Affects HIV-AIDS Susceptibility," *Cell Host & Microbe* 4, no. 1 (2008): 52–62.

51. "Malaria Gene 'Increases HIV Risk,'" BBC News, July 16, 2008.

52. X. Wang, C. Chen, L. Wang, D. Chen, W. Guang, and J. French, "Conception, Early Pregnancy Loss, and Time to Clinical Pregnancy: A Population-Based Prospective Study," *Fertility and sterility* 79, no. 3 (2003): 577–84.

53. Bittles, "Genetic Aspects of Inbreeding and Incest."

54. Garratty, "Blood Groups and Disease: A Historical Perspective"; Mourant, Kopeć, and Domaniewska-Sobczak, *The Distribution of the Human Blood Groups, and Other Polymorphisms*.

55. "Kissing Cousins, Missing Children," *Economist: Demography and Genetics* (February 7, 2008).

56. Bateson, "Optimal Outbreeding"; Mourant, "Associations between Hereditary Blood Factors and Diseases"; M. Adinolfi, "Recurrent Habitual Abortion, HLA Sharing, and Deliberate Immunization with Partner's Cells: A Controversial Topic," *Human Reproduction* 1, no. 1 (1986): 45.

57. Mourant, Kopeć, and Domaniewska-Sobczak, *The Distribution of the Human Blood Groups, and Other Polymorphisms*; Nicholas Bakalar, "Rh Factor, 1944," *New York Times*, January 3, 2011.

58. N. T. N. Ngoc, M. Merialdi, H. Abdel-Aleem, G. Carroli, M. Purwar, N. Zavaleta, L. Campodonico, et al., "Causes of Stillbirths and Early Neonatal Deaths: Data from 7993 Pregnancies in Six Developing Countries," *Bulletin of the World Health Organization* 84, no. 9 (2006): 699–705.

59. A. C. Stevenson, B. Say, S. Ustaoglu, and Z. Durmus, "Aspects of Pre-Eclamptic Toxaemia of Pregnancy, Consanguinity, Twinning in Ankara," *Journal of Medical Genetics* 13, no. 1 (1976): 1.

60. Patrick Bateson, "Inbreeding Avoidance and Incest Taboos," in *Inbreeding, Incest, and the Incest Taboo: The State of Knowledge at the Turn of the Century*, ed.

Arthur P. Wolf and William H. Durham (Stanford, CA: Stanford University Press, 2004), pp. 24–37; Richard Conniff, "Go Ahead, Kiss Your Cousin. Heck, Marry Her If You Want To," *Discover* 24, no. 8 (August 2003).

61. Thiessen and Gregg, "Human Assortative Mating and Genetic Equilibrium: An Evolutionary Perspective"; Bateson, "Optimal Outbreeding."

62. J. R. Alford, P. K. Hatemi, J. R. Hibbing, N. G. Martin, and L. J. Eaves, "The Politics of Mate Choice," *Journal of Politics* 73, no. 2 (2011): 362–79; C. A. Klofstad, R. McDermott, and P. K. Hatemi, "The Dating Preferences of Liberals and Conservatives," *Political Behavior* (2012): 1–20.

63. Bittles, "Genetic Aspects of Inbreeding and Incest."

64. Conniff, "Go Ahead, Kiss Your Cousin. Heck, Marry Her If You Want To."

65. Shields, *Philopatry, Inbreeding, and the Evolution of Sex.*

Chapter 11. How Optimal Mating Happens

a. Thanks to the enormous sample size, the significance of this constant relationship between spousal kinship and fitness is astronomical. The probability that the relationship shown in the first graph occurred by chance is only 1.5 in 1,000,000,000,000,000,000,000, 000,000,000,000,000,000,000,000,000,000,000,000,000,000,000,000,000,000,000,000, 000,000,000,000,000,000,000,000,000,000,000,000,000,000,000,000,000. In other words, the P value for the correlation between spousal kinship and second-generation fertility was 1.5×10^{-129}; for the second graph (the number of children who reproduced) it was 3.6×10^{-66}; and for the number of grandchildren, $P = 7.6 \times 10^{-58}$.

b. This brief description can hardly do justice to the extraordinary diversity of customs that human societies have devised to disperse and trade their reproductive assets. The rigorous study of kinship systems dates back to the late nineteenth century, when it formed the essence of the newly born field of anthropology. Until a relativist fog engulfed the discipline in the 1970s, the study of kinship fascinated virtually all the great anthropologists and social theorists. Those who have made significant contributions to this fundamental social-science area include Lewis Henry Morgan, Bronisław Malinowski, Alfred Radcliffe-Brown, Claude Lévi-Strauss, Sir Edmund Leach, Sir E. E. Evans-Pritchard, and many others.

c. A "locus" is the location of a DNA sequence or gene on a chromosome.

1. P. P. G. Bateson, "Optimal Outbreeding," in *Mate Choice*, ed. P. P. G. Bateson (Cambridge: Cambridge University Press, 1983), pp. 257–77.

2. Clare Jenkins, "Kiss and Patel," BBC News Magazine, June 11, 2009.

3. A. Helgason, S. Palsson, and D. F. Guthbjartsson, "An Association between the Kinship and Fertility of Human Couples," *Science* 319, no. 5864 (2008): 813.

4. The data in this map are based on approximately 170 studies, dating from the 1950s to the first decade of the twenty-first century. A. H. Bittles, "The Role and Significance of Consanguinity as a Demographic Variable," *Population and Development Review* 20, no. 3 (1994): 561–84; Richard Conniff, "Go Ahead, Kiss Your Cousin. Heck, Marry Her If You Want To," *Discover* 24, no. 8 (August 2003).

5. G. O. Tadmouri, P. Nair, T. Obeid, M. T. Al Ali, N. Al Khaja, and H. A. Hamamy, "Consanguinity and Reproductive Health among Arabs," *Reproductive Health* 6, no. 1 (2009): 17.

6. Alan H. Bittles, "Genetic Aspects of Inbreeding and Incest," in *Inbreeding, Incest, and the Incest Taboo: The State of Knowledge at the Turn of the Century*, ed. Arthur P. Wolf and William H. Durham (Stanford, CA: Stanford University Press, 2004), pp. 38–60.

7. Bittles, "The Role and Significance of Consanguinity as a Demographic Variable."

8. These data were collected in the 1980s and 1990s in the cities of Bangalore and Mysore—not in rural areas, where the prevalence of consanguineous marriage was very probably higher. However, some decline in the prevalence of these marriages has probably occurred since then due to smaller family sizes and the consequent reduced availability of these marriageable relatives. Bittles, "Genetic Aspects of Inbreeding and Incest."

9. Bittles, "The Role and Significance of Consanguinity as a Demographic Variable."

10. Ibid.

11. William M. Shields, "Optimal Inbreeding in the Evolution of Philopatry," in *The Ecology of Animal Movement*, ed. Ian R. Swingland and Paul J. Greenwood (Oxford: Clarendon Press, 1983), pp. 132–59; William M. Shields, "The Natural and Unnatural History of Inbreeding and Outbreeding," in *The Natural History of Inbreeding and Outbreeding: Theoretical and Empirical Perspectives*, ed. Nancy Wilmsen Thornhill (Chicago: University of Chicago Press, 1993), pp. 143–69.

12. Bateson, "Optimal Outbreeding."

13. Andrew T. Smith, "The Natural History of Inbreeding and Outbreeding in Small Mammals," in *The Natural History of Inbreeding and Outbreeding: Theoretical and Empirical Perspectives*, ed. Nancy Wilmsen Thornhill (Chicago: University of Chicago Press, 1993), pp. 329–51.

14. Arthur P. Wolf, *Sexual Attraction and Childhood Association: A Chinese Brief for Edward Westermarck* (Stanford, CA: Stanford University Press, 1995).

15. Ibid.

16. Richard W. Wrangham, *Demonic Males: Apes and the Origins of Human Violence*, ed. Dale Peterson (Boston: Houghton Mifflin, 1996).

17. Wolf, *Sexual Attraction and Childhood Association: A Chinese Brief for Edward Westermarck*.

18. R. Labouriau and A. Amorim, "Human Fertility Increases with Marital Radius," *Genetics* 178, no. 1 (2008): 601; R. Labouriau and A. Amorim, "Comment on 'An Association between the Kinship and Fertility of Human Couples,'" *Science* 322, no. 5908 (2008): 1634b.

19. J. R. Alford and J. R. Hibbing, "The Origin of Politics: An Evolutionary Theory of Political Behavior," *Perspectives on Politics* 2, no. 04 (2004): 707–23.

20. J. E. Settle, C. T. Dawes, N. A. Christakis, and J. H. Fowler, "Friendships Moderate an Association between a Dopamine Gene Variant and Political Ideology," *Journal of Politics* 72, no. 04 (2010): 1189–98.

21. C. Chen, M. Burton, E. Greenberger, and J. Dmitrieva, "Population Migration and the Variation of Dopamine D4 Receptor (DRD4) Allele Frequencies around the Globe," *Evolution and Human Behavior* 20, no. 5 (1999): 309–24.

22. D. F. Aberle, U. Bronfenbrenner, E. H. Hess, D. R. Miller, D. M. Schneider, and J. N. Spuhler, "The Incest Taboo and the Mating Patterns of Animals," *American Anthropologist* 65, no. 2 (1963): 253–65.

23. Patrick Bateson, "Inbreeding Avoidance and Incest Taboos," in *Inbreeding, Incest, and the Incest Taboo: The State of Knowledge at the Turn of the Century*, ed. Arthur P. Wolf and William H. Durham (Stanford, CA: Stanford University Press, 2004), pp. 24–37.

24. Bateson, "Optimal Outbreeding."

25. Edward Westermarck, *The History of Human Marriage*, 2nd ed. (London, 1894).

26. Edward Westermarck, *Three Essays on Sex and Marriage* (London: Macmillan, 1934).

27. Arthur P. Wolf, "Explaining the Westermarck Effect, or, What Did Natural Selection Select For," in *Inbreeding, Incest, and the Incest Taboo: The State of Knowledge at the Turn of the Century*, ed. Arthur P. Wolf and William H. Durham (Stanford, CA: Stanford University Press, 2004), pp. 76–92; Edward o. Wilson, *The Social Conquest of Earth* (New York: Liveright, 2012).

28. Patrick Bateson, "Inbreeding Avoidance and Incest Taboos," in *Inbreeding, Incest, and the Incest Taboo: The State of Knowledge at the Turn of the Century*, ed. Arthur P. Wolf and William H. Durham (Stanford, CA: Stanford University Press, 2004), pp. 24-37.

29. Ibid.

30. Sigmund Freud, *Totem and Taboo: Some Points of Agreement between the Mental Lives of Savages and Neurotics* (New York: W. W. Norton & Co., 1950).

31. A. Wiszewska, B. Pawlowski, and L. G. Boothroyd, "Father-Daughter Relationship as a Moderator of Sexual Imprinting: A Facialmetric Study," *Evolution and Human Behavior* 28, no. 4 (2007): 248–52.

32. T. Bereczkei, P. Gyuris, and G. E. Weisfeld, "Sexual Imprinting in Human Mate Choice," *Proceedings of the Royal Society B: Biological Sciences* 271, no. 1544 (2004): 1129.

33. D. Jedlicka, "A Test of the Psychoanalytic Theory of Mate Selection," *Journal of Social Psychology* 112, no. 2 (1980): 295–99.

34. G. Zei, P. Astolfi, and S. D. Jayakar, "Correlation between Father's Age and Husband's Age: A Case of Imprinting?" *Journal of Biosocial Science* 13, no. 04 (1981): 409–18.

35. G. D. Wilson and P. T. Barrett, "Parental Characteristics and Partner Choice: Some Evidence for Oedipal Imprinting," *Journal of Biosocial Science* 19, no. 02 (1987): 157–61.

36. A. J. Golby, J. D. E Gabrieli, J. Y. Chiao, and J. L. Eberhardt, "Differential Responses in the Fusiform Region to Same-Race and Other-Race Faces," *Nature Neuroscience* 4, no. 8 (2001): 845–50.

37. H. A. Elfenbein and N. Ambady, "On the Universality and Cultural Specificity of Emotion Recognition: A Meta-Analysis," *Psychological Bulletin* 128, no. 2 (2002): 203.

38. D. J. Penn and W. K. Potts, "The Evolution of Mating Preferences and Major Histocompatibility Complex Genes," *American Naturalist* 153, no. 2 (1999): 145–64; S. Jacob, M. K. McClintock, B. Zelano, and C. Ober, "Paternally Inherited HLA Alleles Are Associated with Women's Choice of Male Odor," *Nature Genetics* 30, no. 2 (2002): 175–79.

39. C. Ober, "Studies of HLA, Fertility, and Mate Choice in a Human Isolate," *Human Reproduction Update, European Society of Human Reproduction and Embryology* 5, no. 2 (1999): 103.

40. Marcy K. Uyenoyama, "Genetic Incompatibility as a Eugenic Mechanism," in *The Natural History of Inbreeding and Outbreeding: Theoretical and Empirical Perspectives*, ed. Nancy Wilmsen Thornhill (Chicago: University of Chicago Press, 1993), pp. 60–73; J. A. McIntyre and W. P. Faulk, "Recurrent Spontaneous Abortion in Human Pregnancy: Results of Immunogenetical, Cellular, and Humoral Studies," *American Journal of Reproductive Immunology* 4, no. 4 (1983): 165; L. N. Weckstein, P. Patrizio, J. P. Balmaceda, R. H. Asch, and D. W. Branch, "Human Leukocyte Antigen Compatibility and Failure to Achieve a Viable Pregnancy with Assisted Reproductive Technology," *Acta Europaea Fertilitatis* 22, no. 2 (1991): 103; C. Wedekind, T. Seebeck, F. Bettens, and A. J. Paepke, "MHC-Dependent Mate Preferences in Humans," *Proceedings: Biological Sciences* 260, no. 1359 (1995): 245–49.

41. M. F. Reznikoff-Etievant, J. C. Bonneau, D. Alcalay, B. Cavelier, C. Toure, R. Lobet, and A. Netter, "HLA Antigen-Sharing in Couples with Repeated Spontaneous Abortions and the Birthweight of Babies in Successful Pregnancies," *American Journal of Reproductive Immunology* 25, no. 1 (1991): 25.

42. Penn and Potts, "The Evolution of Mating Preferences and Major Histocompatibility Complex Genes."

43. Ober, "Studies of HLA, Fertility, and Mate Choice in a Human Isolate."

44. Penn and Potts, "The Evolution of Mating Preferences and Major Histocompatibility Complex Genes"; K. Yamazaki, G. K. Beauchamp, I. K. Egorov, J. Bard, L. Thomas, and E. A. Boyse, "Sensory Distinction between H-2b and H-2bm1 Mutant Mice," *Proceedings of the National Academy of Sciences of the United States of America* 80, no. 18 (1983): 5685; P. B. Singh, J. Herbert, B. Roser, L. Arnott, D. K. Tucker, and R. E. Brown, "Rearing Rats in a Germ-Free Environment Eliminates Their Odors of Individuality," *Journal of Chemical Ecology* 16, no. 5 (1990): 1667–82; A. G. Singer, G. K. Beauchamp, and K. Yamazaki, "Volatile Signals of the Major Histocompatibility Complex in Male Mouse Urine," *Proceedings of the National Academy of Sciences of the United States of America* 94, no. 6 (1997): 2210.

45. S. C. Roberts, A. C. Little, L. M. Gosling, B. C. Jones, D. I. Perrett, V. Carter, and M. Petrie, "MHC-Assortative Facial Preferences in Humans," *Biology Letters* 1, no. 4 (2005): 400.

46. Wedekind et al., "MHC-Dependent Mate Preferences in Humans"; A. N. Gilbert, K. Yamazaki, G. K. Beauchamp, and L. Thomas, "Olfactory Discrimination of Mouse Strains (Mus Musculus) and Major Histocompatibility Types by Humans (Homo Sapiens)," *Journal of Comparative Psychology* 100, no. 3 (1986): 262–65.

47. J. M. Setchell, M. J. E. Charpentier, K. M. Abbott, E. J. Wickings, and L. A. Knapp, "Opposites Attract: MHC Associated Mate Choice in a Polygynous Primate," *Journal of Evolutionary Biology* 23, no. 1 (2010): 136–48.

48. Jacob et al., "Paternally Inherited HLA Alleles Are Associated with Women's Choice of Male Odor."

49. Wedekind et al., "MHC-Dependent Mate Preferences in Humans."

50. Jacob et al., "Paternally Inherited HLA Alleles Are Associated with Women's Choice of Male Odor."

51. P. K. Hatemi, N. A. Gillespie, L. J. Eaves, B. S. Maher, B. T. Webb, A. C. Heath, S. E. Medland, et al. "A Genome-Wide Analysis of Liberal and Conservative Political Attitudes," *Journal of Politics* 73, no. 1 (2011): 271–85.

Chapter 12. Why Gender Inequality and Fertility Change across Human History

a. Indeed, Charles Darwin observed in the late nineteenth century that differences between a species's sexes that do *not* serve any survival function often aid the animals in sexual selection. For example, since males face stiffer competition in many species of birds, costly ornamentation helps them to compete with other males in the struggle to mate with (the fittest) females.

b. Human women aren't the only beings that pay an unequal cost to reproduce both partners' genes equally. Hermaphroditic flatworms, such as *Pseudoceros bifurcus* and *hancockanus* can become either fathers or mothers (their bodies are equipped with both sperm and ovaries). But it suits them to become fathers because the male invests so much less energy in reproduction. To determine their role, pairs of these flatworms engage in an activity called "penis fencing." It involves a twenty-minute duel in which the marine flatworms rear up and attempt to stab their rival's skin with dagger-like penises to hypodermically inject sperm. The "winner" becomes the father and can continue trying to impregnate other worms—without having to heal the stab wounds and invest in the eggs. See N. K. Michiels and L. J. Newman, "Sex and Violence in Hermaphrodites," *Nature* 391, no. 6668 (1998): 647.

1. Palestinian Media Watch, "Internal Palestinian Issues: Women in PA Society," October 27, 2010.

2. Robert L. Trivers, "Parental Investment and Sexual Selection," in *Sexual Selection and the Descent of Man 1871–1971: The Darwinian Pivot*, ed. Bernard Campbell (Chicago: Aldine Publishing Company, 1972), pp. 136–207.

3. Joann S. Lublin and Kelly Eggers, "More Women Are Primed to Land CEO Roles," *Wall Street Journal*, April 30, 2012.

4. H. Kaplan, "A Theory of Fertility and Parental Investment in Traditional and Modern Human Societies," *American Journal of Physical Anthropology* 101, no. S23 (1996): 91–135.

5. S. Goldberg, *Why Men Rule: A Theory of Male Dominance* (Chicago: Open Court, 1993); M. Harris, "The Evolution of Human Gender Hierarchies: A Trial

Formulation," in *Sex and Gender Hierarchies*, ed. D. Miller (Cambridge: Cambridge University Press, 1993), pp. 57–79; M. Z. Rosaldo, "Woman, Culture, and Society: A Theoretical Overview," in *Woman, Culture, and Society*, ed. M. Z. Rosaldo and L. Lamphere (Stanford, CA: Stanford University Press, 1974), pp. 17–42; P. R. Sanday, "Female Status in the Public Domain," in *Woman, Culture, and Society*, ed. M. Z. Rosaldo and L. Lamphere (Stanford, CA: Stanford University Press, 1974), pp. 189–206.

6. Goldberg, *Why Men Rule: A Theory of Male Dominance*; James Sidanius and Robert Kurzban, "Evolutionary Approaches to Political Psychology," in *Oxford Handbook of Political Psychology*, ed. David O. Sears, Leonie Huddy, and Robert Jervis (New York: Oxford University Press, 2003), pp. 146–81.

7. UNIVEF, "Estimates of Maternal Mortality 2008 (WHO/UNICEF/UNFPA/the World Bank)," *ChildInfo: Monitoring the Situation of Children and Women*, September 2010; "Children Per Woman (Total Fertility Rate) for 2008," Gapminder.org, http://www.gapminder.org/data/documentation/gd008/ (accessed May 21, 2013).

8. F. B. M. de Waal, *Good Natured: The Origins of Right and Wrong in Primates and Other Animals* (Cambridge, MA: Harvard University Press, 1996).

9. Sandra L. Vehrencamp, "A Model for the Evolution of Despotic Versus Egalitarian Societies," *Animal Behaviour* 31, no. 3 (1983): 667–82.

10. Ibid.

11. J. L. Caswell, S. Mallick, D. J. Richter, J. Neubauer, C. Schirmer, S. Gnerre, and D. Reich, "Analysis of Chimpanzee History Based on Genome Sequence Alignments," *PLoS Genetics* 4, no. 4 (2008): e1000057.

12. Y. J. Won and J. Hey, "Divergence Population Genetics of Chimpanzees," *Molecular Biology and Evolution* 22, no. 2 (2005): 297.

13. Richard W. Wrangham, *Demonic Males: Apes and the Origins of Human Violence*, ed. Dale Peterson (Boston: Houghton Mifflin, 1996).

14. Ibid.; F. B. M. de Waal, "Bonobos, Left & Right: Primate Politics Heats up again as Liberals & Conservatives Spindoctor Science," *Skeptic: Promoting Science and Critical Thinking*, August 8, 2007.

15. Wrangham, *Demonic Males: Apes and the Origins of Human Violence*.

16. F. B. M. de Waal, "Bonobo Sex and Society," *Scientific American* 272, no. 3 (1995): 82–88.

17. De Waal, "Bonobos, Left & Right: Primate Politics Heats up again as Liberals & Conservatives Spindoctor Science"; Dinesh D'Souza, "Bonobo Promiscuity? Another Myth Bites the Dust," *AOL Newsbloggers*, August 3, 2007.

18. Hillard Kaplan, Michael Gurven, and Jeffrey Winking, "An Evolutionary Theory of Human Life Span: Embodied Capital and the Human Adaptive Complex," in *Handbook of Theories of Aging*, ed. V. L. Bengtson, M. Silverstein, and N. Putney (New York: Springer, 2009), pp. 39–60; Frank W. Marlow, "Hunter-Gatherers and Human Evolution," *Evolutionary Anthology: Issues, News, and Reviews* 14, no. 2 (2005): 54–67; Hillard Kaplan, Kim Hill, Jane Lancaster, and A. Magdalena Hurtado, "A Theory of Human Life History Evolution: Diet, Intelligence, and Longevity," *Evolutionary Anthology: Issues, News, and Reviews* 9, no. 4 (2000): 156–85.

19. Allen Johnson, "The Political Unconscious: Stories and Politics in Two South

American Cultures," in *Political Psychology: Cultural and Cross-Cultural Foundations*, ed. Stanley Allen Renshon and J. H. Duckitt (New York: New York University Press, 2000), pp. 159–81.

20. B. M. Knauft, T. S. Abler, L. Betzig, C. Boehm, R. K. Dentan, T. M. Kiefer, K. F. Otterbein, J. Paddock, and L. Rodseth, "Violence and Sociality in Human Evolution," *Current Anthropology* 32, no. 4 (1991): 391–428.

21. Paul H. Rubin, *Darwinian Politics: The Evolutionary Origin of Freedom* (New Brunswick, NJ: Rutgers University Press, 2002); Christopher Boehm, *Hierarchy in the Forest: The Evolution of Egalitarian Behavior* (Cambridge, MA: Harvard University Press, 1999).

22. Francis Moran III, "Marxist Gorillas: Nonhuman Primates and the Biopolitics of Class," in *Recent Explorations in Biology and Politics*, ed. Albert Somit and Steven A. Peterson (Greenwich, CN: Jai Press, 1997), pp. 57–70.

23. Nichalos G. Blurton Jones, "A Reply to Dr. Harpending," *American Journal of Physical Anthropology* 93, no. 3 (1994): 391–97.

24. Nichalos G. Blurton Jones, "Bushman Birth Spacing: Direct Tests of Some Simple Predictions," *Ethology and Sociobiology* 8, no. 3 (1987): 183–203.

25. Blurton Jones, "A Reply to Dr. Harpending"; Blurton Jones, "Bushman Birth Spacing: Direct Tests of Some Simple Predictions"; Kim Hill and Hillard Kaplan, "Life History Traits in Humans: Theory and Empirical Studies," *Annual Review of Anthropology* 28 (1999): 397–430; Kim Hill and A. Magdalena Hurtado, *Aché Life History: The Ecology and Demography of a Foraging People* (New York: Aldine de Gruyter, 1996).

26. Johnson, "The Political Unconscious: Stories and Politics in Two South American Cultures."

27. G. A. Schuiling, "Honor Your Father and Your Mother," *Journal of Psychosomatic Obstetrics & Gynecology* 22, no. 4 (2001): 215–19.

28. Jared Diamond, *Guns, Germs, and Steel: The Fates of Human Societies* (New York: W. W. Norton, 1999).

29. Ibid.

30. G. E. Lenski, *Power and Privilege: A Theory of Social Stratification* (Chapel Hill: University of North Carolina Press, 1984).

31. Rubin, *Darwinian Politics: The Evolutionary Origin of Freedom*; Diamond, *Guns, Germs, and Steel: The Fates of Human Societies*; James Sidanius and Felicia Pratto, "Social Dominance Theory: A New Synthesis," in *Political Psychology: Key Readings*, ed. John T. Jost and Jim Sidanius (New York: Psychology Press, 2004), pp. 315–32.

32. Wrangham, *Demonic Males: Apes and the Origins of Human Violence*; Robert Hans Van Gulik, *Sexual Life in Ancient China: A Preliminary Survey of Chinese Sex and Society from Ca. 1500 B. C. Till 1644 A. D.* (Leiden: E. J. Brill, 1974).

33. Kaplan, "A Theory of Fertility and Parental Investment in Traditional and Modern Human Societies."

34. M. Apostolou, "Sexual Selection under Parental Choice in Agropastoral Societies," *Evolution and Human Behavior* 31, no. 1 (2010): 39–47.

35. Ibid.

36. Ibid.

37. J. Greenwood and A. Seshadri, "The U.S. Demographic Transition," *American Economic Review* 92, no. 2 (2002): 153–59.

38. Monique Borgerhoff Mulder, "The Demographic Transition: Are We Any Closer to an Evolutionary Explanation?" *Trends in Ecology & Evolution (TREE)* 13, no. 7 (1998): 266–70.

39. Kaplan, "A Theory of Fertility and Parental Investment in Traditional and Modern Human Societies."

40. Pew Social Trends Staff, "The Decline of Marriage and Rise of New Families," *Social and Demographic Trends* (2010).

41. Apostolou, "Sexual Selection under Parental Choice in Agropastoral Societies."

42. Benjamin I. Page, *The Rational Public: Fifty Years of Trends in Americans' Policy Preferences*, ed. Robert Y. Shapiro (Chicago: University of Chicago Press, 1992).

43. R. L. Cliquet, "Below-Replacement Fertility and Gender Politics," *Research in Biopolitics* 6 (1998): 91–118.

44. P. Jha, M. A. Kesler, R. Kumar, F. Ram, U. Ram, L. Aleksandrowicz, D. G. Bassani, S. Chandra, and J. K. Banthia, "Trends in Selective Abortions of Girls in India: Analysis of Nationally Representative Birth Histories from 1990 to 2005 and Census Data from 1991 to 2011," *Lancet* 377, no. 9781 (June 2011).

Chapter 13. The Biology of War and Genocide

a. Of the remaining 5 percent, 3 percent are avunculocal (newlyweds live with the husband's mother's oldest brother), and 2 percent are neolocal (they move to a new location).

b. The idea that no real difference existed between Tutsis and Hutus before Belgian colonialism is an example of the moralistic fallacy: that there should be no ethnic differences between peoples, so there is no difference (other than that invented by outsiders for self-interested reasons).

1. Carl von Clausewitz, *On War*, ed. Michael Howard and Peter Paret (Princeton, NJ: Princeton University Press, 1976).

2. David Adams, et al. "Appendix: The Seville Statement on Violence," in *Aggression and Peacefulness in Humans and Other Primates*, ed. James Silverberg and J. Patrick Gray (New York: Oxford University Press, 1992), pp. 295–97.

3. Fred C. Alford, "Group Psychology Is the State of Nature," in *Political Psychology*, ed. Kristen R. Monroe (Mahwah, NJ: L. Erlbaum, 2002), pp. 193–205.

4. Ted Robert Gurr and Barbara Harff, *Ethnic Conflict in World Politics* (Boulder, CO: Westview Press, 1994); James Sidanius and Felicia Pratto, "Social Dominance Theory: A New Synthesis," in *Political Psychology: Key Readings*, ed. John T. Jost and Jim Sidanius (New York: Psychology Press, 2004), pp. 315–32.

5. R. B. Zajonc, "The Zoomorphism of Human Collective Violence," in *Understanding Genocide: The Social Psychology of the Holocaust*, ed. Leonard S. Newman and Ralph Erber (Oxford: Oxford University Press, 2002), pp. 222–38; C. K. W. De Dreu,

L. L. Greer, M. J. J. Handgraaf, S. Shalvi, G. A. Van Kleef, M. Baas, F. S. Ten Velden, E. Van Dijk, and S. W. W. Feith, "The Neuropeptide Oxytocin Regulates Parochial Altruism in Intergroup Conflict among Humans," *Science* 328, no. 5984 (2010): 1408.

6. C. R. Ember, "Myths about Hunter-Gatherers," *Ethnology* 17, no. 4 (1978): 439–48.

7. L. H. Keeley, *War before Civilization* (Oxford: Oxford University Press, 1997); J. K. Choi and S. Bowles, "The Coevolution of Parochial Altruism and War," *Science* 318, no. 5850 (2007): 636.

8. J. L. Caswell, S. Mallick, D. J. Richter, J. Neubauer, C. Schirmer, S. Gnerre, and D. Reich, "Analysis of Chimpanzee History Based on Genome Sequence Alignments," *PLoS Genetics* 4, no. 4 (2008): e1000057; J. R. Minkel, "Human-Chimp Gene Gap Widens from Tally of Duplicate Genes," *Scientific American*, December 19, 2006.

9. Richard W. Wrangham, *Demonic Males: Apes and the Origins of Human Violence*, ed. Dale Peterson (Boston: Houghton Mifflin, 1996).

10. Jane Goodall, *The Chimpanzees of Gombe: Patterns of Behavior* (Cambridge, MA: Belknap Press of Harvard University Press, 1986); Johan M. G. Van der Dennen, "The Politics of Peace in Preindustrial Societies: The Adaptive Rationale behind Corroboree and Calumet," *Research in Biopolitics* 6 (1998): 159–92.

11. Melissa J. Ferguson, Travis J. Carter, and Ran R. Hassin, "On the Automaticity of Nationalist Ideology: The Case of the USA," in *Social and Psychological Bases of Ideology and System Justification*, ed. John T. Jost, Aaron C. Kay, and Hulda Thorisdottir (Oxford: Oxford University Press, 2009), pp. 53–82.

12. Wrangham, *Demonic Males: Apes and the Origins of Human Violence.*

13. W. D. Hamilton, "Innate Social Aptitudes of Man: An Approach from Evolutionary Genetics," *Biosocial anthropology* (1975): 133–55.

14. Van der Dennen, "The Politics of Peace in Preindustrial Societies: The Adaptive Rationale behind Corroboree and Calumet."

15. Wrangham, *Demonic Males: Apes and the Origins of Human Violence.*

16. Richard Black, "Species Count Put at 8.7 Million," BBC News Science and Environment, August 23, 2011.

17. Wrangham, *Demonic Males: Apes and the Origins of Human Violence.*

18. Ember, "Myths about Hunter-Gatherers."

19. Wrangham, *Demonic Males: Apes and the Origins of Human Violence*; A. E. Griesse and R. Stites, "Russia: Revolution and War," in *Female Soldiers—Combatants or Noncombatants? Historical and Contemporary Perspectives*, ed. Nancy Loring Goldman (Westport, CN: Greenwood Press, 1982), pp. 61–84.

20. James Sidanius and Robert Kurzban, "Evolutionary Approaches to Political Psychology," in *Oxford Handbook of Political Psychology*, ed. David O. Sears, Leonie Huddy, and Robert Jervis (New York: Oxford University Press, 2003), pp. 146–81; John Keegan, *A History of Warfare* (New York: Alfred A. Knopf, 1993); L. Rodseth, Richard W. Wrangham, A. M. Harrigan, and B. B. Smuts, "The Human Community as a Primate Society," *Current Anthropology* 32 (1991): 221–54.

21. G. P. Murdock and D. R. White, "Standard Cross-Cultural Sample," *Ethnology* 8, no. 4 (1969): 329–69.

22. Benjamin I. Page, *The Rational Public: Fifty Years of Trends in Americans' Policy Preferences*, ed. Robert Y. Shapiro (Chicago: University of Chicago Press, 1992).

23. Wrangham, *Demonic Males: Apes and the Origins of Human Violence*.

24. Ibid.; N. A. Chagnon, "Life Histories, Blood Revenge, and Warfare in a Tribal Population," *Science* 239, no. 4843 (1988): 985.

25. Wrangham, *Demonic Males: Apes and the Origins of Human Violence*.

26. Keeley, *War before Civilization*.

27. Wrangham, *Demonic Males: Apes and the Origins of Human Violence*.

28. Chagnon, "Life Histories, Blood Revenge, and Warfare in a Tribal Population."

29. Paul H. Rubin, *Darwinian Politics: The Evolutionary Origin of Freedom* (New Brunswick, NJ: Rutgers University Press, 2002).

30. Van der Dennen, "The Politics of Peace in Preindustrial Societies: The Adaptive Rationale behind Corroboree and Calumet."

31. J. A. Wilder, Z. Mobasher, and M. F. Hammer, "Genetic Evidence for Unequal Effective Population Sizes of Human Females and Males," *Molecular Biology and Evolution* 21, no. 11 (2004): 2047; Kate Melville, "Ancient Man Spread the Love Around," *Science A Go Go*, September 20, 2004.

32. S. Kanazawa, "Evolutionary Psychological Foundations of Civil Wars," *Journal of Politics* 71, no. 01 (2009): 25–34.

33. Francis Fukuyama, *The Origins of Political Order: From Prehuman Times to the French Revolution* (London: Profile Books, 2011).

34. Wrangham, *Demonic Males: Apes and the Origins of Human Violence*.

35. "9/11 Link to Rise in Male Foetal Death Rate, Study Says," BBC News, May 24, 2010.

36. T. A. Bruckner, R. Catalano, and J. Ahern, "Male Fetal Loss in the U.S. Following the Terrorist Attacks of September 11, 2001," *BMC Public Health* 10, no. 1 (2010): 273.

37. Van der Dennen, "The Politics of Peace in Preindustrial Societies: The Adaptive Rationale behind Corroboree and Calumet"; W. T. Divale, F. Chamberis, and D. Gangloff, "War, Peace, and Marital Residence in Pre-Industrial Societies," *Journal of Conflict Resolution* 20, no. 1 (1976): 57.

38. Daniel Chirot and Clark McCauley, *Why Not Kill Them All? The Logic and Prevention of Mass Political Murder* (Princeton, NJ: Princeton University Press, 2006).

39. Van der Dennen, "The Politics of Peace in Preindustrial Societies: The Adaptive Rationale behind Corroboree and Calumet"; Ragnar Julius Numelin, *The Beginnings of Diplomacy: A Sociological Study of Intertribal and International Relations* (London: Oxford University Press, 1950); E. B. Tylor, "On a Method of Investigating the Development of Institutions; Applied to Laws of Marriage and Descent," *Journal of the Anthropological Institute of Great Britain and Ireland* 18 (1889): 245–72; Robin Fox, *Kinship and Marriage: An Anthropological Perspective* (Baltimore, MD: Penguin, 1967).

40. Gunnar Landtman, *The Kiwai Papuans of British New Guinea, a Nature-Born Instance of Rousseau's Ideal Community* (London: Macmillan and Co., Limited, 1927).

41. R. A. Rappaport, "Ritual Regulation of Environmental Relations among a New

Guinea People," *Ethnology* 6, no. 1 (1967): 17–30; W. Goldschmidt, "Peacemaking and the Institutions of Peace in Tribal Societies," in *The Anthropology of Peace and Nonviolence*, ed. L. E. Sponsel and T. A. Gregor (Boulder, CO: Lynne Rienner, 1994), pp. 109–32.

42. Van der Dennen, "The Politics of Peace in Preindustrial Societies: The Adaptive Rationale behind Corroboree and Calumet"; A. P. Elkin, *The Australian Aborigines; How to Understand Them* (Sydney: Angus & Robertson, 1954).

43. "Facebook Drops 'Intifada' Page for Promoting Violence," BBC News, *Middle East*, March 29, 2011.

44. Palestinian Media Watch, "Religious War: Kill Jews for Allah," http://www.palwatch.org/main.aspx?fi=427 2010 (accessed April 8, 2013).

45. Chirot and McCauley, *Why Not Kill Them All? The Logic and Prevention of Mass Political Murder*.

46. B. Harff and T. R. Gurr, "Victims of the State: Genocides, Politicides, and Group Repression since 1945," *International Review of Victimology* 1, no. 1 (1989): 23–41.

47. Ervin Staub and Daniel Bar-Tal, "Genocide, Mass Killing, and Intractable Conflict: Roots, Evolution, Prevention, and Reconciliation," in *Oxford Handbook of Political Psychology*, ed. David O. Sears, Leonie Huddy, and Robert Jervis (New York: Oxford University Press, 2003), pp. 710–54.

48. B. Harff, "No Lessons Learned from the Holocaust? Assessing Risks of Genocide and Political Mass Murder since 1955," *American Political Science Review* 97, no. 01 (2003): 57–73.

49. Ibid.

50. Zajonc, "The Zoomorphism of Human Collective Violence"; Ervin Staub, "The Psychology of Bystanders, Perpetrators, and Heroic Helpers," in *Understanding Genocide: The Social Psychology of the Holocaust*, ed. Leonard S. Newman and Ralph Erber (Oxford: Oxford University Press, 2002), pp. 11-42; A. L. Des Forges, *Leave None to Tell the Story* (Human Rights Watch, 1999).

51. Harff, "No Lessons Learned from the Holocaust? Assessing Risks of Genocide and Political Mass Murder since 1955."

52. Ibid.

53. Ibid.

54. Staub, "The Psychology of Bystanders, Perpetrators, and Heroic Helpers."

55. "Burma Election: 'Not Going to Vote,'" BBC News Asia-Pacific, November 4, 2010.

56. Sanjoy Majumder, "Ayodhya Verdict: Indian Holy Site 'to Be Divided,'" BBC News South Asia, September 30, 2010; "India on High Alert after Varanasi City Blast," BBC News South Asia, December 8, 2010.

57. Hillel Fendel, "Half the Public Wants to See Holy Temple Rebuilt," *Arutz Sheva, Israel National News*, July 18, 2010.

58. Itamar Marcus and Nan Jacques Zilberdik, "The PA Denies Jewish History in Jerusalem: The Jewish Temple Is 'the Alleged Temple,'" Palestinian Media Watch, August 11, 2011.

59. S. D. Goitein, "The Historical Background of the Erection of the Dome of the Rock," *Journal of the American Oriental Society* 70, no. 2 (1950): 104–108.

60. "Egypt's Pope 'Sorry' for Bishop's Koran Comments," BBC News Middle East, September 27, 2010.

61. Chirot and McCauley, *Why Not Kill Them All? The Logic and Prevention of Mass Political Murder.*

62. D. J. Goldhagen, *Hitler's Willing Executioners: Ordinary Germans and the Holocaust* (New York: Knopf, 1996).

63. Neil Jeffrey Kressel, *Mass Hate: The Global Rise of Genocide and Terror*, updated ed. (Boulder, CO: Westview Press, 2002); Raul Hilberg, *The Destruction of the European Jews*, student ed. (New York Holmes & Meier, 1985).

64. Kressel, *Mass Hate: The Global Rise of Genocide and Terror*; Martha L. Cottam, Beth Dietz-Uhler, Elena Mastors, and Thomas Preston, *Introduction to Political Psychology* (Mahwah, NJ: Lawrence Erlbaum Associates, 2004).

65. H. R. Trevor-Roper, *Hitler's Table Talk 1941–44: His Private Conversations* (London: Weidenfeld and Nicolson, 1973).

66. Cottam et al., *Introduction to Political Psychology.*

67. Kressel, *Mass Hate: The Global Rise of Genocide and Terror.*

68. Ibid.

69. "Libya Protests: UN Security Council Condemns Crackdown," BBC News Middle East, February 23, 2011.

70. "Up to 15,000 Killed in Libya War: U.N. Rights Expert," Reuters, June 9, 2011.

71. R. M. Elgderi, K. S. Ghenghesh, and N. Berbash, "Carriage by the German Cockroach (Blattella Germanica) of Multiple-Antibiotic-Resistant Bacteria That Are Potentially Pathogenic to Humans, in Hospitals and Households in Tripoli, Libya," *Annals of Tropical Medicine and Parasitology* 100, no. 1 (2006): 55–62.

72. Palestinian Media Watch "'Kill a Jew—Go to Heaven,'" December 24, 2002.

73. Translation courtesy of Palestinian Media Watch; Palestinian Media Watch, "Religious War: Kill Jews for Allah."

74. Ethan Bronner, "Poll Shows Most Palestinians Favor Violence over Talks," *New York Times*, March 19, 2008.

75. Chirot and McCauley, *Why Not Kill Them All? The Logic and Prevention of Mass Political Murder.*

76. Staub, "The Psychology of Bystanders, Perpetrators, and Heroic Helpers."

77. Earl Conteh-Morgan, *Collective Political Violence: An Introduction to the Theories and Cases of Violent Conflicts* (New York: Routledge, 2004).

78. Chirot and McCauley, *Why Not Kill Them All? The Logic and Prevention of Mass Political Murder.*

79. Kressel, *Mass Hate: The Global Rise of Genocide and Terror*; Leo Kuper, *Genocide: Its Political Use in the Twentieth Century* (New Haven, CT: Yale University Press, 1981).

80. Kressel, *Mass Hate: The Global Rise of Genocide and Terror.*

81. M. Mamdani, *When Victims Become Killers: Colonialism, Nativism, and the Genocide in Rwanda* (James Currey Publishers, 2001).

82. René Lemarchand, *Burundi: Ethnic Conflict and Genocide*, 1st paperback ed. (Washington, DC: Woodrow Wilson Center Press, 1996).

83. R. J. Hartley, "To Massacre: A Perspective on Demographic Competition," *Anthropological Quarterly* 80, no. 1 (2007): 237–51; Damien De Walque, "Selective Mortality during the Khmer Rouge Period in Cambodia," *Population and Development Review* 31, no. 2 (2005): 351–68.

84. Kressel, *Mass Hate: The Global Rise of Genocide and Terror*.

85. Ibid.

86. Hartley, "To Massacre: A Perspective on Demographic Competition."

87. Ibid.; Binaifer Nowrojee, *Shattered Lives: Sexual Violence during the Rwanda Genocide and Its Aftermath*, ed. Dorothy Q. Thomas and Janet Fleischman (New York: Human Rights Watch, 1996); Alexandra Stiglmayer, "The Rapes in Bosnia-Herzegovina," in *Mass Rape: The War against Women in Bosnia-Herzegovina*, ed. Alexandra Stiglmayer (Lincoln: University of Nebraska Press, 1994), pp. 82–169.

88. Conteh-Morgan, *Collective Political Violence: An Introduction to the Theories and Cases of Violent Conflicts*.

89. Kressel, *Mass Hate: The Global Rise of Genocide and Terror*; Roy Gutman, *A Witness to Genocide: The 1993 Pulitzer Prize–Winning Dispatches on the "Ethnic Cleansing" of Bosnia* (New York: Macmillan, 1993).

90. "UN Official Calls DR Congo 'Rape Capital of the World,'" BBC News, April 28, 2010.

91. "DR Congo Women March against Sexual Violence," BBC News Africa, October 17, 2010.

92. "DR Congo: 48 Rapes Every Hour, US Study Finds," BBC News Africa, May 12, 2011.

93. "'Rape Is Used as a Weapon of War,'" BBC News Africa, June 8, 2011.

94. Pascale Harter, "Libya Rape Victims 'Face Honour Killings,'" BBC News Africa, June 14, 2011.

PART III. DO WE LIVE IN A JUST WORLD?

Chapter 14. Attitudes Toward Inequality and Authority in Society

a. This statement exposes a flaw in the RWA test. Although Altemeyer refined many of the technical snags that plagued its predecessor, he did not fully explain the reasons *why* RWA so successfully predicts left-right orientation and party affiliation. Consequently, some of the test's statements mix important content categories. Specifically, this item about "*the self-righteous forces*" mixes attitudes toward societal authority with attitudes toward religion. An improved test would not mix content categories within a single item. In addition, this item assumes a Western, Cold-War scenario where conservatives worry about radical left-wing dissidents (that is, "godless Communists"); also, the phrasing does not work as well for other parts of the world. In the Middle East, for example, many governments have fought against right-wing Islamist dissidents.

b. For the same reasons, this last item above should simply read: "It is best to treat dissenters with leniency, since their ideas are the lifeblood of change and reform." This revision omits the terms "progressive" and "new ideas," which confound specifically left-wing interests with antigovernment dissidence in general.

1. "Death Penalty Debate," CBC Digital Archives, http://archives.cbc.ca/society/crime_justice/topics/625/2008 (accessed April 8, 2013).

2. S. J. H. McCann, "Societal Threat, Authoritarianism, Conservatism, and US State Death Penalty Sentencing (1977–2004)," *Journal of Personality and Social Psychology* 94, no. 5 (2008): 913.

3. Robert Altemeyer, *Enemies of Freedom: Understanding Right-Wing Authoritarianism*, 1st ed. (San Francisco: Jossey-Bass Publishers, 1988); Robert Altemeyer, *The Authoritarian Specter* (Cambridge, MA: Harvard University Press, 1996).

4. Raphael S. Ezekiel, *The Racist Mind: Portraits of Neo-Nazis and Klansmen* (New York: Viking, 1995).

5. Altemeyer, *Enemies of Freedom: Understanding Right-Wing Authoritarianism*.

6. "George W. Bush Claims UK Lives 'Saved by Waterboarding,'" BBC News UK, November 9, 2010.

7. Peter Singer, *A Darwinian Left: Politics, Evolution, and Cooperation* (London: Weidenfeld & Nicolson, 1999).

8. Robert J. Hoshowsky, "1962: Arthur Lucas and Ronald Turpin," ExecutedToday.com, http://www.executedtoday.com/2007/12/11/1962-arthur-lucas-and-ronald-turpin/ (accessed December 11 2007).

9. George Lakoff, *Moral Politics: How Liberals and Conservatives Think*, 2nd ed. (Chicago, IL: University of Chicago Press, 2002).

10. Carolyn L. Hafer and Becky L. Choma, "Belief in a Just World, Perceived Fairness, and Justification of the Status Quo," in *Social and Psychological Bases of Ideology and System Justification*, ed. John T. Jost, Aaron C. Kay, and Hulda Thorisdottir (Oxford: Oxford University Press, 2009), pp. 107–25.

11. Daniel Chirot and Clark McCauley, *Why Not Kill Them All? The Logic and Prevention of Mass Political Murder* (Princeton, NJ: Princeton University Press, 2006); Robert Chazan, *God, Humanity, and History: The Hebrew First Crusade Narratives* (Berkeley: University of California Press, 2000).

12. Jon Basil Utley, "'Dual Covenant' Christians: Christian Zionists and the Strangest Alliance in History," AntiWar.com, http://antiwar.com/utley/?articleid=9456 (accessed August 2, 2006).

13. "Israel: Haifa Forest Fire 'under Control,'" BBC News Middle East, December 5, 2010.

14. James Davies, Susanna Sandström, Anthony Shorrocks, and Edward Wolff, "Pioneering Study Shows Richest Two Percent Own Half World Wealth," United Nations University, http://www.wider.unu.edu/events/past-events/2006-events/en_GB/05-12-2006/ (accessed December 5, 2006).

15. Francis Fukuyama, "Left Out," *American Interest: Policy, Politics & Culture* (January/February 2011): 22–28.

16. Hafer and Choma, "Belief in a Just World, Perceived Fairness, and Justification of the Status Quo."

17. M. O. Hunt, "Status, Religion, and the 'Belief in a Just World': Comparing African Americans, Latinos, and Whites," *Social Science Quarterly* 81, no. 1 (2000): 325–43.

18. "Bolivia Defends Seizing Foreign Energy Firms," BBC News Business, November 12, 2010.

19. Simon Romero, "In Venezuela, a New Wave of Foreigners," *New York Times*, November 6, 2010.

20. Lakoff, *Moral Politics: How Liberals and Conservatives Think.*

21. Paul M. Sniderman, Philip E. Tetlock, and Laurel Elms, "Public Opinion and Democratic Politics: The Problem of Nonattitudes and the Social Construction of Political Judgment" in *Citizens and Politics: Perspectives from Political Psychology*, ed. James H. Kuklinski (Cambridge: Cambridge University Press, 2001), pp. 254–88.

22. Finlo Rohrer, "What Are You Not Allowed to Say?" BBC News Magazine, November 29, 2010.

23. Ibid.

24. Natalia Sobrevilla Perea, "Peru's Sterilisation Victims Still Await Compensation and Justice," Guardian.co.uk, http://www.guardian.co.uk/commentisfree/cifamerica/2011/jun/17/peru-sterilisation-compensation (accessed June 17, 2011).

25. Charles Sidney Bluemel, *War, Politics, and Insanity*, 2nd rev. ed. (Denver: World Press, 1950).

26. "Russian Calendar Girls in Putin Birthday Battle," BBC News Europe, October 8, 2010.

27. Fred I. Greenstein, *Personality and Politics: Problems of Evidence, Inference, and Conceptualization* (New York: Norton, 1975).

28. Aaron David Miller, *The Much Too Promised Land: America's Elusive Search for Arab-Israeli Peace* (New York: Bantam Books, 2008).

29. David E. Sanger, James Glanz, and Jo Becker, "Around the World, Distress over Iran." *New York Times*, November 28, 2010.

30. Tucker Reals, "Iran: Obama Must 'Unclench' America's Fist." *CBS News, World Watch*, January 28, 2009.

31. Frank Rich, "The Rage Won't End on Election Day," *New York Times*, October 16, 2010.

32. Katie Connolly, "Understanding the 2010 US Election through Campaign Ads," BBC News, November 1, 2010.

33. Mark Leibovich, "Being Glenn Beck," *New York Times Magazine*, September 29, 2010.

34. Eric Zorn, "Arizona Democratic U.S. Rep. Gabrielle Giffords Shot in Rampage at Grocery Store," *Chicago Tribune*, January 8, 2011.

35. Michael Sheridan, "Ex-Klu Klux Klan Leader and White Supremacist Uses Ad to Run for Congress," *New York Daily News*, March 9, 2010.

Chapter 15. Attitudes toward Inequality and Authority Within the Family

a. Pinker's criticisms include Lakoff's failing to cite scientific studies or recognize his intellectual predecessors, a "cognitive relativism" that puts too much faith in the power of metaphors as the root cause of political conflict, and a bias that belittles conservatives while self-congratulating progressives.

b. To a lesser extent, the connotation of the word "motherland" in modern English evokes the far-left regime of the former Soviet Union.

c. As for the other Big Five traits, later-borns are, on average, more extraverted and agreeable, while first-borns have sometimes been shown to be more neurotic.

1. Marc J. Hetherington and Jonathan Daniel Weiler, *Authoritarianism and Polarization in American Politics*, ed. Jonathan Daniel Weiler (Cambridge: Cambridge University Press, 2009).

2. Nicholas D. Kristof, "Our Politics May Be All in Our Head," *New York Times*, February 13, 2010.

3. Robert Altemeyer, *Enemies of Freedom: Understanding Right-Wing Authoritarianism*, 1st ed. (San Francisco: Jossey-Bass Publishers, 1988).

4. J. Graham, J. Haidt, and B. A. Nosek, "Liberals and Conservatives Rely on Different Sets of Moral Foundations," *Journal of Personality and Social Psychology* 96, no. 5 (2009): 1029–46; Jonathan Haidt and Jesse Graham, "Planet of the Durkheimians, Where Community, Authority, and Sacredness Are Foundations of Morality," in *Social and Psychological Bases of Ideology and System Justification*, ed. John T. Jost, Aaron C. Kay, and Hulda Thorisdottir (Oxford: Oxford University Press, 2009), pp. 371–401; Nicholas D. Kristof, "Would You Slap Your Father? If So, You're a Liberal," *New York Times*, May 27, 2009.

5. George Lakoff, *Moral Politics: How Liberals and Conservatives Think*, 2nd ed. (Chicago, IL: University of Chicago Press, 2002).

6. Ibid.

7. R. D. Smither, "Authoritarianism, Dominance, and Social Behavior: A Perspective from Evolutionary Personality Psychology," *Human Relations* 46, no. 1 (1993): 23.

8. Martha L. Cottam, Beth Dietz-Uhler, Elena Mastors, and Thomas Preston, *Introduction to Political Psychology* (Mahwah, NJ: Lawrence Erlbaum Associates, 2004).

9. K. Demyttenaere, R. Bruffaerts, J. Posada-Villa, I. Gasquet, V. Kovess, J. P. Lepine, M. C. Angermeyer, et al. "Prevalence, Severity, and Unmet Need for Treatment of Mental Disorders in the World Health Organization World Mental Health Surveys," *Journal of the American Medical Association* 291, no. 21 (2004): 2581.

10. R. Chris Fraley, Brian N. Griffin, Jay Belsky, and Glenn I. Roisman, "Developmental Antecedents of Political Ideology: A Longitudinal Investigation from Birth to Age 18 Years," *Psychological Science* 20, no. 10 (2012): 1–7.

11. Allen Johnson, "The Political Unconscious: Stories and Politics in Two South

American Cultures," in *Political Psychology: Cultural and Cross-Cultural Foundations*, ed. Stanley Allen Renshon and J. H. Duckitt (New York: New York University Press, 2000), pp. 159–81.

12. K. Keniston, "The Sources of Student Dissent," *Journal of Social Issues* 23 (1967): 108–37.

13. R. Flacks, "The Liberated Generation: An Exploration of the Roots of Student Protest," *Journal of Social Issues* 23, no. 3 (1967): 52–75.

14. Daniel Valella, "Award-Winning 'White Ribbon' Does Not Dress up the Beginnings of Modern Warfare," *Columbia Spectator*, February 1, 2010.

15. Michael Omasta and Michael Pekler, "In Jedem Meiner Filme Muss Ich Laut Lachen." Falter.at, http://www.falter.at/falter/kategorie/autoren/michael-pekler/ (accessed September 16, 2009).

16. Neil Jeffrey Kressel, *Mass Hate: The Global Rise of Genocide and Terror*, updated ed. (Boulder, CO: Westview Press, 2002).

17. Ervin Staub, "The Psychology of Bystanders, Perpetrators, and Heroic Helpers," *Understanding Genocide: The Social Psychology of the Holocaust*, ed. Leonard S. Newman and Ralph Erber (Oxford: Oxford University Press, 2002), pp. 11–42; A. Miller, *For Your Own Good: Hidden Cruelty in Child-Rearing and the Roots of Violence* (Farrar, Straus and Giroux, 2002); E. D. Devereux, "Authority and Moral Development among German and American Children: A Cross-National Pilot Experiment," *Journal of Comparative Family Studies* 3, no. 1 (1972): 99–124.

18. David M. Levy, "Anti-Nazis: Criteria of Differentiation," in *Personality and Political Crisis: New Perspectives from Social Science and Psychiatry for the Study of War and Politics*, ed. Alfred H. Stanton (Glencoe, IL: Free Press, 1951), pp. 151–227.

19. R. E. Money-Kyrle, *Psychoanalysis and Politics: A Contribution to the Psychology of Politics and Morals* (New York: W. W. Norton, 1951), p. 13.

20. Mark Leibovich, "Being Glenn Beck," *New York Times Magazine*, September 29, 2010.

21. Raphael S. Ezekiel, *The Racist Mind: Portraits of Neo-Nazis and Klansmen* (New York: Viking, 1995).

22. D. S. Kalmuss and M. A. Straus, "Wife's Marital Dependency and Wife Abuse," *Journal of Marriage and Family* 44, no. 2 (1982): 277–86.

23. Hetherington and Weiler, *Authoritarianism and Polarization in American Politics*.

24. Lakoff, *Moral Politics: How Liberals and Conservatives Think*.

25. Steven Pinker, "Block That Metaphor!" *New Republic*, October 2, 2006.

26. Christopher Lasch, "Family and Authority," in *Capitalism and Infancy: Essays on Psychoanalysis and Politics*, ed. Barry Richards (London: Free Association Books, 1984), pp. 22–37.

27. Daniel Rancour-Laferriere, *Signs of the Flesh: An Essay on the Evolution of Hominid Sexuality* (Berlin: Mouton, 1985).

28. Harold Dwight Lasswell, *Psychopathology and Politics* (New York: Viking Press, 1960).

29. John Carl Flügel, *Man, Morals and Society: A Psycho-Analytical Study* (New York: International Universities Press, 1945).

30. Gabriella Klein, "Language Policy during the Fascist Period: The Case of Language Education," in *Language, Power, and Ideology: Studies in Political Discourse*, ed. Ruth Wodak (Amsterdam/Philadelphia: John Benjamins Publishing Company, 1989), pp. 39–55.

31. Frank J. Sulloway, *Born to Rebel: Birth Order, Family Dynamics, and Creative Lives* (New York: Pantheon Books, 1996).

32. Flügel, *Man, Morals, and Society: A Psycho-Analytical Study*.

33. Sigmund Freud, *The Ego and the Id*, ed. James Strachey and trans. Joan Riviere (New York: W. W. Norton, 1960).

34. Alfred Adler, *The Individual Psychology of Alfred Adler: A Systematic Presentation in Selections from His Writings*, 1st ed. (New York: Basic Books, 1967).

35. Everette Hall and Ben Barger, "Attitudinal Structures of Older and Younger Siblings," *Journal of Individual Psychology* 20 (1964): 59–68.

36. Allison Aubrey, "How Much Does Birth Order Shape Our Lives?" National Public Radio, November 23, 2010.

37. "First-Borns Have Higher IQ Scores." BBC News, June 22, 2007.

38. P. J. Weber, "The Birth Order Oddity in Supreme Court Appointments," *Presidential Studies Quarterly* 14, no. 4 (1984): 561–68.

39. Aubrey, "How Much Does Birth Order Shape Our Lives?"; Frank J. Sulloway, "Why Siblings Are Like Darwin's Finches: Birth Order, Sibling Competition, and Adaptive Divergence within the Family," in *The Evolution of Personality and Individual Differences* (New York: Oxford University Press, 2010), pp. 86–119; Albert Somit, Alan Arwine, and Steven A. Peterson, *Birth Order and Political Behavior* (New York: University Press of America, Inc., 1996).

40. Peterson, *Birth Order and Political Behavior*.

41. R. Bagudu, *Judging Annan* (AuthorHouse, 2007).

42. Somit, Arwine, and Peterson, *Birth Order and Political Behavior*; I. Harris, "Who Would Kill a President: Little Brother," *Psychology Today* (1976): 48–53.

43. Sulloway, "Why Siblings Are Like Darwin's Finches: Birth Order, Sibling Competition, and Adaptive Divergence within the Family"; F. J. Sulloway and R. L. Zweigenhaft, "Birth Order and Risk Taking in Athletics: A Meta-Analysis and Study of Major League Baseball," *Personality and Social Psychology Review* 14, no. 4 (2010): 402.

44. Sulloway, *Born to Rebel: Birth Order, Family Dynamics, and Creative Lives*.

45. Ibid.; Sulloway, "Why Siblings Are Like Darwin's Finches: Birth Order, Sibling Competition, and Adaptive Divergence within the Family"; D. L. Paulhus, P. D. Trapnell, and D. Chen, "Birth Order Effects on Personality and Achievement within Families," *Psychological Science* 10, no. 6 (1999): 482.

46. Aubrey, "How Much Does Birth Order Shape Our Lives?"

47. G. J. Manaster, C. Rhodes, M. B. Marcus, and J. C. Chan, "The Role of Birth Order in the Acculturation of Japanese Americans," *Psychologia* 41 (1998): 155–70.

48. Sulloway, *Born to Rebel: Birth Order, Family Dynamics, and Creative Lives*.

49. Somit, Arwine, and Peterson, *Birth Order and Political Behavior*.

50. Alix Spiegel, "Siblings Share Genes, but Rarely Personalities," National Public Radio, November 22, 2010.

51. Sulloway, *Born to Rebel: Birth Order, Family Dynamics, and Creative Lives*.

52. Weber, "The Birth Order Oddity in Supreme Court Appointments"; Somit, Arwine, and Peterson, *Birth Order and Political Behavior*.

53. Somit, Arwine, and Peterson, *Birth Order and Political Behavior*.

54. R. L. Zweigenhaft, "Birth Order Effects and Rebelliousness: Political Activism and Involvement with Marijuana," *Political Psychology* 23, no. 2 (2002): 219–33.

55. R. L. Zweigenhaft and J. Von Ammon, "Birth Order and Civil Disobedience: A Test of Sulloway's 'Born to Rebel' Hypothesis," *Journal of Social Psychology* 140, no. 5 (2000): 624–27.

56. Sulloway, "Why Siblings Are Like Darwin's Finches: Birth Order, Sibling Competition, and Adaptive Divergence within the Family"; Manaster et al., "The Role of Birth Order in the Acculturation of Japanese Americans."

57. Sulloway, "Why Siblings Are Like Darwin's Finches: Birth Order, Sibling Competition, and Adaptive Divergence within the Family."

58. Ibid.; F. J. Sulloway, "Birth Order, Sibling Competition, and Human Behavior," in *Conceptual Challenges in Evolutionary Psychology, Innovative Research Strategies: Studies in Cognitive Systems*, vol. 27, ed. J. H. Fetzer (series) and H. R. Holcomb III (volume) (Dordrecht: Kluwer, 2001): 39–83.

59. Jim Wyss, "At Ecuador's Top, It's Brother vs. Brother," *Miami Herald*, December 11, 2010.

PART IV. THE BIOLOGY OF FAMILY CONFLICT

Chapter 16. Why Sibling Conflict Occurs and Polarizes Political Personalities

a. Trivers's genetic explanation of parent-offspring conflict coincides with an observation that Freud had made about sibling rivalry fifty years earlier. In 1922, Freud wrote: "The elder child would certainly like to put his successor jealously aside, to keep it away from the parents, and to rob it of all its privileges"; however, "this younger child (like all that come later) is loved by the parents as much as he himself is." (See Sigmund Freud, "Group Psychology and the Analysis of the Ego," in *The Standard Edition of the Complete Psychological Works of Sigmund Freud*, vol. 18 [London: Institute of Psychoanalysis/Hogarth Press, 1922], pp. 69–143.)

b. The abolishment of primogeniture likely had egalitarian appeal to many early converts to Islam in areas of the Middle East where the custom of primogeniture disfavored later-borns. Under Islamic inheritance rules, property is divided between male heirs (with half portions for female heirs). Over time, however, it becomes difficult to maintain large concentrations of wealth within a family. Coupled with the resource scarcity of much of the pre-oil-age Islamic world, partible inheritance increased the incentive for consanguineous unions. Marrying a first cousin prevents wealth from escaping from families.

1. Robert L. Trivers, "Parental Investment and Sexual Selection," in *Sexual Selection and the Descent of Man 1871–1971: The Darwinian Pivot*, ed. Bernard Campbell (Chicago: Aldine Publishing Company, 1972), pp. 136–207.

2. R. L. Trivers, "Parent-Offspring Conflict," *Integrative and Comparative Biology* 14, no. 1 (1974): 249.

3. Ibid.

4. Hillard Haplan, Michael Gurven, and Jeffrey Winking, "An Evolutionary Theory of Human Life Span: Embodied Capital and the Human Adaptive Complex," in *Handbook of Theories of Aging*, ed. V. L. Bengtson, M. Silverstein, and N. Putney (New York: Springer Publishing Company, 2009), pp. 39–60.

5. Frank J. Sulloway, *Born to Rebel: Birth Order, Family Dynamics, and Creative Lives* (New York: Pantheon Books, 1996); Frank J. Sulloway, "Why Siblings Are Like Darwin's Finches: Birth Order, Sibling Competition, and Adaptive Divergence within the Family," in *The Evolution of Personality and Individual Differences* (New York: Oxford University Press, 2010), pp. 86–119.

6. C. R. Badcock, *Psychodarwinism: The New Synthesis of Darwin & Freud* (London: Harper Collins, 1994).

7. N. G. Blurton Jones and E. da Costa, "A Suggested Adaptive Value of Toddler Night Waking: Delaying the Birth of the Next Sibling," *Ethology and Sociobiology* 8, no. 2 (1987): 135–42.

8. G. A. Schuiling, "Honor Your Father and Your Mother," *Journal of Psychosomatic Obstetrics & Gynecology* 22, no. 4 (2001): 215–19.

9. Sulloway, "Why Siblings Are Like Darwin's Finches: Birth Order, Sibling Competition, and Adaptive Divergence within the Family."

10. Ibid.

11. F. Trillmich and J. B. W. Wolf, "Parent-Offspring and Sibling Conflict in Galápagos Fur Seals and Sea Lions," *Behavioral Ecology and Sociobiology* 62, no. 3 (2008): 363–75.

12. Kim Hill and Hillard Kaplan, "Life History Traits in Humans: Theory and Empirical Studies," *Annual Review of Anthropology* 28 (1999): 397–430.

13. Daniel Chirot and Clark McCauley, *Why Not Kill Them All? The Logic and Prevention of Mass Political Murder* (Princeton, NJ: Princeton University Press, 2006); Harold Nicolson, *Kings, Courts, and Monarchy* (New York: Simon and Schuster, 1962); John V. A Fine Jr., *The Late Medieval Balkans: A Critical Survey from the Late Twelfth Century to the Ottoman Conquest* (Ann Arbor: University of Michigan Press, 1994); Francis Fukuyama, *The Origins of Political Order: From Prehuman Times to the French Revolution* (London: Profile Books, 2011).

14. Hill and Kaplan, "Life History Traits in Humans: Theory and Empirical Studies."

15. H. Kaplan, "A Theory of Fertility and Parental Investment in Traditional and Modern Human Societies," *American Journal of Physical Anthropology* 101, no. S23 (1996): 91–135.

16. Badcock, *Psychodarwinism: The New Synthesis of Darwin & Freud.*

17. Trivers, "Parent-Offspring Conflict."

18. Frans B. M. de Waal, "Aggression as a Well-Integrated Part of Primate Social Relationships: A Critique of the Seville Statement on Violence," in *Aggression and Peacefulness in Humans and Other Primates*, ed. James Silverberg and J. Patrick Gray (New York: Oxford University Press, 1992), pp. 37–56.

19. Badcock, *Psychodarwinism: The New Synthesis of Darwin & Freud.*

20. Blurton Jones and da Costa, "A Suggested Adaptive Value of Toddler Night Waking: Delaying the Birth of the Next Sibling."

21. Alan Dundes, "The Hero Pattern and the Life of Jesus," in *Protocol of the 25th Colloquy of the Center for Hermeneutical Studies in Hellenistic and Modern Culture* (Berkeley, CA: Graduate Theological Union/University of California, 1977).

22. Frans B. M. de Waal, *Chimpanzee Politics: Power and Sex among Apes* (Baltimore: Johns Hopkins University Press, 2007).

23. Desmond Morris, *Manwatching: A Field Guide to Human Behaviour* (New York: H. N. Abrams, 1977).

24. Sulloway, *Born to Rebel: Birth Order, Family Dynamics, and Creative Lives.*

25. Sulloway, "Why Siblings Are Like Darwin's Finches: Birth Order, Sibling Competition, and Adaptive Divergence within the Family"; Alix Spiegel, "Siblings Share Genes, but Rarely Personalities," National Public Radio, November 22, 2010.

26. Sulloway, "Why Siblings Are Like Darwin's Finches: Birth Order, Sibling Competition, and Adaptive Divergence within the Family."

27. Everette Hall and Ben Barger, "Attitudinal Structures of Older and Younger Siblings," *Journal of Individual Psychology* 20 (1964): 59–68.

28. J. T. Jost, "The End of the End of Ideology," *American Psychologist* 61, no. 7 (2006): 651.

29. R. Hertwig, J. N. Davis, and F. J. Sulloway, "Parental Investment: How an Equity Motive Can Produce Inequality," *Psychological Bulletin* 128, no. 5 (2002): 728.

30. Sulloway, "Why Siblings Are Like Darwin's Finches: Birth Order, Sibling Competition, and Adaptive Divergence within the Family."

PART V. ARE PEOPLE BY NATURE COOPERATIVE OR COMPETITIVE?

Chapter 17. Sages through the Ages

a. By identifying a source of evil that is outside of the individual, these good/blank-slate philosophers are similar to people who attribute the world's inequalities to outer, structural injustices (i.e., believers in an unjust world). Believers in a *just* world, in contrast, attribute inequalities (such as poverty) to the character and *internal* flaws of individuals. Those who have a high BJW are more likely to believe in a selfish human nature (as do the political philosophers reviewed later in this chapter).

b. The following line of the US Constitution (1776) reflects Locke's philosophy:

"We hold these truths to be self-evident, that all men are created equal, that they are endowed by their Creator with certain unalienable Rights, that among these are Life, Liberty and the pursuit of Happiness."

c. Hobbes published *Leviathan*, his influential work on the social contract and governance, in 1651. On the frontispiece of the book is an image of a mighty ruler. The Latin writing above the image reads, "There is no power on earth to be compared to him." This phrase refers to the biblical passage in the Book of Job that describes the Leviathan, which was a terrifying sea monster:

> His sneezings flash forth light, and his eyes are like the eyelids of the morning. Out of his mouth go burning torches, and sparks of fire leap forth. Out of his nostrils goeth smoke, as out of a seething pot and burning rushes. His breath kindleth coals, and a flame goeth out of his mouth. . . . When he raiseth himself up, the mighty are afraid; by reason of despair they are beside themselves. If one lay at him with the sword, it will not hold; nor the spear, the dart, nor the pointed shaft. He esteemeth iron as straw, and brass as rotten wood. The arrow cannot make him flee; slingstones are turned with him into stubble. Clubs are accounted as stubble; he laugheth at the rattling of the javelin. . . . Upon earth there is not his like, who is made to be fearless (Job 41:10–25)

Like the biblical Leviathan, Hobbes's state is a beast—a sovereign that no lesser powers can hope to defeat. The left side of the book's frontispiece displays symbols of military and political power (a castle, a crown, a cannon, weapons, and a battlefield); the right side shows symbols of religious power (a church, a mitre, excommunication, a logic diagram, and a religious court). Thus, the absolutist power of the Hobbesian state is a giant that integrates earthly and celestial authority. (NB: In modern Hebrew, the word "leviathan" לִוְיָתָן has come to mean "whale.")

1. Robert H. Blank, *Biology and Political Science*, ed. Samuel M. Hines (London: Routledge, 2001).

2. Aristotle, *Politics*, ed. H. Rackham (Cambridge, MA: Harvard University Press, 1944).

3. Roger D. Masters, *The Nature of Politics* (New Haven, CT: Yale University Press, 1989).

4. Kwong Loi Shun, "Mencius," in *The Stanford Encyclopedia of Philosophy*, ed. Edward N. Zalta, http://plato.stanford.edu/archives/win2010/entries/mencius/ (accessed April 8, 2013).

5. Masters, *The Nature of Politics*; Peter Singer, *A Darwinian Left: Politics, Evolution, and Cooperation* (London: Weidenfeld & Nicolson, 1999).

6. Alex Tuckness, "Locke's Political Philosophy," in *The Stanford Encyclopedia of Philosophy*, ed. Edward N. Zalta, http://plato.stanford.edu/archives/fall2011/entries/locke-political/ (accessed April 8, 2013; William Uzgalis, "John Locke," in *The Stanford Encyclopedia of Philosophy*, ed. Edward N. Zalta, http://plato.stanford.edu/archives/win2010/entries/locke/ (accessed April 8, 2013).

7. Christopher Bertram, "Jean Jacques Rousseau," in *The Stanford Encyclopedia of Philosophy*, ed. Edward N. Zalta, http://plato.stanford.edu/archives/spr2011/entries/rousseau/ (accessed April 8, 2013).

8. Paul H. Rubin, *Darwinian Politics: The Evolutionary Origin of Freedom* (New Brunswick, NJ: Rutgers University Press, 2002).

9. Singer, *A Darwinian Left: Politics, Evolution, and Cooperation.*

10. Masters, *The Nature of Politics.*

11. Rachel Barney, "Callicles and Thrasymachus," in *The Stanford Encyclopedia of Philosophy*, ed. Edward N. Zalta, http://plato.stanford.edu/archives/spr2011/entries/callicles-thrasymachus/ (accessed April 8, 2013.

12. Chanakya, *Sri Chanakya Niti-Sastra: The Political Ethics of Chanakya Pandit*, compiled by Miles Davis. Lucknow, India, http://www.philosophy.ru/library/asiatica/indica/authors/kautilya/canakya_niti_sastra.html (accessed April 8, 2013).

13. Shun, "Mencius."

14. B. Watson, Di Mo, Xunzi, F. Han, and Columbia University Department of History, *Basic Writings of Mo Tzu, Hsun Tzu, and Han Fei Tzu* (New York: Columbia University, 1967).

15. Dan Robins, "Xunzi," in *The Stanford Encyclopedia of Philosophy*, ed. Edward N. Zalta, http://plato.stanford.edu/archives/fall2008/entries/xunzi/ (accessed April 8, 2013).

16. Thomas Hobbes, *Leviathan*. ed. C. B. Macpherson (Baltimore: Penguin books, 1968).

17. James Madison, *The Federalist*, ed. Jacob Ernest Cooke, 1st ed. (Middletown, CN: Wesleyan University Press, 1961).

18. Barry Goldwater, *The Conscience of a Conservative* (New York: Hillman Books, 1960).

19. David O. Sears and Carolyn L. Funk, "The Role of Self-Interest in Social and Political Attitudes," *Advances in Experimental Social Psychology* 24, no. 1 (1991): 1–91.

20. R. Flacks, "The Liberated Generation: An Exploration of the Roots of Student Protest," *Journal of Social Issues* 23, no. 3 (1967): 52–75.

21. Daniel Dombey and Tobias Buck, "Obama Urges Israel to Open Gaza Borders," *Financial Times*, January 22, 2009.

22. Roger Hardy, "Obama's New Bid to Engage the Muslim World," BBC News Middle East, November 8, 2010.

23. Rubin, *Darwinian Politics: The Evolutionary Origin of Freedom.*

24. J. C. Flugel, *Man, Morals, and Society: A Psycho-Analytical Study* (New York: International Universities Press, 1945); Andrew Samuels, *The Political Psyche* (London: Routledge, 1993); George Lakoff, *Moral Politics: How Liberals and Conservatives Think*, 2nd ed. (Chicago, IL: University of Chicago Press, 2002).

25. Peter Railton, "Moral Camouflage or Moral Monkeys?" *New York Times*, July 18, 2010.

26. Singer, *A Darwinian Left: Politics, Evolution, and Cooperation.*

27. Else Frenkel-Brunswik, "Sex, People, and Self as Seen through the Interviews,"

in *The Authoritarian Personality*, ed. T. W. Adorno, Else Frenkel-Brunswik, Daniel J. Levinson, and R. Nevitt Sanford (New York: Harper & Row, 1950), pp. 390–441.

28. Lakoff, *Moral Politics: How Liberals and Conservatives Think*; R. E. Money-Kyrle, *Psychoanalysis and Politics: A Contribution to the Psychology of Politics and Morals* (New York: W. W. Norton, 1951); Fred I. Greenstein, *Personality and Politics: Problems of Evidence, Inference, and Conceptualization* (New York: W. W. Norton, 1975).

29. Neil Jeffrey Kressel, *Mass Hate: The Global Rise of Genocide and Terror*, updated ed. (Boulder, CO: Westview Press, 2002).

30. Alan Bullock, *Hitler, a Study in Tyranny*, completely rev. ed. (New York: Harper & Row, 1962).

31. Raphael S. Ezekiel, *The Racist Mind: Portraits of Neo-Nazis and Klansmen* (New York: Viking, 1995).

32. Mark Leibovich, "Being Glenn Beck," *New York Times Magazine*, September 29, 2010.

33. Jacob M. Vigil, "Facial Expression Processing Varies with Political Affiliation," available at Nature Precedings, http://precedings.nature.com/documents/2414/version/1 (accessed April 8, 2013); Jacob M. Vigil, "Political Leanings Vary with Facial Expression Processing and Psychosocial Functioning," *Group Processes & Intergroup Relations* 13, no. 5 (2010): 547.

34. Karl Evan Giuseffi, "Processing Facial Emotions: An EEG Study of the Differences between Conservatives and Liberals and across Political Participation," University of Nebraska, 2012.

35. D. R. Oxley, K. B. Smith, J. R. Alford, M. V. Hibbing, J. L. Miller, M. Scalora, P. K. Hatemi, and J. R. Hibbing, "Political Attitudes Vary with Physiological Traits," *Science* 321, no. 5896 (2008): 1667–70.

36. K. Bulkeley, "Dream Content and Political Ideology," *Dreaming* 12, no. 2 (2002): 61–77; J. T. Jost, J. Glaser, A. W. Kruglanski, and F. J. Sulloway, "Political Conservatism as Motivated Social Cognition," *Psychological Bulletin* 129, no. 3 (2003): 339.

37. Richard P. Eibach and Lisa K. Libby, "Ideology of the Good Old Days: Exaggerated Perceptions of Moral Decline and Conservative Politics," in *Social and Psychological Bases of Ideology and System Justification*, ed. John T. Jost, Aaron C. Kay, and Hulda Thorisdottir (Oxford: Oxford University Press, 2009), pp. 402–23; P. M. Sniderman and H. Brady, "Multi-Investigator Study, Cumulative Datafile and Codebook 1998–1999 (Computer File). Berkeley, CA: Survey Research Center," University of California, Berkeley (distributor), retrieved August 2001; G. LaFree, "Declining Violent Crime Rates in the 1990s: Predicting Crime Booms and Busts," *Annual Review of Sociology* (1999).

38. "Gingrich Calls Crimes 'Artifacts of Bad Policy,'" *Boston Globe*, May 20, 1995, p. 5.

39. Newt Gingrich, *To Save America: Stopping Obama's Secular-Socialist Machine* (Washington, DC: Regnery Press, 2011).

40. Leibovich, "Being Glenn Beck."

41. Ezekiel, *The Racist Mind: Portraits of Neo-Nazis and Klansmen.*

Chapter 18. Do Perceptions of Human Nature Change as We Age?

a. This quotation comes from the 1903 play *Man and Superman*. Actually, the line is from a political treatise "written" by the play's fictitious protagonist, John Tanner (in reality, of course, Shaw himself authored and attached the work to the play). In this fifty-eight-page appendix, titled "The Revolutionist's Handbook and Pocket Companion," Shaw combines his socialist political leanings with his support for eugenics.

Here, Shaw argues that the way to produce the genetic Superman is through good breeding. And good breeding can occur when motivated by natural *attraction*. To find their suitable objects, according to Shaw's manifesto, human desires must be free to transcend traditional ethnic, social, and economic barriers. "The Revolutionist's Handbook" asserts: "Equality is essential to good breeding; and equality, as all economists know, is incompatible with property." Thus, Shaw supported the far left and its explicit agenda to eradicate private property in the service of equality—even until he was an old man. (See George Bernard Shaw, "The Revolutionist's Handbook and Pocket Companion," in *Man and Superman* (S.1.), the Limited Editions Club, 1962 [1903].)

b. The original quotation, in fact, was uttered by Jack Weinberg, a leader of the free-speech movement at Berkeley. His immediate purpose was to refute a reporter's insinuation that the students were being coopted by older, Communist infiltrators.

c. Just before Marisol Valles fled Mexico, she gave birth to her first child. The dangers she paid attention to at the time were extreme—threats from the drug cartels on her life and that of her newborn son.

1. Luis Chaparro, "Mejor Estar Viva En Eu Que Muerta En México: Marisol Valles," *El Diario de El Paso*, May 11, 2011; Robin Yapp, "Mexican Student Police Chief Will Live in Fear for Rest of Life," *Telegraph*, June 3, 2011; Arthur Brice, "20-Year-Old Woman Becomes Top Cop in Violent Mexican Municipality," CNN, October 20, 2010; Nicholas Casey, "Mexican Cop Flees to U.S.: Young Woman Hailed for Leading Town's Force Amid Drug Violence Seeks Asylum," *Wall Street Journal*, March 9, 2011.

2. Aristotle, *Rhetoric*, ed. W. Rhys Roberts, Ingram Bywater, and Friedrich Solmsen, 1st Modern Library ed. (New York: Modern Library, 1954).

3. Fred R. Shapiro and Joseph Epstein, ed. *The Yale Book of Quotations* (New Haven, CT: Yale University Press, 2006).

4. E. E. Sampson, "Student Activism and the Decade of Protest," *Journal of Social Issues* 23, no. 3 (1967): 1–33.

5. R. R. McCrae, P. T. Costa Jr., M. P. Lirna, A. Simoes, F. Ostendorf, A. Angleitner, I. Marusic, et al. "Age Differences in Personality across the Adult Life Span: Parallels in Five Cultures," *Developmental Psychology* 35, no. 2 (1999): 466–77.

6. J. Allik and R. R. McCrae, "Toward a Geography of Personality Traits: Patterns of Profiles across 36 Cultures," *Journal of Cross Cultural Psychology* 35, no. 1 (2004): 13–28.

7. Patricia Cohen, "Professor Is a Label That Leans to the Left," *New York Times*, January 17, 2010.

8. Robert Altemeyer, *Enemies of Freedom: Understanding Right-Wing Authoritarianism*, 1st ed. (San Francisco: Jossey-Bass Publishers, 1988).

9. Richard P. Eibach and Lisa K. Libby, "Ideology of the Good Old Days: Exaggerated Perceptions of Moral Decline and Conservative Politics," in *Social and Psychological Bases of Ideology and System Justification*, ed. John T. Jost, Aaron C. Kay, and Hulda Thorisdottir (Oxford: Oxford University Press, 2009), pp. 402–23.

10. R. P. Eibach, L. K. Libby, and T. D. Gilovich, "When Change in the Self Is Mistaken for Change in the World," *Journal of Personality and Social Psychology* 84, no. 5 (2003): 917–31.

11. McCrae et al., "Age Differences in Personality across the Adult Life Span: Parallels in Five Cultures."

PART VI. ILLUMINATING OUR TRUE HUMAN NATURE

1. Paul H. Rubin, *Darwinian Politics: The Evolutionary Origin of Freedom* (New Brunswick, NJ: Rutgers University Press, 2002).

2. Glendon Schubert, "Politics as a Life Science," in *Biology and Politics: Recent Explorations*, ed. Albert Somit (The Hague: Mouton, 1976), pp. 155–96.

3. Stephen K. Sanderson, *The Evolution of Human Sociality: A Darwinian Conflict Perspective* (Lanham, MD: Rowman & Littlefield, 2001).

Chapter 19. The Conservative Altruism: Kin Selection

1. R. L. Trivers, "Parent-Offspring Conflict," *Integrative and Comparative Biology* 14, no. 1 (1974): 249; Edward O. Wilson, *Sociobiology: The New Synthesis* (Cambridge, MA: Belknap Press of Harvard University Press, 1975); M. A. Nowak and K. Sigmund, "Evolution of Indirect Reciprocity," *Nature* 437, no. 7063 (2005): 1291–98.

2. J. P. Rushton, D. W. Fulker, M. C. Neale, D. K. B. Nias, and H. J. Eysenck, "Altruism and Aggression: The Heritability of Individual Differences," *Journal of Personality and Social Psychology* 50, no. 6 (1986): 1192.

3. P. W. Sherman, "Nepotism and the Evolution of Alarm Calls," *Science* 197, no. 4310 (1977): 1246–53; Susan Milius, "The Science of Eeeeek: What a Squeak Can Tell Researchers about Life, Society, and All That," *Science News*, September 12, 1998.

4. S. A. West, I. Pen, and A. S. Griffin, "Cooperation and Competition between Relatives," *Science* 296, no. 5565 (2002): 72; Timothy Shanahan, *The Evolution of Darwinism: Selection, Adaptation, and Progress in Evolutionary Biology* (Cambridge: Cambridge University Press, 2004).

5. W. D. Hamilton, "The Genetical Evolution of Social Behavior, Parts 1 and 2," *Journal of Theoretical Biology* 7, no. 1 (1964): 1–52.

6. Nowak and Sigmund, "Evolution of Indirect Reciprocity."

7. Trivers, "Parent-Offspring Conflict."

8. Iver Mysterud, Thomas Drevon, and Tore Slagsvold, "An Evolutionary Interpretation of Gift-Giving Behavior in Modern Norwegian Society," *Evolutionary Psychology* 4 (2006): 406–25.

9. Martin S. Smith, Bradley J. Kish, and Charles B. Crawford, "Inheritance of Wealth as Human Kin Investment," *Ethology and Sociobiology* 8, no. 3 (1987): 171–82.

10. D. S. Judge and S. B. Hrdy, "Allocation of Accumulated Resources among Close Kin: Inheritance in Sacramento, California, 1890–1984," *Ethology and Sociobiology* 13 (1992): 495–522.

11. G. D. Webster, "Human Kin Investment as a Function of Genetic Relatedness and Lineage," *Evolutionary Psychology* 2 (2004): 129–41.

12. E. A. Madsen, R. J. Tunney, G. Fieldman, H. C. Plotkin, R. I. M. Dunbar, J. M. Richardson, and D. McFarland, "Kinship and Altruism: A Cross-Cultural Experimental Study," *British Journal of Psychology* 98, no. 2 (2007): 339–59.

13. L. M. DeBruine, "Facial Resemblance Enhances Trust," *Proceedings of the Royal Society of London. Series B: Biological Sciences* 269, no. 1498 (2002): 1307.

14. J. N. Bailenson, S. Iyengar, N. Yee, and N. A. Collins, "Facial Similarity between Voters and Candidates Causes Influence," *Public Opinion Quarterly* 72, no. 5 (2009): 935–61; Adam Gorlick, "Researchers Say Voters Swayed by Candidates Who Share Their Looks," *Stanford Report*, October 22, 2008.

15. G. K. Leak and S. B. Christopher, "Freudian Psychoanalysis and Sociobiology: A Synthesis," *American Psychologist* 37, no. 3 (1982): 313–22.

16. Frans B. M. de Waal, "For Goodness' Sake," *New York Times*, July 9, 2010; C. K. W. De Dreu, L. L. Greer, M. J. J. Handgraaf, S. Shalvi, G. A. Van Kleef, M. Baas, F. S. Ten Velden, E. Van Dijk, and S. W. W. Feith, "The Neuropeptide Oxytocin Regulates Parochial Altruism in Intergroup Conflict among Humans," *Science* 328, no. 5984 (2010): 1408.

17. Ibid.; Nicholas Wade, "Depth of the Kindness Hormone Appears to Know Some Bounds," *New York Times*, January 10, 2011.

18. De Dreu et al., "The Neuropeptide Oxytocin Regulates Parochial Altruism in Intergroup Conflict among Humans."

19. J. L. Eberhardt, "Imaging Race," *American Psychologist* 60, no. 2 (2005): 181–90; A. J. Golby, J. D. E Gabrieli, J. Y. Chiao, and J. L. Eberhardt, "Differential Responses in the Fusiform Region to Same-Race and Other-Race Faces," *Nature Neuroscience* 4, no. 8 (2001): 845–50.

20. R. Kanai, T. Feilden, C. Firth, and G. Rees, "Political Orientations Are Correlated with Brain Structure in Young Adults," *Current Biology*, accepted manuscript, 2011.

21. W. A. Cunningham, M. K. Johnson, C. L. Raye, J. C. Gatenby, J. C. Gore, and M. R. Banaji, "Separable Neural Components in the Processing of Black and White Faces," *Psychological Science* 15, no. 12 (2004): 806; J. A. Richeson, A. A. Baird, H. L. Gordon, T. F. Heatherton, C. L. Wyland, S. Trawalter, and J. N. Shelton, "An fMRI Investigation of the Impact of Interracial Contact on Executive Function," *Nature Neuroscience* 6, no. 12 (2003): 1323–28.

22. Kanai et al., "Political Orientations Are Correlated with Brain Structure in Young Adults."

23. D. R. Oxley, K. B. Smith, J. R. Alford, M. V. Hibbing, J. L. Miller, M. Scalora,

P. K. Hatemi, and J. R. Hibbing, "Political Attitudes Vary with Physiological Traits," *Science* 321, no. 5896 (2008): 1667–70.

24. A. Knafo, S. Israel, A. Darvasi, R. Bachner Melman, F. Uzefovsky, L. Cohen, E. Feldman, et al. "Individual Differences in Allocation of Funds in the Dictator Game Associated with Length of the Arginine Vasopressin 1a Receptor RS3 Promoter Region and Correlation between RS3 Length and Hippocampal mRNA," *Genes, Brain and Behavior* 7, no. 3 (2008): 266–75.

Chapter 20. The Liberal Altruism: Reciprocity

1. "Zebra Snapped Putting Head in Hippopotamus's Mouth," BBC News, March 12, 2010.

2. "Crocodile Crazy: The Man Who Enjoys Giving His Dangerous 'Companion' Kisses and Cuddles," *Daily Mail Online*, http://www.dailymail.co.uk/news/article-1206872/Crocodile-crazy-The-man-enjoys-giving-dangerous-companion-cuddle.html (accessed August 17, 2009); "Fisherman's 'Inseparable Relationship' with Crocodile," BBC News Latin America and Caribbean, September 28, 2010.

3. R. L. Trivers, "The Evolution of Reciprocal Altruism," *Quarterly Review of Biology* 46 (1971): 35–57.

4. Peter Railton, "Moral Camouflage or Moral Monkeys?" *New York Times*, July 18, 2010.

5. Trivers, "The Evolution of Reciprocal Altruism."

6. W. D. Hamilton, "Innate Social Aptitudes of Man: An Approach from Evolutionary Genetics," *Biosocial Anthropology* (1975): 133–55.

7. "Palestinians Get Saddam Funds," BBC News, March 13, 2003.

8. Itamar Marcus and Nan Jacques Zilberdik, "Martyrs Get 'a Full Life with Allah and 72 Wives,'" Palestinian Media Watch, February 7, 2010.

9. "Mother about Terrorist 'Martyr' Son: He Wanted 70 Wives," Palestinian Media Watch, August 2, 2011.

10. Marcus and Zilberdik, "Martyrs Get 'a Full Life with Allah and 72 Wives'"; Palestinian TV (Fatah), "Beautiful Virgins Await Muslim Martyrs in Paradise," Palestinian Media Watch, February 7, 2010.

11. M. A. Nowak and K. Sigmund, "Evolution of Indirect Reciprocity," *Nature* 437, no. 7063 (2005): 1291–98.

12. James Sidanius and Robert Kurzban, "Evolutionary Approaches to Political Psychology," in *Oxford Handbook of Political Psychology*, ed. David O. Sears, Leonie Huddy, and Robert Jervis (New York: Oxford University Press, 2003), pp. 146–81.

13. Robert M Axelrod, *The Evolution of Cooperation* (New York: Basic Books, 1984).

14. J. R. Alford and J. R. Hibbing, "The Origin of Politics: An Evolutionary Theory of Political Behavior," *Perspectives on Politics* 2, no. 04 (2004): 707–23.

15. Trivers, "The Evolution of Reciprocal Altruism."

16. Frans B. M. de Waal, "Aggression as a Well-Integrated Part of Primate Social

Relationships: A Critique of the Seville Statement on Violence," in *Aggression and Peacefulness in Humans and Other Primates*, ed. James Silverberg and J. Patrick Gray (New York: Oxford University Press, 1992), pp. 37–56.

17. F. Warneken and M. Tomasello, "Varieties of Altruism in Children and Chimpanzees," *Trends in Cognitive Sciences* 13, no. 9 (2009): 397–402.

18. S. F. Brosnan and F. B. M. de Waal, "Monkeys Reject Unequal Pay," *Nature* 425, no. 6955 (2003): 297–99.

19. Iver Mysterud, Thomas Drevon, and Tore Slagsvold, "An Evolutionary Interpretation of Gift-Giving Behavior in Modern Norwegian Society," *Evolutionary Psychology* 4 (2006): 406–25.

20. A. Zahavi, "Mate Selection—a Selection for a Handicap," *Journal of Theoretical Biology* 53, no. 1 (1975): 205–14; E. Fehr and U. Fischbacher, "The Nature of Human Altruism," *Nature* 425, no. 6960 (2003): 785–91; R. Kurzban, P. DeScioli, and E. O'Brien, "Audience Effects on Moralistic Punishment," *Evolution and Human Behavior* 28, no. 2 (2007): 75–84.

21. Nowak and Sigmund, "Evolution of Indirect Reciprocity."

22. De Waal, "Aggression as a Well-Integrated Part of Primate Social Relationships: A Critique of the Seville Statement on Violence."

23. M. Gurven, W. Allen-Arave, K. Hill, and M. Hurtado, "'It's a Wonderful Life': Signaling Generosity among the Ache of Paraguay," *Evolution and Human Behavior* 21, no. 4 (2000): 263–82.

24. Fehr and Fischbacher, "The Nature of Human Altruism."

25. M. Milinski, D. Semmann, and H. Krambeck, "Donors to Charity Gain in Both Indirect Reciprocity and Political Reputation," *Proceedings of the Royal Society of London. Series B: Biological Sciences* 269, no. 1494 (2002): 881.

26. Nowak and Sigmund, "Evolution of Indirect Reciprocity"; H. Gintis, S. Bowles, R. Boyd, and E. Fehr, "Explaining Altruistic Behavior in Humans," *Evolution and Human Behavior* 24, no. 3 (2003): 153–72.

27. R. Boyd, H. Gintis, S. Bowles, and P. J. Richerson, "The Evolution of Altruistic Punishment," *Proceedings of the National Academy of Sciences of the United States of America* 100, no. 6 (2003): 3531.

28. Fehr and Fischbacher, "The Nature of Human Altruism."

29. Kurzban, DeScioli, and O'Brien, "Audience Effects on Moralistic Punishment."

30. B. Gobin, J. Billen, and C. Peeters, "Policing Behaviour Towards Virgin Egg Layers in a Polygynous Ponerine Ant," *Animal Behaviour* 58, no. 5 (1999): 1117–22.

31. M. D. Hauser and P. Marler, "Food-Associated Calls in Rhesus Macaques (Macaca Mulatta): II. Costs and Benefits of Call Production and Suppression," *Behavioral Ecology* 4, no. 3 (1993): 206.

32. M. E. Price, "Punitive Sentiment among the Shuar and in Industrialized Societies: Cross-Cultural Similarities," *Evolution and Human Behavior* 26 (2005): 279–87.

33. M. Okuno-Fujiwara and A. Postlewaite, "Social Norms and Random Matching Games," *Games and Economic Behavior* 9, no. 1 (1995): 79–109.

34. Gintis et al., "Explaining Altruistic Behavior in Humans."

35. Ibid.; J. Henrich, R. McElreath, A. Barr, J. Ensminger, C. Barrett, A. Bolyanatz, J. C. Cardenas, et al. "Costly Punishment across Human Societies," *Science* 312, no. 5781 (2006): 1767.

36. Allen Johnson, "The Political Unconscious: Stories and Politics in Two South American Cultures," in *Political Psychology: Cultural and Cross-Cultural Foundations*, ed. Stanley Allen Renshon and J. H. Duckitt (New York: New York University Press, 2000), pp. 159–81.

37. J. Henrich, "Does Culture Matter in Economic Behavior? Ultimatum Game Bargaining among the Machiguenga of the Peruvian Amazon," *American Economic Review* 90, no. 4 (2000): 973–79.

38. Gintis et al., "Explaining Altruistic Behavior in Humans."

39. Henrich, "Does Culture Matter in Economic Behavior? Ultimatum Game Bargaining among the Machiguenga of the Peruvian Amazon."

40. L. A. Cameron, "Raising the Stakes in the Ultimatum Game: Experimental Evidence from Indonesia," *Economic Inquiry* 37, no. 1 (1999): 47–59.

41. Henrich et al., "Costly Punishment across Human Societies."

42. J. K. Rilling, D. A. Gutman, T. R. Zeh, G. Pagnoni, G. S. Berns, and C. D. Kilts, "A Neural Basis for Social Cooperation," *Neuron* 35, no. 2 (2002): 395–405.

43. D. J. F. De Quervain, U. Fischbacher, V. Treyer, M. Schellhammer, U. Schnyder, A. Buck, and E. Fehr, "The Neural Basis of Altruistic Punishment," *Science* 305, no. 5688 (2004): 1254.

Chapter 21. Altruism across the Lifespan: The Neurological Development of Cynicism

1. Aristotle, *Rhetoric*, ed. W. Rhys Roberts, Ingram Bywater, and Friedrich Solmsen, 1st Modern Library ed. (New York: Modern Library, 1954).

2. H. Kaplan, "A Theory of Fertility and Parental Investment in Traditional and Modern Human Societies," *American Journal of Physical Anthropology* 101, no. S23 (1996): 91–135.

3. Ibid.

4. Polly Wiessner, "Parent-Offspring Conflict in Marriage: Implications for Social Evolution and Material Culture among the Ju/'Hoansi Bushmen," in *Pattern and Process in Cultural Evolution*, ed. Stephen Shennan (Berkeley: University of California Press, 2009), pp. 251–63.

5. Lindon Eaves, Nicholas Martin, Andrew Heath, Richard Schieken, Joanne Meyer, Judy Silberg, Michael Neale, and Linda Corey, "Age Changes in the Causes of Individual Differences in Conservatism," *Behavior Genetics* 27, no. 2 (1997): 121–24.

6. J. N. Giedd, "The Teen Brain: Insights from Neuroimaging," *Journal of Adolescent Health* 42, no. 4 (2008): 335–43.

7. Sheena Iyengar, *The Art of Choosing*, 1st ed. (New York, NY: Twelve, 2010).

8. Giedd, "The Teen Brain: Insights from Neuroimaging"; Robin Marantz Henig, "What Is It about 20-Somethings?" *New York Times Magazine*, August 18, 2010.

9. E. R. McAnarney, "Adolescent Brain Development: Forging New Links?" *Journal of Adolescent Health* 42, no. 4 (2008): 321–23.

10. R. R. McCrae, P. T. Costa Jr., M. P. Lirna, A. Simoes, F. Ostendorf, A. Angleitner, I. Marusic, et al. "Age Differences in Personality across the Adult Life Span: Parallels in Five Cultures," *Developmental Psychology* 35, no. 2 (1999): 466–77.

11. Aristotle, *Rhetoric*.

Chapter 22. The Altruism That Isn't: Self-Deception among People and Politicians

a. Clearly, definitions of morality (and therefore of altruism) vary considerably between groups. Liberal church congregations would not consider Rekers's antigay activism as altruistic, as did his colleagues at NARTH.

b. These words were originally stated by the English Catholic writer Lord Acton in 1887 to express his opposition to the doctrine of papal infallibility.

c. "Jamahiriya" is a play on the Arabic word for republic, which translates roughly as "peopledom," "state of the masses," or "people's republic."

1. R. L. Trivers, "The Evolution of Reciprocal Altruism," *Quarterly Review of Biology* 46 (1971): 35–57.

2. Peter Railton, "Moral Camouflage or Moral Monkeys?" *New York Times*, July 18, 2010.

3. R. Trivers, "Sociobiology and Politics," in *Sociobiology and Human Politics*, ed. Elliott White (Lexington, MA: Lexington Books, 1981), pp. 1–43.

4. M. O. Slavin, "The Dual Meaning of Repression and the Adaptive Design of the Human Psyche," *Journal of the American Academy of Psychoanalysis and Dynamic Psychiatry* 18, no. 2 (1990): 307–41; Robert Trivers, *The Folly of Fools: The Logic of Deceit and Self-Deception in Human Life* (New York: Basic Books, 2012), pp. 40–48.

5. Robert L. Trivers, "Foreword," in Richard Dawkins, *The Selfish Gene* (New York: Oxford University Press, 1976).

6. R. M. Nesse, "The Evolutionary Functions of Repression and the Ego Defenses," *Journal of the American Academy of Psychoanalysis and Dynamic Psychiatry* 18, no. 2 (1990): 260–85.

7. R. M. Nesse and A. T. Lloyd, "The Evolution of Psychodynamic Mechanisms," in *The Adapted Mind: Evolutionary Psychology and the Generation of Culture*, ed. Leda Cosmides and John Tooby (New York: Oxford University Press, 1992), pp. 601–24.

8. C. R. Badcock, *Psychodarwinism: The New Synthesis of Darwin & Freud* (London: Harper Collins, 1994).

9. Nesse, "The Evolutionary Functions of Repression and the Ego Defenses."

10. Heinz Hartmann, *Ego Psychology and the Problem of Adaptation* (New York: International Universities Press, 1964).

11. Sigmund Freud, "Thoughts for the Times on War and Death: The Disillusionment

of the War," in *The Standard Edition of the Complete Psychological Works of Sigmund Freud*, vol. 14 (London: Institute of Psychoanalysis and Hogarth Press, 1918), pp. 273–88.

12. Penn Bullock and Brandon K Thorp, "Christian Right Leader George Rekers Takes Vacation with 'Rent Boy,'" *Miami New Times*, May 6, 2010.

13. Frank J. Sulloway, *Freud, Biologist of the Mind: Beyond the Psychoanalytic Legend*, paperback ed. (New York: Basic Books, 1983).

14. Jean Laplanche, *The Language of Psycho-Analysis* (New York: Norton, 1974).

15. Nina Bernstein, "Foes of Sex Trade Are Stung by the Fall of an Ally," *New York Times*, March 12, 2008.

16. Nesse, "The Evolutionary Functions of Repression and the Ego Defenses."

17. Nesse and Lloyd, "The Evolution of Psychodynamic Mechanisms."

18. Harold Dwight Lasswell, "Propaganda and Mass Insecurity," in *Personality and Political Crisis: New Perspectives from Social Science and Psychiatry for the Study of War and Politics*, ed. Alfred H. Stanton and Stewart E. Perry (Glencoe, IL: Free Press, 1951), pp. 15–43.

19. Chris Rodda, "Michele Bachmann Revises Her Own Family History to Sound More Iowan," *Huffington Post*, April 11, 2011.

20. Ezra Deutsch-Feldman, Michael Fitzgerald, Matthew McKnight, and Laura Stampler, "Slideshow: The Worst Excuses in Politics," *New Republic*, January 12, 2011.

21. J. Lammers, D. A. Stapel, and A. D. Galinsky, "Power Increases Hypocrisy," *Psychological Science* 21, no. 5 (2010): 737.

22. Railton, "Moral Camouflage or Moral Monkeys?"

23. Robert Klitgaard, *Addressing Corruption in Haiti*, (Claremont, CA: Claremont Graduate University, 2010).

24. A. Dundes, "Laughter behind the Iron Curtain: A Sample of Rumanian Political Jokes," *Ukrainian Quarterly* 27 (1971): 50–59.

25. Ervin Staub, "The Psychology of Bystanders, Perpetrators, and Heroic Helpers," in *Understanding Genocide: The Social Psychology of the Holocaust*, ed. Leonard S. Newman and Ralph Erber (Oxford: Oxford University Press, 2002), pp. 11–42.

26. Peter Singer, *A Darwinian Left: Politics, Evolution, and Cooperation* (London: Weidenfeld & Nicolson, 1999).

27. Nasser Karimi, "Hugo Chavez Receives Iran's Highest Honor," *Washington Post*, July 30, 2006.

28. Franklin Briceno, "Ex-Peru Rebels, out Prison, Run for Elected Office," *Miami Herald*, September 30, 2010.

29. Singer, *A Darwinian Left: Politics, Evolution, and Cooperation*.

30. James Sidanius and Felicia Pratto, "Social Dominance Theory: A New Synthesis," in *Political Psychology: Key Readings*, ed. John T. Jost and Jim Sidanius (New York: Psychology Press, 2004), pp. 315–32.

31. Alastair Leithead, "Communist Vietnam's $35 Bowl of Noodle Soup," BBC Radio 4, January 21, 2011.

32. Tony Halpin, "'Father of All Turkmen' Toppled under Orders of Successor," *Times* (UK), January 20, 2010.

33. "Read My Words, Go to Heaven, Leader Says," *Los Angeles Times*, March 21, 2006; Macy Halford, "Shadow of the Ruhnama," *New Yorker*, April 27, 2010; Rory Mulholland, "The Cult of the Turkmen Leader," BBC News Asia-Pacific, November 2, 2001.

34. Will Grant, "Venezuela Bans Unauthorised Use of Hugo Chavez's Image," BBC News Latin America and Caribbean, November 23, 2010.

35. Jim Wyss, "Hemisphere Loser: Venezuela," *Miami Herald*, October 27, 2010.

36. "The Rise and Rise of the Cognitive Elite: Brains Bring Ever Larger Rewards," *Economist*, January 20, 2011.

37. Yŏng-sŏn Chŏn, *Tasi Koch'yo SsŭN Pukhan ŬI Sahoe Wa Munhwa / ChŏN YŏNg-SŏN Chŏ*. 다시 고쳐 쓴 북한 의 사회 와 문화 / 전 영선 저 (Sŏul: Yŏngnak, 2006).

38. Jerrold M. Post, *Leaders and Their Followers in a Dangerous World: The Psychology of Political Behavior*, ed. Alexander George, 1st ed. (Ithaca, NY: Cornell University Press, 2004).

39. "North Korea Leader Lauded in Giant Hillside Slogan," BBC News Asia, November 23, 2012.

40. Peter Suedfeld and Mark Schaller, "Authoritarianism and the Holocaust: Some Cognitive and Affective Implications," in *Understanding Genocide: The Social Psychology of the Holocaust*, ed. Leonard S. Newman and Ralph Erber (Oxford: Oxford University Press, 2002), pp. 68–90.

41. Daniel Rancour-Laferriere, *Signs of the Flesh: An Essay on the Evolution of Hominid Sexuality* (Berlin: Mouton, 1985).

42. "Law on Sex Equality," *Korea News Service* (DPRK), July 30, 2011.

43. Post, *Leaders and Their Followers in a Dangerous World: The Psychology of Political Behavior*.

44. Malachi Martin, *The Decline and Fall of the Roman Church* (New York: Putnam, 1981); J. P. Migne, ed. *Patrologia Latina Database*, version 5.0b, Alexandria, VA; Hans Küng, *The Catholic Church: A Short History* (New York: Modern Library, 2001).

45. *Monumenta Germaniae Historica* (Hannoverae: impensis bibliopolii Hahniani, 1891).

46. Peter Damian, *Liber Gomorrhianus: Omosessualità Ecclesiastica E Riforma Della Chiesa*, ed. Edoardo D'Angelo (Alessandria: Edizioni dell'Orso, 2001)

47. "Rush Limbaugh Arrested on Drug Charges," CBSNews.com, http://www.cbsnews.com/2100-201_162-1561324.html (accessed March 5, 2009).

48. R. A. Friedman, J. W. House, W. M. Luxford, S. Gherini, and D. Mills, "Profound Hearing Loss Associated with Hydrocodone/Acetaminophen Abuse," *Otology & Neurotology* 21, no. 2 (2000): 188; "Did Drugs Cause Limbaugh's Hearing Loss?" ABC, *Good Morning America*, October 13, 2003.

49. Martha Raddatz, "Porn Found in Osama Bin Laden Evidence Trove," ABC News, May 13, 2011.

50. "Nobel Peace Prize Faces Boycotts over Liu Xiaobo," BBC News Europe, December 7, 2010.

51. Karimi, "Hugo Chavez Receives Iran's Highest Honor."

52. "Bolivia's President Lands in Iran," BBC News, September 1, 2008; Helen Coster, "Iranian Cash Building Bonds with Bolivia," *Washington Post*, December 5, 2010.

53. "Iran and Venezuela Deepen 'Strategic Alliance,'" BBC News Latin America and Caribbean, October 20, 2010.

54. "China 'Trying to Block Publication of UN Darfur Report,'" BBC News Africa, October 21, 2010.

Chapter 23. The Enigmatic Altruism of Heroic Rescuers

1. "Times Topics: Wesley Autrey," March 19, 2009.

2. Cara Buckley, "Man Is Rescued by Stranger on Subway Tracks," *New York Times*, January 3, 2007.

3. "Times Topics: Liviu Librescu," *New York Times*, September 10, 2011.

4. Judith Lichtenberg, "Is Pure Altruism Possible?" *New York Times*, October 19, 2010.

5. "Irena Sendler, Saviour of Children in the Warsaw Ghetto, Died on May 12th, Aged 98," *Economist*, May 22, 2008.

6. Ibid.

7. Ervin Staub, "The Psychology of Bystanders, Perpetrators, and Heroic Helpers," in *Understanding Genocide: The Social Psychology of the Holocaust*, ed. Leonard S. Newman and Ralph Erber (Oxford: Oxford University Press, 2002), pp. 11–42; P. London, "The Rescuers: Motivational Hypotheses about Christians Who Saved Jews from the Nazis," *Altruism and Helping Behavior: Social Psychological Studies of Some Antecedents and Consequences* (1970): 241–50; N. Tec, *When Light Pierced the Darkness: Christian Rescue of Jews in Nazi-Occupied Poland* (New York: Oxford University Press, 1987).

8. K. R. Monroe, "John Donne's People: Explaining Differences between Rational Actors and Altruists through Cognitive Frameworks," *Journal of Politics* 53, no. 02 (1991): 394–433.

CONCLUDING THOUGHTS

a. For example, Hurricane Katrina—and the Bush administration's weak response to this national emergency in 2005—tipped the president's approval rating to a clear minority (see Richard Morin, "Bush Approval Rating at All-Time Low," *Washington Post*, September 12, 2005).

Climate change over time can have numerous adverse effects on the environment. These include a scarcity of water, food, and fishery stocks, conflicts over these diminished resources, poverty, migrations, urbanization, and disease. So greater climate change is likely to cause political instability.

b. Just like natural disasters, wars can erupt quite quickly. Public opinion judges critical leadership decisions made in the course of fast-moving, uncertain military developments. For instance, rapid change in American public opinion occurred in early 1968, just after the North Vietnamese launched the Tet Offensive. In this attack, the North Vietnamese brought fighting to the courtyard of the US Embassy in Saigon. Many political elites perceived the development as disastrous for the United States. Consequently, the number of Americans who described themselves as "hawks" about the war dropped precipitously by 22 percent in only two months (see Benjamin I. Page, *The Rational Public: Fifty Years of Trends in Americans' Policy Preferences*, ed. Robert Y. Shapiro (Chicago: University of Chicago Press, 1992).

c. This is the margin by which Franklin D. Roosevelt beat Alf Landon in 1936.

1. "US Census: Hispanic Children Now Majority in California," BBC News US and Canada, March 9, 2011.

2. "Non-Hispanic US White Births Now the Minority in US," BBC News US and Canada, May 17, 2012.

3. Ted C. Fishman, "As Populations Age, a Chance for Younger Nations," *New York Times Magazine*, October 14, 2010.

4. Benjamin I. Page, *The Rational Public: Fifty Years of Trends in Americans' Policy Preferences*, ed. Robert Y. Shapiro (Chicago: University of Chicago Press, 1992).

5. "Interracial Marriage at New US High," BBC News US and Canada, February 17, 2012; Daniel Chirot and Clark McCauley, *Why Not Kill Them All? The Logic and Prevention of Mass Political Murder* (Princeton, NJ: Princeton University Press, 2006); John A. Hall, *Is America Breaking Apart?* ed. Charles Lindholm (Princeton, NJ: Princeton University Press, 1999); Oscar Handlin, *The Uprooted*, 2nd ed. (Boston: Little, Brown, 1973); Orlando Patterson, *The Ordeal of Integration: Progress and Resentment in America's "Racial" Crisis* (Washington, DC: Civitas/Counterpoint, 1997); Mary C. Waters, *Ethnic Options: Choosing Identities in America* (Berkeley: University of California Press, 1990).

6. Page, *The Rational Public: Fifty Years of Trends in Americans' Policy Preferences*.

7. "Mid-East: Will There Be a Domino Effect?" BBC News Middle East, February 3, 2011.

8. Fishman, "As Populations Age, a Chance for Younger Nations."

9. Albert Somit, Alan Arwine, and Steven A. Peterson, *Birth Order and Political Behavior* (New York: University Press of America, 1996).

10. F. J. Sulloway, Philip T. Starks, and Michael B. Shermer, "The Adaptive Significance of Religion: A Mutualistic Relationship between Genes and Memes" (Berkeley: University of California, Berkeley; Tufts University; Claremont Graduate University, 2011).

11. Moisés Naím, *The End of Power: From Boardrooms to Battlefields and Churches to States, Why Being in Charge Isn't What It Used to Be* (New York: Basic Books, 2013).

12. Yilmaz Esmer and Thorleif Pettersson, "The Effects of Religion and Religiosity on Voting Behavior," in *The Oxford Handbook of Political Behavior*, ed. Russell J. Dalton and Hans-Dieter Klingemann (Oxford: Oxford University Press, 2007), pp. 481–503.

13. "Jihadists Use Mobiles as Propaganda Tools," BBC News Technology, April 21, 2011.

14. "Was Al Qaeda Magazine the Blueprint for Boston Bombers' Pressure Cooker Explosions?" CBSNews.com, April 23, 2013, http://www.cbsnews.com/8301-505263_162-57580874/was-al-qaeda-magazine-the-blueprint-for-boston-bombers-pressure-cooker-explosions/ (accessed May 15, 2013).

15. Page, *The Rational Public: Fifty Years of Trends in Americans' Policy Preferences*.

16. Scott Shane and Andrew W. Lehren, "Cables Obtained by WikiLeaks Shine Light into Secret Diplomatic Channels," *New York Times*, November 28, 2010.

17. Hulda Thorisdottir, John T. Jost, and Aaron C. Kay, "On the Social and Psychological Bases of Ideology and System Justification," in *Social and Psychological Bases of Ideology and System Justification*, ed. John T. Jost, Aaron C. Kay, and Hulda Thorisdottir (Oxford: Oxford University Press, 2009), pp. 3–23.

18. Page, *The Rational Public: Fifty Years of Trends in Americans' Policy Preferences*.

19. David O. Sears, "Long-Term Psychological Consequences of Political Events," in *Political Psychology*, ed. Kristen R. Monroe (Mahwah, NJ: L. Erlbaum, 2002), pp. 249–69.

20. Kauṭilīya, *Kauṭilīya Arthaśāstra: Sanskrit, Transliteration, and English Translation with an Exhaustive Introduction*, ed. R. Shama Sastri and V. Narain (Delhi: Chaukhamba Sanskrit Pratishthan, 2005).

21. Larry Diamond, *The Spirit of Democracy: The Struggle to Build Free Societies throughout the World* (New York: Times Books, 2008).

22. C. Fred Alford, "Group Psychology Is the State of Nature," in *Political Psychology*, ed. Kristen R. Monroe (Mahwah, NJ: L. Erlbaum, 2002), pp. 193–205.

23. Rabindranath Tagore, "Poem #35, Where the Mind Is without Fear," in *The Complete Poems of Rabindranath Tagore's Gitanjali: Texts and Critical Evaluation*, edited by Samiran Kumar Paul (New Delhi: Sarup & Sons, 2006), p. 162.

APPENDIX A: A BRIEF WORD ON CORRELATIONS AND LEVEL OF SIGNIFICANCE FOR LAYPEOPLE

a. The slope of the flat line at the bottom in the middle is zero, but the correlation itself is undefined because the dependent variable doesn't change at all (it has a variance of zero).

1. Edward Glaeser and Bryce A. Ward, "Myths and Realities of American Political Geography," National Bureau of Economic Research, 2006.

APPENDIX B: JOE THE PLUMBER, POLITICAL ELITES, AND THE CONTROVERSIAL EXISTENCE OF PUBLIC OPINION

1. Guillermo C. Jiménez, *Red Genes, Blue Genes: Exposing Political Irrationality* (Brooklyn, NY: Autonomedia, 2009).

2. Martha L. Cottam, Beth Dietz-Uhler, Elena Mastors, and Thomas Preston, *Introduction to Political Psychology* (Mahwah, NJ: Lawrence Erlbaum Associates, 2004); John Zaller, *The Nature and Origins of Mass Opinion* (Cambridge: Cambridge University Press, 1992).

3. Cottam et al., *Introduction to Political Psychology.*

4. Benjamin I. Page, *The Rational Public: Fifty Years of Trends in Americans' Policy Preferences*, ed. Robert Y. Shapiro (Chicago: University of Chicago Press, 1992).

5. Zaller, *The Nature and Origins of Mass Opinion.*

6. Angus Campbell, P. E. Converse, W. E. Miller, and E. Donald, *The American Voter*, unabridged ed. (Chicago: University of Chicago Press, 1980).

7. Ibid.

8. Robert Reich, "Reading America's Tea Leaves," *American Interest: Policy, Politics & Culture* (November/December 2010): 6–17.

9. Nate Silver, "What Do Economic Models Really Tell Us about Elections?" *FiveThirtyEight: Nate Silver's Political Calculus*, http://fivethirtyeight.blogs.nytimes.com/2011/06/03/what-do-economic-models-really-tell-us-about-elections/ (accessed June 3, 2011).

10. Cottam et al., *Introduction to Political Psychology.*

11. Zaller, *The Nature and Origins of Mass Opinion.*

12. Richard G. Niemi and Herbert F. Weisberg, ed. *Controversies in Voting Behavior*, 3rd ed. (Washington, DC: CQ Press, 1993); Kathleen Knight and Robert Erikson, "Ideology in the 1990s," in *Understanding Public Opinion*, ed. Barbara Norrander and Clyde Wilcox (Washington, DC: CQ Press, 1997), pp. 89–110; Robert S. Erikson, Kent L. Tedin, and Norman R. Luttbeg, *American Public Opinion: Its Origins, Content, and Impact* (New York: Longman, 2011), p. 77.

13. Zaller, *The Nature and Origins of Mass Opinion.*

14. Paul M. Sniderman, Gretchen C. Crosby, and William G. Howell, "The Politics of Race," in *Racialized Politics: The Debate about Racism in America*, ed. David O. Sears, Jim Sidanius, and Lawrence Bobo (Chicago: University of Chicago Press, 2000), pp. 236–79.

15. Zaller, *The Nature and Origins of Mass Opinion*; Thomas F. Pettigrew, "Systematizing the Predictors of Prejudice," in *Racialized Politics: The Debate about Racism in America*, ed. David O. Sears, Jim Sidanius, and Lawrence Bobo (Chicago: University of Chicago Press, 2000), pp. 280–301.

FIGURE CREDITS

Figure 3: Correlation between the Size of Two Brain Regions and Political Ideology. Reprinted from Ryota Kanai, Tom Feilden, Colin Firth, and Geraint Rees, "Political Orientations Are Correlated with Brain Structure in Young Adults," *Current Biology* 21, no. 8 (2011): 677–80, with permission from Elsevier.

Figure 5: Correlation between Political Moderation and GDP per Capita for Seventy-One Countries (1981–2000). Reprinted from Carl-Johan Dalgaard and Ola Olsson, "Why Are Rich Countries More Politically Cohesive?" In Department of Economics, University of Copenhagen: Discussion Papers, No. 09-23, 2009, as arranged with author.

"Liberals!" Cartoon by Dana Fradon/The *New Yorker* Collection/www.cartoon bank.com, published January 10, 1970.

Figure 17: The Human Family Tree (not drawn in proportion to time periods). Courtesy of *Nature*/BBC.

Figure 18: Dispersal Distribution for Female Great Tits (*Parus major*), Wytham Woods, United Kingdom (1964–2007). Reprinted from Marta Szulkin, Ben C. Sheldon, "Dispersal as a Means of Inbreeding Avoidance in a Wild Bird Population," *Proceedings of the Royal Society B* 275, no. 1635 (2008): 703–11, by permission of the Royal Society. Photo by Luc Viatour/www.Lucnix.be.

Figure 19: Charles II of Spain (1661–1700), by Juan Carreño de Miranda (1614–1685). Courtesy of *Wikipedia*.

Figure 20: Correlation between Historical Pathogen Prevalence and Individualism/Collectivism (93 regions). Reprinted from Corey L. Fincher, Randy Thornhill, Damian R. Murray, Mark Schaller, "Pathogen Prevalence Predicts Human Cross-Cultural Variability in Individualism/Collectivism," *Proceedings of the Royal Society B* 275, no. 1640 (2008): 1279–85, by permission of the Royal Society.

Figure 21: Rh Blood Group System. Distribution of the *C* Gene in the Indigenous Populations of the World. Reprinted from A. E. Mourant et al., ed. *The Distribution of the Human Blood Groups, and Other Polymorphisms*, 2nd ed. (London: Oxford University Press, 1976), by permission of Oxford University Press.

Figure 23: Holy Roman Emperor Maximilian I (1459–1519), *left*, painted by Albrecht Dürer; and Philip IV of Spain (1605–1665), *right*, painted by Diego Valázquez. Courtesy of *Wikipedia.*

Figure 24: Average Standardized Number of Children, Children Who Reproduce, and Grandchildren, by Relatedness of Couples (Iceland, 1800–1965); the Bars Are 95 Percent Confidence Intervals. Reprinted from A. Helgason, S. Palsson, and D. F. Guthbjartsson, "An Association between the Kinship and Fertility of Human Couples," *Science* 319, no. 5864 (2008): 813. Reprinted with permission from AAAS.

Figure 25: Probability of Marrying Second Cousin or Closer. Courtesy of Alan Bittles.

Figure 26: The Islamic World. Courtesy of *Wikipedia.*

Figure 27: Barren Areas of the World, Seen from Space. Courtesy of NASA Goddard Space Flight Center.

Figure 28: Relationship between Marital Radius and Lifetime Fertility for Danish Women Born in 1954. Reprinted from R. Labouriau, and A. Amorim, "Comment on 'an Association between the Kinship and Fertility of Human Couples,'" *Science* 322, no. 5908 (2008): 1634b, with permission from AAAS.

Figure 29: Measurements Taken on Faces of Female Subjects' Fathers and Stimulus Males. Reprinted from Agnieszka Wiszewska, Boguslaw Pawlowski, and Lynda G. Boothroyd, "Father-Daughter Relationship as a Moderator of Sexual Imprinting: a Facialmetric Study," *Evolution and Human Behavior* 28, no. 4 (2007): 248–52, with permission from Elsevier.

Figure 30: The Location of the Human Leukocyte Antigen Genes on Chromosome 6. Courtesy of Philip Deitiker/*Wikipedia.*

Figure 31: Lifetime Risk of Maternal Death, by Total Children per Woman. Made with free material from http://www.gapminder.org.

Figure 34: Total Fertility Rates, by GDP per Capita, in 2009. Courtesy of *CIA World Factbook*/*Wikipedia*.

Figure 36: Percentage of Deaths Caused by War. Adaptation of figure 6.1, "War Fatality Rates for Various Prestate and Civilized Societies," in "The Harvest of Mars: The Casualties of War," ch. 6 from Lawrence H. Keeley, *War before Civilization: The Myth of the Peaceful Savage* (Oxford: Oxford University Press, 1997), by permission of Oxford University Press, Inc.

Figure 37: Arthur Lucas (*left*) and Ronald Turpin (*right*). Photo of Arthur Lucas, ca. 1962. © Government of Canada. Reproduced with the permission of the Minister of Public Works and Government Services Canada (2013), from Library and Archives Canada/Department of Justice fonds/RG13-B-1, volume 1839, file CC902, C-131142. Photo of Ronald Turpin, ca. 1962 © Government of Canada. Reproduced with the permission of the Minister of Public Works and Government Services Canada (2013), from Library and Archives Canada/Department of Justice fonds/RG13-B-1, volume 1842, file CC904, C-131148.

"There is no justice in the world." Cartoon by Robert Mankoff/The *New Yorker* Collection/www.cartoonbank.com, published February 9, 1981.

Figure 39: Child Suffering from *Kwashiorkor*. Courtesy of Centers for Disease Control and Prevention/*Wikipedia*.

"My God! I went to sleep a Democrat and I've awakened a Republican." Cartoon by Dana Fradon/The *New Yorker* Collection/www.cartoonbank.com, published December 24, 1984.

Figure 41: Presidential Candidates, and Subliminally Morphed Photographs of Subjects and Candidates. Reprinted from Jeremy N. Bailenson, Shanto Iyengar, Nick Yee, and Nathan A. Collins, "Facial Similarity between Voters and Candidates Causes Influence," *Public Opinion Quarterly* 72, no. 5 (2008): 935–61, by permission of Oxford University Press.

Figure 42: Pocho and Chito. Courtesy of Adam C. Smith Photography, adamc smithphotography.com.

"All right, I lied to you. All governments lie." Cartoon by Dana Fradon/The *New Yorker* Collection/www.cartoonbank.com, published June 21, 1993.

Figure 44: Map Showing the Partition of Poland, according to the Soviet-Nazi Agreement Made in 1939. From the Collection of the Archives of Modern Records in Warsaw (AAN); oval added for emphasis.

Figure 45: Correlation Coefficients and Graphs of the Two Variables Compared. Courtesy of Denis Boigelot/*Wikipedia*.

INDEX

Note: Page numbers in *italic type* refer to photographs, graphs, or illustrations.